Lecture Notes in Computer Science 15369

Founding Editors

Gerhard Goos
Juris Hartmanis

Editorial Board Members

Elisa Bertino, *Purdue University, West Lafayette, IN, USA*
Wen Gao, *Peking University, Beijing, China*
Bernhard Steffen , *TU Dortmund University, Dortmund, Germany*
Moti Yung , *Columbia University, New York, NY, USA*

The series Lecture Notes in Computer Science (LNCS), including its subseries Lecture Notes in Artificial Intelligence (LNAI) and Lecture Notes in Bioinformatics (LNBI), has established itself as a medium for the publication of new developments in computer science and information technology research, teaching, and education.

LNCS enjoys close cooperation with the computer science R & D community, the series counts many renowned academics among its volume editors and paper authors, and collaborates with prestigious societies. Its mission is to serve this international community by providing an invaluable service, mainly focused on the publication of conference and workshop proceedings and postproceedings. LNCS commenced publication in 1973.

Ruber Hernández-García · Ricardo J. Barrientos · Sergio A. Velastin

Editors

Progress in Pattern Recognition, Image Analysis, Computer Vision, and Applications

27th Iberoamerican Congress, CIARP 2024
Talca, Chile, November 26–29, 2024
Proceedings, Part II

Editors
Ruber Hernández-García ⓘ
Universidad Católica del Maule
Talca, Chile

Ricardo J. Barrientos ⓘ
Universidad Católica del Maule
Talca, Chile

Sergio A. Velastin ⓘ
Queen Mary University of London
London, UK

ISSN 0302-9743 ISSN 1611-3349 (electronic)
Lecture Notes in Computer Science
ISBN 978-3-031-76603-9 ISBN 978-3-031-76604-6 (eBook)
https://doi.org/10.1007/978-3-031-76604-6

© The Editor(s) (if applicable) and The Author(s), under exclusive license
to Springer Nature Switzerland AG 2025, corrected publication 2025

This work is subject to copyright. All rights are solely and exclusively licensed by the Publisher, whether the whole or part of the material is concerned, specifically the rights of translation, reprinting, reuse of illustrations, recitation, broadcasting, reproduction on microfilms or in any other physical way, and transmission or information storage and retrieval, electronic adaptation, computer software, or by similar or dissimilar methodology now known or hereafter developed.
The use of general descriptive names, registered names, trademarks, service marks, etc. in this publication does not imply, even in the absence of a specific statement, that such names are exempt from the relevant protective laws and regulations and therefore free for general use.
The publisher, the authors and the editors are safe to assume that the advice and information in this book are believed to be true and accurate at the date of publication. Neither the publisher nor the authors or the editors give a warranty, expressed or implied, with respect to the material contained herein or for any errors or omissions that may have been made. The publisher remains neutral with regard to jurisdictional claims in published maps and institutional affiliations.

This Springer imprint is published by the registered company Springer Nature Switzerland AG
The registered company address is: Gewerbestrasse 11, 6330 Cham, Switzerland

If disposing of this product, please recycle the paper.

Preface

The 27th Iberoamerican Congress on Pattern Recognition (CIARP) was the 2024 edition of the annual international conference CIARP, focusing on all aspects of pattern recognition, computer vision, artificial intelligence, data mining, and related areas, with contributions covering a broad spectrum of theory and applications to foster international collaboration and knowledge. Over the years, CIARP has become a key research event and one of the most important in Pattern Recognition for the Iberoamerican community. We are pleased to acknowledge the endorsement of CIARP 2024 by the International Association for Pattern Recognition (IAPR) and the support of the Chilean Association for Pattern Recognition (ACHIRP).

As in previous editions, CIARP 2024 brought together worldwide researchers and experts to showcase and disseminate ongoing research on mathematical methods and computing techniques for Artificial Intelligence and Pattern Recognition, particularly in Bioinformatics, Biometrics, Cognitive and Humanoid Vision, Computer Vision, Image Analysis, and Intelligent Data Analysis, as well as their application in several diverse areas such as industry, health, robotics, data mining, opinion mining and sentiment analysis, telecommunications, document analysis, and natural language processing. Moreover, CIARP 2024 was a platform for the global scientific community to share their research experiences, disseminate novel insights, and foster collaborations among research groups specializing in artificial intelligence, pattern recognition, and related fields.

CIARP has always prided itself on its international character, and this edition received contributions from 16 countries. Among the Iberoamerican contributors were Chile, Brazil, Ecuador, Mexico, Portugal, Colombia, Uruguay, Peru, Costa Rica, and Spain. Other notable submissions came from France, the UK, the Netherlands, Hungary, Egypt, and South Africa.

After a rigorous double-blind reviewing process, where three highly qualified reviewers spent significant time and effort reviewing each of the 61 submissions, 38 papers were accepted for inclusion in these proceedings, reflecting an acceptance rate of 62%. All accepted papers achieved scientific quality scores exceeding the overall mean rating. The selection of reviewers was guided by their expertise, ensuring representation from diverse countries and institutions worldwide. We want to extend our heartfelt gratitude to all the members of the Program Committee for their work, which undoubtedly enhanced the quality of the selected papers.

The conference was held at Universidad Católica del Maule, Talca - Chile, from November 26–29, 2024, and comprised four days of engaging single-track paper sessions, tutorials, and keynotes. The keynotes were delivered by distinguished lecturers Josep LLADOS, Julian FIERREZ, Angel D. SAPPA, and Domingo MERY. Moreover, the 1st IAPR LATAM School on Advanced Biometrics Techniques (Hybrid Event) took place during the conference with the participation of leading international researchers, organized by the ANID FOVI230126 Project and held in conjunction with CIARP 2024, aiming to provide up-to-date skills to participating students, professionals, academics, and researchers from the Latin American region in technical, regulatory, and ethical aspects of advanced biometric systems.

CIARP 2024 also awarded the Aurora Pons-Porrata Medal, honoring an Iberoamerican woman with a prestigious career in Pattern Recognition and related fields, the IAPR Best Paper and the IAPR Best Student Paper, whose authors were invited to submit an extended paper for publication in the Pattern Recognition Letters journal.

CIARP 2024 was jointly organized by the Faculty of Engineering Sciences of Universidad Católica del Maule and the Chilean Association for Pattern Recognition (ACHIRP). We express our sincere gratitude for their valuable contributions to its success. Furthermore, we wish to acknowledge the dedication of all members of the Organizing and Local Committees in organizing an outstanding conference and proceedings.

We especially thank the EquinOCS and LNCS teams at Springer for their invaluable support and guidance throughout the preparation of this volume.

Finally, our deepest gratitude goes to all authors who submitted their work to CIARP 2024, including those whose papers could not be accepted. We trust these proceedings will serve as a valuable reference for the global pattern recognition research community.

November 2024
Ruber Hernández-García
Ricardo J. Barrientos
Sergio A. Velastin

Organization

Conference Chairs

Ruber Hernández-García Universidad Católica del Maule, Chile
Ricardo J. Barrientos Universidad Católica del Maule, Chile

Program Chair

Sergio A. Velastin Queen Mary University of London, UK

CIARP Steering Committee

Álvaro Pardo APRU, Uruguay
César A. Astudillo ACHIRP, Chile
César Beltrán-Castañón APeRP, Peru
Joan A. Sánchez AERFAI, Spain
João Paulo Papa SIGPR-BR, Brazil
José F. Martínez-Trinidad MACVNR, Mexico
José Ruiz-Shulcloper ACRP, Cuba
Hélder Oliveira APRP, Portugal
Manuel G. Forero Vargas ACORP, Colombia
Marta Mejail SARP, Argentina

Local Committee

César A. Astudillo Universidad de Talca, Chile
Cristian A. Martínez Universidad Nacional de Salta, Argentina
Felipe Tirado Universidad Católica del Maule, Chile
Felipe J. Valencia Universidad Católica del Maule, Chile
Ingrid M. López Universidad Católica del Maule, Chile
Juan Bekios-Calfa Universidad Católica del Norte, Chile
Marco Mora Universidad Católica del Maule, Chile
Viviana Torres Universidad Católica del Maule, Chile
Wladimir E. Soto-Silva Universidad Católica del Maule, Chile
Xaviera A. López-Cortés Universidad Católica del Maule, Chile

Technical Support

Luis Morán — Universidad Católica del Maule, Chile

Program Committee

Abel Díaz Berenguer	Vrije Universiteit Brussel, Belgium
Adrián Pérez-Suay	University of Valencia, Spain
Agustín Moreno Cañadas	Universidad Nacional de Colombia, Colombia
Alessandro Bof	Universidade Federal do Pampa, Brazil
Alexei Machado	Pontifícia Universidade Católica de Minas Gerais, Brazil
Alfonso Estudillo-Romero	Université de Rennes I, France
Amel Tuama	Northern Technical University, Iraq
Ana María Bernardos	Universidad Politécnica de Madrid, Spain
Ana María Mendonça	Universidade do Porto, Portugal
Ana Sequeira	Institute for Systems and Computer Engineering, Technology and Science, Portugal
Annette Morales-González	CENATAV, Cuba
Antoine Manzanera	ENSTA-ParisTech, France
Antoni Grau	Universitat Politècnica de Catalunya, Spain
Antonio José Sanchez-Salmeron	Universitat Politècnica de València, Spain
Attallah Bilal	Mohamed Boudiaf University of M'Sila, Algeria
Aurelio Lopez-Lopez	National Institute of Astrophysics, Optics and Electronics, Mexico
Barbara Benato	Universidade Estadual de Campinas, Brazil
Billy Peralta	Universidad Andres Bello, Chile
Carlo Sansone	University of Naples Federico II, Italy
Carlos Valle	Universidad de Playa Ancha, Chile
Catarina Silva	Universidade de Coimbra, Portugal
César A. Astudillo	Universidad de Talca, Chile
Clovis Tauber	UMR U1253 iBrain, Université de Tours, Inserm, France
Cristian Martínez	Universidad Nacional de Salta, Argentina
Domingo Mery	Pontificia Universidad Católica de Chile, Chile
Edgar Roman Rangel	Instituto Tecnológico Autónomo de México, Mexico
Fabricio Lopes	Universidade Tecnológica Federal do Paraná, Brazil

Felipe Belém	ESIEE, France
Felipe Tirado	Universidad Católica del Maule, Chile
Francesc J. Ferri	Universitat de València, Spain
Gaurav Jaswal	IIT Mandi, India
Giorgio Fumera	Università degli Studi di Cagliari, Italy
Guillermo Sanchez-Diaz	Universidad Autónoma de San Luis Potosí, Mexico
Gustavo Fernandez Dominguez	Austrian Institute of Technology, Austria
Hasan Aljabbouli	New York University, USA
Heber Ivan Mejia Cabrera	Universidad Señor de Sipán, Peru
Helio Pedrini	Universidade Estadual de Campinas, Brazil
Jacques Facon	Universidade Federal do Espírito Santo, Brazil
Jesús Ariel Carrasco-Ochoa	National Institute of Astrophysics, Optics and Electronics, Mexico
João Paulo Papa	Universidade Estadual Paulista, Brazil
Joel Arrais	Universidade de Coimbra, Portugal
José Eladio Medina Pagola	Universidad de las Ciencias Informáticas, Cuba
José Francisco Martínez-Trinidad	National Institute of Astrophysics, Optics and Electronics, Mexico
José García Rodríguez	Universidad de Alicante, Spain
Jose M. Molina	Universidad Carlos III de Madrid, Spain
José Ruiz-Shulcloper	Universidad de las Ciencias Informáticas, Cuba
Juan Carlos Briñez de Leon	Universidad Nacional de Colombia, Colombia
Juan Zamora	Pontificia Universidad Católica de Valparaíso, Chile
Julio Madera	Universidad de Camagüey, Cuba
Kalman Palagyi	University of Szeged, Hungary
Larbi Boubchir	University of Paris 8, France
Laurent Heutte	Université de Rouen, France
Lazaro Bustio	Universidad Iberoamericana, Mexico
Leopoldo Altamirano	National Institute of Astrophysics, Optics and Electronics, Mexico
Luis Enrique Sucar	National Institute of Astrophysics, Optics and Electronics, Mexico
Luis Gomez Deniz	Universidad de Las Palmas de Gran Canaria, Spain
M. Angelica Pinninghoff	Universidad de Concepción, Chile
Manuel S. Lazo Cortés	Tecnológico Nacional de México, Mexico
Marcelo Mendoza	Universidad Técnica Federico Santa María, Chile
Marcos Antonio Levano	Universidad Católica de Temuco, Chile
Marie Beurton-Aimar	University of Bordeaux, France
Mario Bruno	Universidad de Playa Ancha, Chile
Marjory da Costa Abreu	Sheffield Hallam University, UK

Martha R. Ortiz-Posadas	Universidad Autónoma Metropolitana-Iztapalapa, Mexico
Martin Kampel	Technische Universität Wien, Austria
Matilde Santos Peñas	Universidad Complutense de Madrid, Spain
Michal Haindl	Institute of Information Theory and Automation, Czech Republic
Miguel Moctezuma-Flores	Universidad Nacional Autónoma de México, Mexico
Mohit Dua	National Institute of Technology Kurukshetra, India
Nguyen Anh Minh Mai	Valeo, France
Nicolas Torres	Universidad Técnica Federico Santa María, Chile
Pedro Bugatti	Universidade Federal de São Carlos, Brazil
Pedro Couto	Universidade de Trás-os-Montes e Alto Douro, Portugal
Pedro Real	Universidad de Sevilla, Spain
Pilar Gómez-Gil	National Institute of Astrophysics, Optics and Electronics, Mexico
Priscila Saito	Universidade Federal de São Carlos, Brazil
Qiao Wang	Southeast University, China
Ricardo Contreras	Universidad de Concepción, Chile
Ripudaman Singh Arora	Blue River Technology, John Deere, USA
Rodrigo Salas	Universidad de Valparaíso, Chile
Rosana Matuk	Universidad Nacional de Luján, Argentina
Samuel Silva	Universidade de Aveiro, Portugal
Sebastian Moreno	Universidad Adolfo Ibañez, Chile
Shridhar Devamane	KLE Institute of Technology, India
Sonia Gouveia	Universidade de Aveiro, Portugal
Teresa Gonçalves	Universidade de Évora, Portugal
Vinay Kumar Venkataramana	IVIS LABS Pvt Ltd., India
Vitaly Kober	Centro de Investigación Científica y de Educación Superior de Ensenada, Mexico
Vladimir Milián Núñez	Universidad de las Ciencias Informáticas, Cuba
Walter Gómez	Universidad de La Frontera, Chile
Wilson Rivera	University of Puerto Rico, Puerto Rico
Xaviera A. López-Cortés	Universidad Católica del Maule, Chile
Yaima Filiberto Cabrera	AMV Soluciones, Spain
Yanio Hernandez Heredia	Universidad de las Ciencias Informáticas, Cuba
Yunia Reyes González	Universidad de las Ciencias Informáticas, Cuba

Contents – Part II

Unmasking Phishing Attempts: A Study on Detection in Spanish Emails 1
    Vitali Herrera-Semenets, Lázaro Bustio-Martínez,
    Yamel Pérez-Guadarramas, Jorge Ángel González-Ordiano,
    and Jan van den Berg

Comparative Analysis of Spatial and Spectral Methods in GNN for Power
Flow in Electrical Power Systems 16
    Paulo A. Espinoza and Gonzalo A. Ruz

An Effective Artificial Intelligence Pipeline for Automatic Manatee Count
Using Their Tonal Vocalizations 30
    Fabricio Quirós-Corella, Priscilla Cubero-Pardo, Athena Rycyk,
    Beth Brady, César Castro-Azofeifa, Sebastián Mora-Ramírez,
    and Juan Pablo Ureña-Madrigal

Exploring Neural Joint Activity in Spiking Neural Networks for Fraud
Detection 45
    Dylan Perdigão, Francisco Antunes, Catarina Silva,
    and Bernardete Ribeiro

Rethinking the Quality of Synthetic Palm Vein Images from Spectral
Analysis 60
    Colton Clarke, Edwin H. Salazar-Jurado, and Ruber Hernández-García

An Uncertainty-Driven ScaledYOLOv4 for Open-Pit Mining Helmet
Detection 74
    Roger Calle and Eduardo Aguilar

A Generative Algorithm to Compute NanoFingerprints 90
    Francesc Serratosa

Impact of Agricultural Production on Climate Change in South America:
Comparative Analysis Between 1990 and 2020 104
    Carlos Miguel Aizaga and Rafael Melgarejo-Heredia

VAVnets: Retinal Vasculature Segmentation in Few-Shot Scenarios 120
    Idris Dulau, Benoit Recur, Catherine Helmer, Cecile Delcourt,
    and Marie Beurton-Aimar

Remote-Sensing Based Precipitation Detection Using Conditional GAN
and Recurrent Neural Networks 135
   Pablo Negri, Alejo Silvarrey, Sergio Gonzalez, Juan Ruiz,
   and Luciano Vidal

Data-Driven Genetic Algorithm for the Optimization of Water Distribution
Networks: A New Surrogate Model for Estimating Investment
and Operational Costs in Pumping Stations 151
   Nicolás Gajardo-Sepúlveda, Thalía Faúndez-Lizama,
   Jimmy H. Gutiérrez-Bahamondes, Daniel Mora-Melia,
   and César A. Astudillo

Gene Regulatory Network for the Tryptophanase Operon Under
the Threshold Boolean Network Model 161
   Felipe Encina-Chacana and Gonzalo A. Ruz

Multilabel Classification of Intracranial Hemorrhages Using Deep
Learning and Preprocessing Techniques on Non-contrast CT Images 175
   Rodrigo Salas, Juan Sebastian Castro, Marvin Querales,
   Carolina Saavedra, Claudia Prieto, and Steren Chabert

Segmentation of Brain Tumor Parts from Multi-spectral MRI Records
Using Deep Learning and U-Net Architecture 191
   Szabolcs Csaholczi, Ágnes Györfi, Levente Kovács, and László Szilágyi

Exploiting the *Segment Anything Model* (SAM) for Lung Segmentation
in Chest X-ray Images 205
   Gabriel Bellon de Carvalho and Jurandy Almeida

Predicting Next Phases of Multi-Stage Network Attacks: A Comparative
Study of Statistical and Deep-Learning Models 219
   Antonia Severín, Claudio Canales, Romina Torres, César Roudergue,
   and Rodrigo Salas

Improving Suicide Ideation Screening with Machine Learning
and Questionnaire Optimization Through Feature Analysis 233
   Ignacio Martínez, César Astudillo, and Daniel Núñez

Aquila Optimizer for Hyperparameter Metaheuristic Optimization in ELM 244
   Philip Vasquez-Iglesias, David Zabala-Blanco, Amelia E. Pizarro,
   Juan Fuentes-Concha, and Paulo Gonzalez

Mixture of LSTM Experts for Sales Prediction with Diverse Features 259
Matías Soto, Felipe Cortés, Tímar Contreras, and Billy Peralta

Correction to: Impact of Agricultural Production on Climate Change
in South America: Comparative Analysis Between 1990 and 2020 .... C1
Carlos Miguel Aizaga and Rafael Melgarejo-Heredia

Author Index 275

Contents – Part I

Towards a Lightweight CNN for Semantic Food Segmentation 1
  Bastián Muñoz, Beatriz Remeseiro, and Eduardo Aguilar

Ensemble Approach to Adaptable Behavior Cloning for a Fighting Game AI ... 16
  José García, Carlos Castro, and Carlos Valle

TeleoWatch: Pose-Transformer-Based Advanced Action Recognition 31
  Hanno Jacobs and Thambo Nyathi

Fruit Deformity Classification Through Single-Input and Multi-input
Architectures Based on CNN Models Using Real and Synthetic Images 46
  Tommy D. Beltran, Raul J. Villao, Luis E. Chuquimarca,
  Boris X. Vintimilla, and Sergio A. Velastin

CNN Sensitivity Analysis for Land Cover Map Models Using Sparse
and Heterogeneous Satellite Data 63
  Sebastián Moreno, Javier Lopatin, Diego Corvalán,
  and Alejandra Bravo-Diaz

Video Game Joystick by Recognizing Breathing Patterns 78
  Diego Robles, Andrea Lira, Carla Taramasco, and Jorge Mauro

Recovering Latent Hierarchical Relationships in Image Datasets Through
Hyperbolic Embeddings 92
  Ian Roberts, Mauricio Araya, Ricardo Ñanculef, and Mario Mallea

SwinDehazing: Haze Removal Using U-Net and Swin Transformer 104
  Percy Maldonado-Quispe and Helio Pedrini

A Proposal for Explainable Fruit Quality Recognition Using Multimodal
Models 118
  Felipe Nuñez, Billy Peralta, Orietta Nicolis, Luis Caro, and Marco Mora

Negative Sampling for Triplet-Based Loss: Improving Representation
in Self-supervised Representation Learning 133
  Manuel Alejandro Goyo and Mauricio Hidalgo

Seed-Based Superpixel Re-Segmentation for Improving Object Delineation 148
*Lucca S. P. Lacerda, Felipe C. Belém,
Zenilton Kleber Gonçalves do Patrocínio Júnior, Alexandre X. Falcão,
and Silvio J. F. Guimarães*

Towards Interactive Video Segmentation by Dynamic and Iterative
Spanning Forest 162
*Danielle Vieira, Isabela Borlido Barcelos, Zenilton K. G. Patrocínio Jr,
Alexandre Falcão, and Silvio Jamil F. Guimarães*

Data-Driven Evolutionary Algorithms for Optimizing Pumping Stations
in Water Distribution Networks: Classifier-Guided Search Space Reduction 178
*Thalía Faúndez-Lizama, Nicolás Gajardo-Sepúlveda,
Jimmy H. Gutiérrez-Bahamondes, Daniel Mora-Melia,
and César A. Astudillo*

Depth Map Completion Using a Specific Graph Metric and Balanced
Infinity Laplacian for Autonomous Vehicles 187
Vanel Lazcano

Beta Distribution Approach for Outlier Exposure in Multi-class Text
Classification 198
Camilo Maldonado, Carlos Valle, and Héctor Allende

Impact of Quantization on Large Language Models for Portuguese
Classification Tasks 213
*Danilo Samuel Jodas, Gabriel Lino Garcia, Pedro Henrique Paiola,
João Renato Ribeiro Manesco, and João Paulo Papa*

GemBode and PhiBode: Adapting Small Language Models to Brazilian
Portuguese 228
*Gabriel Lino Garcia, Pedro Henrique Paiola, Eduardo Garcia,
João Renato Ribeiro Manesco, and João Paulo Papa*

Hate Speech Detection in Portuguese Using BERTimbau 244
*João Otávio Rodrigues Ferreira Frediani, Gabriel Lino Garcia,
Pedro Henrique Paiola, Leandro Aparecido Passos, João Paulo Papa,
and Aparecido Nilceu Marana*

An Effective Approach to Text Detection and Recognition in Degraded
Historical Documents 256
Percy Maldonado-Quispe and Helio Pedrini

Author Index 271

Unmasking Phishing Attempts: A Study on Detection in Spanish Emails

Vitali Herrera-Semenets[1], Lázaro Bustio-Martínez[2(✉)], Yamel Pérez-Guadarramas[1], Jorge Ángel González-Ordiano[3], and Jan van den Berg[4]

[1] Advanced Technologies Application Center (CENATAV), La Habana, Cuba
{vherrera,yperez}@cenatav.co.cu
[2] Departamento de Estudios en Ingeniería la Innovaciniversidad Iberoamericana Ciudad de México, Mexico City, Mexico
lazaro.bustio@ibero.mx
[3] Instituto de Investigación Aplicada y Tecnología Universidad Iberoamericana Ciudad de México, Mexico City, Mexico
jorge.gonzalez@ibero.mx
[4] Intelligent Systems Department, Delft University of Technology, Delft, The Netherlands
j.vandenberg@tudelft.nl

Abstract. Phishing, a pervasive cybersecurity issue, involves fraudulent attempts to obtain sensitive information and to provoke unintentional money transfers or malware downloads, among others, by disguising as trustworthy entities in electronic communications. This paper presents an innovative approach to phishing detection in Spanish emails using patterns represented as rules. Through a comprehensive, still efficient analysis of emails, we identify interpretable recurring patterns and relevant phrases used in phishing attempts. These phrases and words often aim to persuade victims into revealing personal or financial information. These patterns are translated into a set of rules that are applied to evaluate incoming emails. Additionally, a proof-of-concept is carried out using a phishing data set of Spanish emails created for this study. Our method achieved promising results in identifying phishing attempts, providing an additional layer of security for email users. Moreover, this approach can be adapted to detect phishing in other languages, making it a potentially global solution to this persistent cybersecurity issue.

Keywords: Machine learning · Phishing detection · Rule-based systems · Spanish emails

1 Introduction

Phishing is a pervasive cybersecurity issue that continues to plague individuals and organizations worldwide. It involves fraudulent attempts to obtain sensitive information such as usernames, passwords, credit card details and to provoke unintentional money transfers or malware downloads by disguising as a trustworthy entity in an electronic communication.

In 2022, Europol reported the arrest of a phishing gang that was behind losses worth several million euros in Belgium and the Netherlands [11]. The modus operandi involved sending emails and text messages containing phishing links to fake banking websites. Thinking they were viewing their own bank accounts through this website, the victims were duped into providing their banking credentials to the suspects. According to Proofpoint's annual "State of Phish" report published in 2023 [20], the 90% of Spanish companies surveyed experienced a successful phishing attack via email. Consequently, a quarter of them recorded financial losses.

The problem is further exacerbated when the phishing attempts are made in languages other than English, such as Spanish, where detection tools are less developed and public databases for phishing detection are scarce. This lack of resources presents a significant challenge in developing effective phishing detection systems for Spanish emails, leaving a large population of internet users vulnerable to these attacks.

Motivated by this gap, this paper introduces an innovative approach to tackle this issue. We present a method for phishing detection in Spanish emails using patterns represented as rules. Through a comprehensive analysis of emails, we identify recurring patterns and key phrases used in phishing attempts. These phrases and words often aim to persuade victims into revealing sensitive information or perform an unintentional action. To extract the key phrases, 212 samples of different phishing attempts, in Spanish emails, were collected. The patterns identified from the data set are then translated into a set of rules that are applied to evaluate incoming emails.

Our method demonstrates high efficacy in identifying phishing attempts. Additionally, the proposed method can provide an additional layer of security for email users by integrating it into existing defensive tools. Moreover, our approach can be adapted to detect phishing attempts in emails written in other languages, making it a potentially global solution to this persistent cybersecurity issue.

The remainder of this paper is structured as follows. First, the background about the motivation underlying this research, phishing detection in Spanish emails, is described in Sect. 2. Second, the proposed strategy is presented in Sect. 3. Third, in Sect. 4, the experimental results using different settings are discussed. Finally, the conclusions and future work are outlined in Sect. 5.

2 Background

A wide variety of proposals for phishing detection based on machine learning techniques are reported in the literature [18]. Figure 1 shows a taxonomy that covers the most addressed approaches.

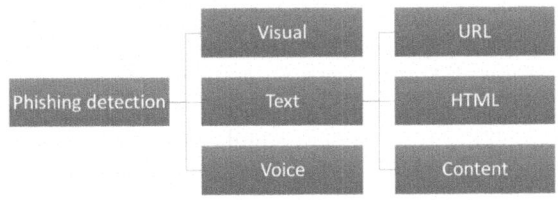

Fig. 1. Taxonomy of the most addressed approaches for phishing detection.

The analysis of visual information can range from the processing of different types of images, including logos or even favicons[1] [6,15]. These approaches are often expensive, in terms of temporal and spatial complexity, since it requires tracing the images to identify visual elements that contribute to extract patterns that describe a phishing attempt. Furthermore, large storage capacities are required and large-scale use is not feasible.

Voice Phishing, also called vishing or phone call scam, is another way to steal people's data and information. The proposals in this area have focused mostly on sentiment analysis, the content of the conversation and verification of the call originating number. To do this, the use of techniques based on natural language processing [5], machine learning [16], and deep learning [17,30], has been more common. Vishing detection is an area in which more promising results are beginning to be seen in recent years. However, there are still several limitations that make people very vulnerable to vishing. For example, difficulty tracking and verifying numbers, emotional manipulation, use of deepfakes, and the ability to detect in real time. The latter is very important, since being faced with a text that may be suspicious, where a person can take time to decide whether to respond or not, is not the same as being in the middle of a phone call, where the attacker permanently insists on obtaining information immediately.

The analysis of the information available in text has been approached from multiple perspectives, the most popular being: URL[2], HTML and content-based. URL and HTML analysis has a wide variety of proposals, mostly focused on detecting phishing websites. Wang et al. [28] uses Internationalized Domain Names (IDN) analysis to detect phishing attempts. To do this, they transform domain names into images and calculate their similarity by Siamese neural networks. This way they can identify whether a domain name is IDN homograph or not. A lightweight data representation for phishing URLs detection in IoT environments was proposed in [7]. Here, the combination of a novel feature selection algorithm and a classifier based on decision trees, led them to achieve an efficacy above 99% in the detection of phishing URLs. There is another large number of reported works, which are based on neural networks [10,22,29].

---

[1] A favicon, or "favorite icon", is a small 16 × 16 pixel icon used in web browsers to represent a site or web page.
[2] URL is an acronym for Uniform Resource Locator and is a reference to a unique resource on the Internet.

A content-based feature extraction was proposed by Bountakas and Xenakis [4]. The authors argued that the contents of emails contain information that can be employed for phishing email detection. More specifically, the proposed methodology supports that phishing emails follow a particular format, where their contents differ from benign emails. Based on the format of emails as well as the phishing email traits, 4 feature categories have been used:

- Phishing emails are designed to perform particular malicious actions (e.g., deceiving the victim to share personal information or click a malicious attachment). To accomplish these actions, phishing emails employ <u>Body Features</u> that are related to the body structure of emails.
- The text of phishing emails is comprised of strong persuasion traits that are represented by <u>Syntactic Features</u> from the emails body and subject fields.
- The emails include a header field that generates <u>Header Features</u>, which provide useful information about the emails type and the number of recipients.
- <u>URL Features</u> obtained from the hyperlinks and the domains of emails to indicate whether the email contains malicious hyperlinks or is generated by peculiar domains.

Content analysis is a widely addressed approach, mainly in electronic messages or emails. In these cases, it is usually common to use methods based on semantic analysis and natural language processing to analyze text and detect inappropriate statements that are indicative of phishing attacks [1,3]. The use of deep neural networks is also noticeable in this area [26]. Hiransha et al. [14] utilized Keras Word Embedding and Convolutional Neural Networks (CNN) to construct their model, which aims to differentiate phishing emails from legitimate ones. The experimental results reported in [21], show that deep learning word embedding, specifically Long Short-Term Memory (LSTM), is appropriate for the email anti-phishing task.

Regardless of the area addressed, the marked tendency to use techniques based on deep learning that lead to good results is notable. Specifically, the CNN together with the LSTM networks are the most used in these scenarios [2,27,31]. However, people often face phishing attempts almost daily and there is not always a mechanism at hand, whether based on deep learning or not, that helps in making decisions in a situation of this type. The interpretability of specific patterns or characteristics that can lead a person to intuit that they are in the presence of a phishing attempt is very limited when deep learning techniques are used.

On the other hand, there are approaches, such as rule-based ones, that are easy to understand and explain. Additionally, rule-based methods do not require large volumes of data for their development, unlike deep learning systems that often need large data sets to train effectively. This aspect is very important when we face phishing attempts in languages other than English, where detection is usually more complex due to the lack of available data sets that allow training models. According to [24], Spanish is the 4th most spoken language in the world and the second with the largest number of native speakers (see Fig. 2). This situation represents a substantial number of individuals who may be exposed to or at risk of becoming victims of phishing, making this group a common target for phishing attempts.

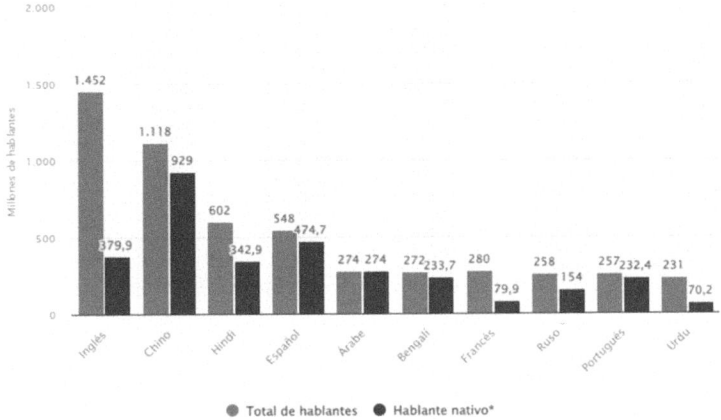

Fig. 2. The most spoken languages worldwide in 2023 [24].

From the previous analysis it is possible to formulate the following reasons that will guide our work:

- **Language coverage:** Although many phishing detection systems are developed and tested primarily in English, phishing is not limited to any particular language. A system that can effectively detect phishing in Spanish could protect a large number of Spanish-speaking users.
- **Cultural adaptability:** Phishing attacks are often adapted to local cultures and contexts to be more convincing. A system that understands the specific rules and patterns of Spanish could be more effective in detecting these culturally adapted attacks.
- **Speed and efficiency:** As mentioned before, rules-based systems can be faster and more efficient than deep learning-based systems. This could allow for faster detection and response to phishing attempts.
- **Interpretability:** A rules-based system can provide a clear explanation of why a message was flagged as phishing. This can be useful for user education and improving trust in the system.

Taking the above into account, we can deduce that it would be very beneficial to have a rule-based mechanism to detect phishing in Spanish. Therefore, in this work a rule-based strategy is proposed to identify, with a high accuracy, the presence of a phishing attempt in Spanish. The purpose of having a rules-based strategy is not only to detect phishing, but also to contribute to user education and awareness in preventing phishing.

3 Proposal

The proposed strategy consists of four fundamental steps (see Fig. 3).

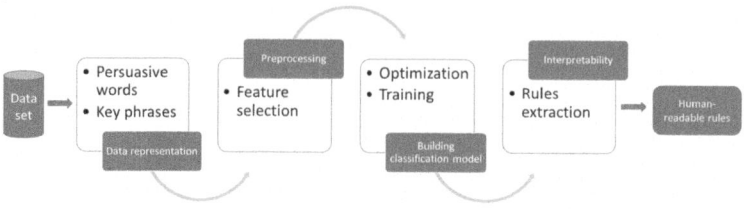

Fig. 3. General scheme of the proposed strategy.

The first step is to represent the data as persuasive words and key phrases. Persuasive words are those used to convince, influence or induce someone to take an action or change their point of view. These words can provoke specific emotions and push potential victims to certain actions. These actions may have various purposes, such as political, marketing and, especially, phishing. These words are powerful tools in the art of copywriting, which is writing texts in such a way that they provoke a direct reaction in the reader. Figure 4 shows some examples of persuasive words grouped into five categories.

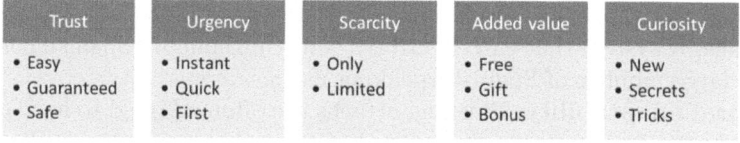

Fig. 4. Examples of persuasive words.

The key phrases are those commonly used in phishing emails to trick users. These phrases may include, for example, urgent requests for personal information, promises of financial rewards, threats of negative consequences if immediate action is not taken, among others. The key phrases extraction algorithm used in this strategy operates in 5 stages: Text-Preprocessing, Candidate Phrases Extraction, Topics Identification, Topics Ranking Construction and Key-phrases Selection. For more details about this algorithm, we refer to the work described in [19]. In this way, each message is represented as an n-dimensional vector, where n is the total number of persuasive words and key phrases. Each position of the vector is assigned a binary value, indicating the presence (1) or not (0) of the word or phrase. This representation captures the essence of phishing content, which often includes persuasive and relevant language to trick users. A characteristic of this data representation is that a high-dimensional data set is obtained, where not all the features provide useful information for the training process.

Therefore, the second step involves feature selection. For this, an algorithm that has been shown to be feasible for intrusion detection task was used [13]. In essence, the algorithm involves using three measures: Chi-Squared, Information Gain, and ReliefF. Each measure leverages different types of qualitative information to select features. After obtaining a ranking of scores with each measure, the mean score is computed for each one. This mean serves as a threshold for selecting features above that value. Then, by combining the selected feature subsets, the final feature set is obtained.

The third step concerns the building of the classification model. Here three classifiers were evaluated, from which rule-based patterns can be obtained. Such classifiers are RIPPER [9], IREP [12] and DecisionTreeClassifier [23]. Section 4 shows that the latter best suited this context.

Algorithm 1: Building classification model

Input: τ: Threshold, ρ: hyper-parameters, ζ: classifier, D: preprocessed dataset
Output: ζ_M: Classification Model

1 $scores = [Accuracy, Recall, Precision]$
2 $models = GridSearch(\zeta, \rho, scores, D, cv = 10)$
3 **foreach** $model$ **in** $models$ **do**
4   **if** $model.Accuracy > \tau$ **then**
5   selectedModels.Add(model)
6   **end**
7 **end**
8 **foreach** $model$ **in** $selectedModels$ **do**
9   **if** $model.Recall > (selectedModels.MaxRecall() - selectedModels.StdRecall())$ **then**
10   filteredModels.Add(model)
11   **end**
12 **end**
13 $bestModel$ = filteredModels.Get_Model_by($Max\_Precision$)
14 ζ_M = TrainModel($\zeta, D, bestModel.Optimal\_Values(\rho)$)
15 **return** ζ_M

As can be seen in line 2 of Algorithm 1 shown below, the building classification model step involves applying Grid Search to optimize the model hyper-parameters based on three quality measures: Accuracy (see Eq. 1), Recall (see Eq. 2) and Precision (see Eq. 3). The values of the quality measures reported in this process were obtained by also applying a 10-fold cross-validation during the Grid Search process. To select the best combination of hyper-parameters, a strategy was established based on a defined threshold for the Accuracy measure. Those built models, during Grid Search, that report an accuracy above the threshold are selected as candidates (see lines 3–7 of Algorithm 1). Then, those models whose Recall is greater than the maximum reported Recall minus the standard deviation are filtered (see lines 8–12 of Algorithm 1). From the filtered models, the one whose precision is the highest is finally selected (see line 13 of Algorithm 1). Once the model with the best combination of hyper-parameters is selected, such hyper-parameters are used to train the classifier using the pre-processed dataset (see line 14 of Algorithm 1).

$$\text{Accuracy} = \frac{\text{True Positives} + \text{True Negatives}}{\text{Total Predictions}} \quad (1)$$

$$\text{Recall} = \frac{\text{True Positives}}{\text{True Positives} + \text{False Negatives}} \quad (2)$$

$$\text{Precision} = \frac{\text{True Positives}}{\text{True Positives} + \text{False Positives}} \quad (3)$$

The fourth and final step involves rule extraction. In this step, human-readable rules are extracted from the classification model. Although there are models that allow patterns represented as rules to be directly obtained, there are others, such as those based on decision trees, of which it is also possible to represent them in rule structures. For this case, the idea of representing a rule as a path from the root node to a leaf node was followed. This allowed us to obtain a set of rules made up of all possible paths from the root node to each of the leaf nodes of the decision tree. These rules help to understand how the model makes decisions and provide transparency, which is especially important in security applications such as phishing detection.

This proposal offers a systematic and understandable approach to detecting phishing in emails, from the initial representation of the data to the interpretation of the model results.

4 Experimental Results

The non-availability of a data set of phishing emails in Spanish led us to create a first approximation of what could be a reference data set to evaluate approaches aimed at detecting phishing in Spanish. It is important to note that there is a tendency in the literature to assume that a spam email is the same as a phishing email. This assumption is wrong because phishing attacks involve targeted attempts to trick people into revealing sensitive information or taking specific actions, while spam typically encompasses unsolicited mass messages. Combining both types of emails can lead to inaccuracies in detection, as phishing emails often mimic legitimate communications and rely heavily on psychological manipulation to attempt to promote certain emotions in their victims using Principles of Persuasion [8], as opposed to spam. Additionally, the different characteristics and intent of phishing attacks require customized detection methods that go beyond traditional spam filters.

Taking the above into account, 212 phishing emails in Spanish were collected. To balance the data set, 212 legitimate emails were also collected. Finally, the data set was made up of a total of 424 emails. This data set was collected through the joint work of 2 institutions: IBERO and CENATAV. The phishing messages collected primarily include attempts to obtain sensitive personal and financial information, such as account numbers, usernames and passwords. A part of these emails seek to get the victim to access spoofing websites by posing as bank officials, legal authorities, or online site administrators. In other cases, they request the information directly by email. However, a common aspect in these emails is the use of certain persuasive words and relevant phrases that seduce the victim to follow the scammer's game.

By applying the first step of the proposed strategy, 687 key phrases were extracted from the training set. Additionally, 145 persuasive words were used to represent the data. This adds up to a total of 832 words and phrases, of which 13 are repeated, which is why they were removed. Therefore, each email was represented as a vector of 819 dimensions, where each dimension is associated with a phrase or word.

It is important to note that not all features used to represent data are necessarily useful or provide relevant information. Therefore, it is essential to apply a feature selection process. This process allows us to identify and select those features that significantly contribute to our analysis, thereby improving the efficiency and accuracy of our data models. In the next step, the feature selection was carried out. This allowed us to reduce the dimensionality of the data to 25 features.

The next step consists of applying a Grid Search process to obtain the optimal values of the hyper-parameters. This procedure was implemented on three distinct classifiers, enabling the acquisition of rule-based pattern representations: RIPPER, IREP and Decision Tree Classifier. For the latter, the fourth step of the proposed strategy was applied, which allows building rules from the created decision tree.

As described in the previous section, a heuristic to select the optimal combination of values for the hyper-parameters was designed. In this sense, for each classifier, the model built that reported the highest accuracy, during the Grid Search process, was selected. In the case of IREP and RIPPER, the efficacy achieved, in terms of accuracy, was 0.65 and 0.66 respectively; somewhat lower than the 0.74 achieved with Decision Tree Classifier. This result led to using Decision Tree Classifier as the model for building rules from the proposed data representation.

In order to evaluate the rules obtained with Decision Tree Classifier, the data set was divided into a training set (85%) and a test set (15%). Once the classifier was trained, from the decision tree built, it was possible to create 20 rules (12 legitimate rules and 8 phishing rules). Below is an example of a phishing rule obtained.

- if (comunidad = 0) and (saludo = 0) and (investigador = 0) and (caso = 0) and (correo = 1) and (cuenta = 1) and (clic = 1) then class: **phishing** (proba: 100.0%) | based on 14 samples

This rule evaluates the presence, or absence, of certain words or phrases in a text, in such a way that they define a pattern associated with a phishing attempt. The rule shown has a probability, or confidence, of 100% based on 14 samples that were covered during training and that matched the phishing class that this rule represents.

The classification model demonstrated a robust performance with an accuracy of 0.81. Notably, it achieved a promising recall of 0.90 for the phishing class, contributing to an average recall of 0.82 (see Fig. 5). Although the data set used is still small to be able to estimate the behavior of the rules obtained in a real-world context, the results are encouraging.

```
                 precision    recall  f1-score   support

        legit        0.90      0.74      0.81        35
     phishing        0.74      0.90      0.81        29

     accuracy                            0.81        64
    macro avg        0.82      0.82      0.81        64
 weighted avg        0.83      0.81      0.81        64
```

Fig. 5. Achieved results with Decision Tree Classifier.

In Fig. 6 it can be seen that only 3 emails were false negatives and 9 false positives. A false negative occurs when a real phishing attack goes undetected and is allowed to continue. This can lead to serious consequences, such as the theft of personal or financial information. Therefore, it is essential that phishing detection systems are able to correctly identify most, if not all, phishing attacks. On the other hand, a false positive occurs when a legitimate email is incorrectly flagged as phishing. While this can be annoying and can lead to the loss of important information if the email is ignored, the consequences of a false positive are generally not as severe as those of a false negative. Therefore, in the fight against phishing, it is preferable to have more false positives than false negatives. However, the ultimate goal is to minimize both to improve the accuracy of the phishing detection system. In general, the results achieved are acceptable, showing a false positive rate of 0.26 and a false negative rate of 0.10.

4.1 Data Representation Comparison

The approach proposed in this study utilizes a data representation grounded in key phrases and persuasive words. However, recent work by Bountakas and Xenakis [4] introduces a content-based representation that encompasses four distinct categories, as outlined in the related work section. Each of these categories comprises a set of features, totaling 22 in all. These features were employed to represent the dataset presented in this study. Subsequently, we applied three

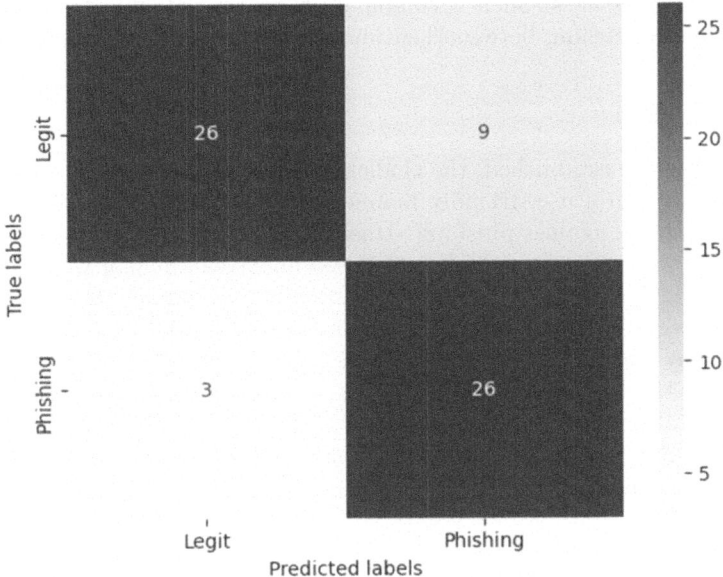

Fig. 6. Confusion matrix obtained using Decision Tree Classifier.

classifiers from different families: Support Vector Machine (SVM), k-Nearest Neighbors (KNN), and Decision Tree Classifier. The resulting outcomes are summarized in Table 1.

Table 1. Comparison of Classifier Performance using different Data Representation

Classifier (Data Representation)	Precision	Recall	F-score	Accuracy
SVM (key phrases and persuasive words)	0.71	0.83	0.76	0.77
SVM (Bountakas and Xenakis [4])	0.70	0.80	0.75	0.75
KNN (key phrases and persuasive words)	0.74	0.84	0.79	0.78
KNN (Bountakas and Xenakis [4])	0.71	0.83	0.76	0.77
Decision Tree (key phrases and persuasive words)	0.74	0.90	0.81	0.81
Decision Tree (Bountakas and Xenakis [4])	0.71	0.86	0.78	0.78

Overall, the results obtained highlight the effectiveness of the data representation proposed in this study for the evaluated classifiers. Specifically, the Decision Tree classifier, when utilizing this data representation, achieved an impressive accuracy of 0.81 and a recall of 0.90. These metrics indicate its strong performance in correctly identifying positive instances. This finding underscores the significance of combining persuasive language cues with key phrases extracted from phishing messages. Such a combination proves valuable for classifiers in effectively distinguishing between legitimate messages and phishing attempts.

4.2 Demo Tool

Once the rules are established, the challenge lies in conveying this knowledge to the average user in a user-friendly manner, making them aware of the need to protect themselves against phishing attacks.

For this purpose, a demo web tool based on Streamlit [25] was created. As depicted in Fig. 7, this tool allows users to interact by inputting text or uploading a file for its content to be analyzed. Upon processing the user-provided information, if any phishing or legitimate rule is met, the tool will alert the user. Additionally, the tool highlights the words within the user's text that triggered the alert. It also informs the user about the type of rule that was triggered and the probability that the email is phishing or legitimate (based on the rule's probability or confidence).

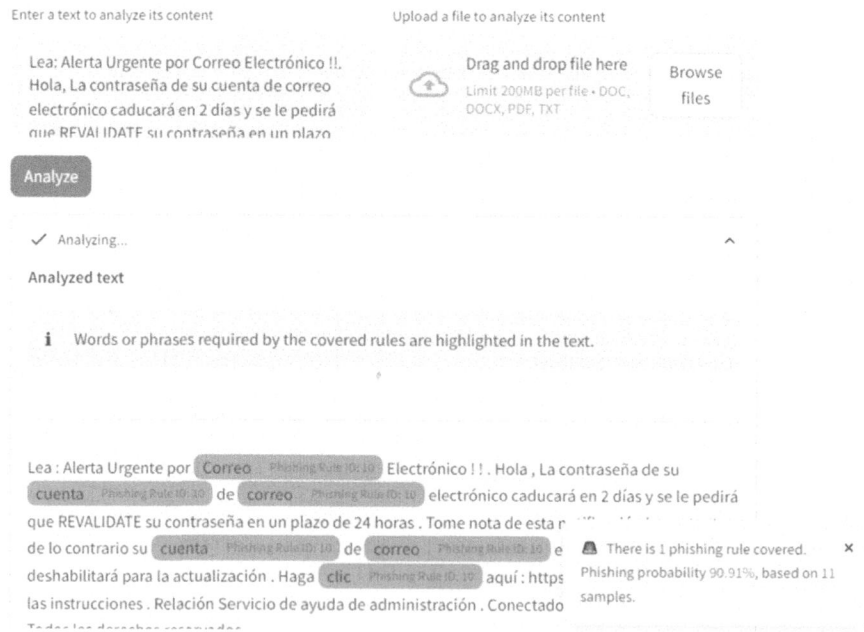

Fig. 7. Demo web tool created to interact with users.

Although it is currently only a demo tool, work is underway to develop this concept further and enable other options to make it more appealing to users and subtly contribute to their cybersecurity culture.

5 Conclusions

Based on the evaluated data set, this study developed an effective method for detecting phishing attempts in Spanish emails. The results obtained have shown that our approach has an accuracy of 81%, indicating an acceptable accuracy in identifying phishing emails. Such accuracy rate provides significant protection to Spanish email users, minimizing the risk of falling into phishing traps.

The study findings emphasize the importance of leveraging persuasive language cues alongside key phrases from phishing messages, as demonstrated by the Decision Tree classifier remarkable accuracy and recall. This integrated approach holds promise for robustly distinguishing between legitimate messages and phishing attempts.

In addition, as part of this work, we have created a demonstration tool that allows users to interactively evaluate a text and determine its likelihood of being a phishing attempt. This tool not only serves as a practical application of our research but also provides users with a tangible way to understand and apply our findings.

In summary, our work has not only proven to be effective in detecting phishing in Spanish emails, but also has the potential to be adapted to other languages. The proof-of-concept conducted showed promising results, but there is still a lot of development work to be done to create a high-accurate detection tool. The gradual incorporation of new phishing emails to the data set will allow adding more useful information for the construction of new rules that help to improve the results achieved so far. Our ultimate goal is to contribute to a safer and more reliable digital environment for all.

References

1. Alhogail, A., Alsabih, A.: Applying machine learning and natural language processing to detect phishing email. Comput. Secur. **110**, 102414 (2021)
2. Ariyadasa, S., Fernando, S., Fernando, S.: Detecting phishing attacks using a combined model of LSTM and CNN. Int. J. Adv. Appl. Sci **7**(7), 56–67 (2020)
3. Banu, R., Anand, M., Kamath, A., Ashika, S., Ujwala, H., Harshitha, S.: Detecting phishing attacks using natural language processing and machine learning. In: 2019 International Conference on Intelligent Computing and Control Systems (ICCS), pp. 1210–1214. IEEE (2019)
4. Bountakas, P., Xenakis, C.: Helphed: hybrid ensemble learning phishing email detection. J. Netw. Comput. Appl. **210**, 103545 (2023)
5. Boussougou, M.K.M., Jin, S., Chang, D., Park, D.J.: Korean voice phishing text classification performance analysis using machine learning techniques. In: Proceedings of the Korea Information Processing Society Conference, pp. 297–299. Korea Information Processing Society (2021)
6. Bozkir, A.S., Aydos, M.: LogoSense: a companion hog based logo detection scheme for phishing web page and e-mail brand recognition. Comput. Secur. **95**, 101855 (2020)
7. Bustio-Martínez, L., Álvarez-Carmona, M.A., Herrera-Semenets, V., Feregrino-Uribe, C., Cumplido, R.: A lightweight data representation for phishing URLs detection in IoT environments. Inf. Sci. **603**, 42–59 (2022)

8. Bustio-Martínez, L., et al.: Towards automatic principles of persuasion detection using machine learning approach. In: Hernández Heredia, Y., Milián Núñez, V., Ruiz Shulcloper, J. (eds.) IWAIPR 2023. LNCS, vol. 14335, pp. 155–166. Springer, Cham (2023). https://doi.org/10.1007/978-3-031-49552-6_14
9. Cohen, W.W.: Fast effective rule induction. In: Machine Learning Proceedings 1995, pp. 115–123. Elsevier (1995)
10. Dilhara, B.: Phishing URL detection: a novel hybrid approach using long short-term memory and gated recurrent units. Int. J. Compu. Appl. **975**, 8887 (2021)
11. Europol: Phishing gang behind several million euros worth of losses busted in Belgium and the Netherlands (2022). https://www.europol.europa.eu/media-press/newsroom/news/phishing-gang-behind-several-million-euros-worth-of-losses-busted-in-belgium-and-netherlands. Accessed 31 Jan 2024
12. Fürnkranz, J., Widmer, G.: Incremental reduced error pruning. In: Machine Learning Proceedings 1994, pp. 70–77. Elsevier (1994)
13. Herrera-Semenets, V., Bustio-Martínez, L., Hernández-León, R., van den Berg, J.: A multi-measure feature selection algorithm for efficacious intrusion detection. Knowl.-Based Syst. **227**, 107264 (2021)
14. Hiransha, M., Unnithan, N.A., Vinayakumar, R., Soman, K., Verma, A.: Deep learning based phishing e-mail detection. In: Proceedings of 1st AntiPhishing Shared Pilot 4th ACM International Workshop Security Privacy Analysis (IWSPA), pp. 1–5. Tempe, AZ, USA (2018)
15. Lee, J., Xin, Z., See, M.N.P., Sabharwal, K., Apruzzese, G., Divakaran, D.M.: Attacking logo-based phishing website detectors with adversarial perturbations. In: Tsudik, G., Conti, M., Liang, K., Smaragdakis, G. (eds.) ESORICS 2023. LNCS, vol. 14346, pp. 162–182. Springer, Cham (2023). https://doi.org/10.1007/978-3-031-51479-1_9
16. Lee, M., Park, E.: Real-time Korean voice phishing detection based on machine learning approaches. J. Ambient. Intell. Humaniz. Comput. **14**(7), 8173–8184 (2023)
17. Moussavou Boussougou, M.K., Park, D.J.: Attention-based 1D CNN-BiLSTM hybrid model enhanced with fasttext word embedding for Korean voice phishing detection. Mathematics **11**(14), 3217 (2023)
18. Naqvi, B., Perova, K., Farooq, A., Makhdoom, I., Oyedeji, S., Porras, J.: Mitigation strategies against the phishing attacks: a systematic literature review. Comput. Secur. 103387 (2023)
19. Pérez-Guadarramas, Y., Simón-Cuevas, A., Romero, F.P., Olivas, J.A.: Topic modeling based on OWA aggregation to improve the semantic focusing on relevant information extraction problems. In: Rivera, G., Cruz-Reyes, L., Dorronsoro, B., Rosete, A. (eds.) Data Analytics and Computational Intelligence: Novel Models, Algorithms and Applications. Studies in Big Data, vol. 132, pp. 17–42. Springer, Cham (2023). https://doi.org/10.1007/978-3-031-38325-0_2
20. Proofpoint: 2023 state of the phish: Europe and the middle east (2023). https://www.proofpoint.com/uk/resources/threat-reports/state-of-phish. Accessed 31 Jan 2024
21. Ra, V., HBa, B.G., Ma, A.K., KPa, S., Poornachandran, P., Verma, A.: Deepantiphishnet: applying deep neural networks for phishing email detection. In: Proceedings of 1st AntiPhishing Shared Pilot 4th ACM Int. Workshop Security Privacy Analysis (IWSPA), pp. 1–11. Tempe, AZ, USA (2018)
22. Sahingoz, O.K., Buber, E., Demir, O., Diri, B.: Machine learning based phishing detection from URLs. Expert Syst. Appl. **117**, 345–357 (2019)

23. sklearn: Decisiontreeclassifier (2024). https://scikit-learn.org/stable/modules/generated/sklearn.tree.DecisionTreeClassifier.html. Accessed 5 Feb 2024
24. Statista: Los idiomas mas hablados en el mundo en 2023 (2024). https://es.statista.com/estadisticas/635631/los-idiomas-mas-hablados-en-el-mundo/. Accessed 31 Jan 2024
25. Streamlit: Api reference (2024). https://docs.streamlit.io/library/api-reference. Accessed 5 Feb 2024
26. Thakur, K., Ali, M.L., Obaidat, M.A., Kamruzzaman, A.: A systematic review on deep-learning-based phishing email detection. Electronics **12**(21), 4545 (2023)
27. Vazhayil, A., Vinayakumar, R., Soman, K.: Comparative study of the detection of malicious URLs using shallow and deep networks. In: 2018 9th International Conference on Computing, Communication and Networking Technologies (ICCCNT), pp. 1–6. IEEE (2018)
28. Wang, M., Zang, X., Cao, J., Zhang, B., Li, S.: Phishhunter: detecting camouflaged IDN-based phishing attacks via Siamese neural network. Comput. Secur. **138**, 103668 (2024)
29. Wei, W., Ke, Q., Nowak, J., Korytkowski, M., Scherer, R., Woźniak, M.: Accurate and fast URL phishing detector: a convolutional neural network approach. Comput. Netw. **178**, 107275 (2020)
30. Yang, J., Lee, C., Kim, S.: Development and utilization of voice phishing prevention service through koBERT-based voice call analysis. KIISE Trans. Comput. Pract **29**, 205–213 (2023)
31. Zhang, Q., Bu, Y., Chen, B., Zhang, S., Lu, X.: Research on phishing webpage detection technology based on CNN-BiLSTM algorithm. J. Phys. Conf. Ser. **1738**, 012131 (2021). IOP Publishing

Comparative Analysis of Spatial and Spectral Methods in GNN for Power Flow in Electrical Power Systems

Paulo A. Espinoza[1(✉)] and Gonzalo A. Ruz[1,2,3]

[1] Faculty of Engineering and Sciences, Universidad Adolfo Ibáñez, Santiago, Chile
pauloandrese@gmail.com
[2] Center of Applied Ecology and Sustainability (CAPES), Santiago, Chile
[3] Data Observatory Foundation, Santiago, Chile

Abstract. This paper explores the application of Graph Neural Networks (GNNs) to power flow problems, comparing several spectral and spatial methods. The research reveals that spatial methods generally outperform their spectral counterparts, which do not rely on spectral theory, eigenvalues, or eigenvectors. GraphSAGE [9] demonstrates the best performance among the spatial methods tested, achieving a Mean Absolute Percentage Error (MAPE) of 0.79% on the test set in an experiment with 14-buses and 0.53% in the experiment with 30-buses. These findings suggest that for power flow problems, it is beneficial to consider at least hybrid or predominantly spatial models that leverage information from non-immediate neighbors. This research highlights the potential of spatial GNN methods in accurately capturing the complexities of power systems, paving the way for more robust and efficient solutions in the domain.

Keywords: Graph Neural Networks · Power Flow · Power Transmission Systems · Graph Convolutional Networks

1 Introduction

Accurately estimating and optimizing power flows is crucial for ensuring operational stability and efficiency in power systems. Historically, these tasks have been carried out using deterministic methods and physical models, such as the Newton-Raphael method and metaheuristic optimization algorithms. However, these approaches have limitations regarding scalability and adaptability to the changing topologies of electrical networks and real-time analysis.

Graph Neural Networks (GNNs) are increasingly recognized as a robust tool for modeling and analyzing complex systems that are organized as graphs, such as electrical networks. GNNs are adept at capturing intricate relationships and dependencies among the graph's nodes, enabling the amalgamation of topological information and nodal attributes for inference and predictions. This capability holds significant value in the field of power systems, where network structure and electrical characteristics play a pivotal role in power flow analysis and control.

According to [18] have demonstrated that GNNs can optimize power flow in networks with complex topologies, highlighting their ability to offer local and scalable solutions. However, the authors identified the need to improve the scalability of these solutions to large networks and their ability to react to topological changes in real-time. Similarly, [16] emphasized the importance of developing GNNs that can understand and adapt to the topological complexities of electrical networks, thereby improving the generalization of these models to different types of networks and situations.

In the same sense [12,21], show that the use of Machine Learning (ML) in power systems offers advantages in terms of complexity, efficiency, and resource intensity compared to conventional methods and approaches. In [17], the use of ML and physics-based models is suggested, additionally highlighting the importance of combining these with (GNNs) models. According to [6,11,16,18,26], the adaptability of these networks to different configurations and their ability to handle real-time data are emphasized, along with the good results that can be achieved with GNNs.

Based on [25], GNNs are categorized into four groups: recurrent graph neural networks, convolutional graph neural networks, graph autoencoders, and spatial-temporal graph neural networks. This paper focuses on analyzing spatial and spectral methods in the convolutional graph neural networks group and deep into the power flow problem in power systems. Spatial methods perform convolutions directly on the graph structure, considering the characteristics of the nodes and the connections between them. On the other hand, spectral methods rely on graph theory and use the graph Laplacian matrix to perform convolutions. The Laplacian matrix captures the graph structure through its eigenvalues and eigenvectors, allowing Fourier transforms to be applied to the graph.

In [14], provides a comprehensive review of GNN applications in power systems, highlighting their advantages in handling graph-structured data for tasks like fault detection and power flow estimation and emphasizing the need for improved scalability and real-time adaptability.

On the other hand, [15] introduces a topology-aware GNN for AC-OPF solutions, demonstrating enhanced generalizability and feasibility by incorporating grid topology and a physics-aware regularization approach and validating its effectiveness through numerical tests on various systems. In addition, [23] presents a GNN model that outperforms traditional DC power flow models and other ML approaches in predicting power flow outcomes, showcasing its accuracy and efficiency, especially in renewable energy contexts.

Finally [27] offer an in-depth review of GNN methods and their applications across various domains, highlighting the strengths and limitations of spectral and spatial approaches and their practical implementations in complex networked systems

The contribution of this paper lies in taking the most well-known GNN methods and conducting a comprehensive performance comparison of each. Unlike existing state-of-the-art studies that typically focus on specific methods without addressing the different taxonomical characteristics, this paper seeks to fill that gap by providing a broader analysis.

This paper is structured as follows: Sect. 2 provides an overview of the current state of Graph Neural Networks (GNN) and their application to the Energy Flow Problem. Section 3 outlines the methodology employed in this study. Section 4 presents the experiments and primary results obtained across various electrical networks. Section 5 discusses the main results. Finally, Sect. 6 concludes the paper with final remarks and future research directions.

2 Background

In the context of GNNs, a graph is represented as $G = (V, E)$, where V denotes the set of nodes (or vertices) and E denotes the set of edges connecting the nodes. The cardinality of V, denoted as $|V|$, is N, which is the number of nodes in the graph. The set E consists of edges and the adjacency matrix $A \in \mathbb{R}^{N \times N}$. The adjacency matrix contains a_{ij}, which is equal to 0 if $e_{ij} \notin E$, and a_{ij} is equal to 1 if $e_{ij} \in E$. This section will mainly focus on Convolutional Graph Neural Networks (ConvGNN) in two categories: spectral and spatial models. Since both categories of GCNs have many variants, only a few classic models are listed to illustrate the principle and structure.

2.1 Spectral Approaches

Spectral approaches operate in the graph's spectral domain, leveraging the graph Laplacian's properties. The graph Laplacian matrix L is defined as:

$$L = D - A, \tag{1}$$

where D is the degree matrix and A is the graph's adjacency matrix. The normalized graph Laplacian is defined as:

$$\tilde{L} = I - D^{-1/2} A D^{-1/2}. \tag{2}$$

The eigenvalues and eigenvectors of the normalized Laplacian $\tilde{L}$ are used to perform spectral graph convolutions. If $\tilde{L} = U \Lambda U^T$, where U is the matrix of eigenvectors and Λ is the diagonal matrix of eigenvalues, the graph Fourier transform of a signal x is given by:

$$F(x) = U^T x, \tag{3}$$

and the inverse graph Fourier transform is:

$$F^{-1}(x) = U x. \tag{4}$$

The convolution operation in the spectral domain is expressed as:

$$g \star x = F^{-1}(F(g) \cdot F(x)) = U(U^T g \cdot U^T x), \tag{5}$$

where g is the filter in the spectral domain. Among the most representative spectral methods are Chebyshev Networks (ChebNet) [10], which approximate the convolution operation using Chebyshev polynomials. The graph convolution operation can be approximated as:

$$g \star x \approx \sum_{k=0}^{K} \theta_k T_k(\tilde{L}) x, \tag{6}$$

where $\tilde{L} = \frac{2}{\lambda_{\max}} L - I$ is the rescaled Laplacian, $\lambda_{\max}$ is the largest eigenvalue of L, θ_k are the Chebyshev coefficients, and $T_k(\tilde{L})$ are the Chebyshev polynomials recursively defined as:

$$T_0(\tilde{L}) = I, \quad T_1(\tilde{L}) = \tilde{L}, \quad T_k(\tilde{L}) = 2\tilde{L} T_{k-1}(\tilde{L}) - T_{k-2}(\tilde{L}). \tag{7}$$

This method reduces the computational complexity by avoiding the need for eigenvector computation.

Graph Convolutional Networks (GCN) [13] simplify the Chebyshev polynomial expansion to a first-order approximation. The convolution operation is defined as:

$$H^{(l+1)} = \sigma(\tilde{D}^{-1/2} \tilde{A} \tilde{D}^{-1/2} H^{(l)} W^{(l)}), \tag{8}$$

where $\tilde{A} = A + I$ is the adjacency matrix with added self-loops, $\tilde{D}$ is the degree matrix of $\tilde{A}$, $H^{(l)}$ is the activation matrix of the l-th layer, $W^{(l)}$ is the weight matrix, and σ is a nonlinear activation function (e.g., ReLU).

2.2 Spatial Approaches

Spatial approaches directly operate on the graph structure, defining convolution operations based on the local neighborhood of nodes.

Graph Sample and Aggregation (GraphSAGE) [9] is an inductive framework that generates node embeddings by sampling and aggregating features from a node's local neighborhood. The layer-wise propagation rule is defined as:

$$h_v^{(l+1)} = \sigma\left(W^{(l)} \cdot \text{AGGREGATE}^{(l)}\left(\{h_u^{(l)}, \forall u \in \mathcal{N}(v)\}\right)\right), \tag{9}$$

where $h_v^{(l)}$ is the representation of node v at layer l, $\mathcal{N}(v)$ denotes the set of neighbors of v, and $\text{AGGREGATE}^{(l)}$ is a function that aggregates the neighbor representations (e.g., mean, LSTM, pooling).

Graph Attention Networks (GAT) [24] incorporate the attention mechanism into the propagation step, assigning different weights to different neighbors. The attention coefficient α_{ij} between nodes i and j is computed as:

$$\alpha_{ij} = \frac{\exp\left(\text{LeakyReLU}\left(a^T[Wh_i\|Wh_j]\right)\right)}{\sum_{k\in\mathcal{N}(i)} \exp\left(\text{LeakyReLU}\left(a^T[Wh_i\|Wh_k]\right)\right)}, \tag{10}$$

where a^T is a learnable weight vector, W is a weight matrix, and $\|$ denotes concatenation. The node representation is then updated as:

$$h_i^{(l+1)} = \sigma\left(\sum_{j\in\mathcal{N}(i)} \alpha_{ij} W h_j\right). \tag{11}$$

Message-passing Neural Networks (MPNN) [8] generalize several GNN variants by defining message-passing and readout phases. During the message passing phase, the hidden state of each node is updated based on messages from its neighbors:

$$h_v^{(t+1)} = \text{UPDATE}\left(h_v^{(t)}, \sum_{u\in\mathcal{N}(v)} \text{MESSAGE}(h_v^{(t)}, h_u^{(t)}, e_{vu})\right), \tag{12}$$

where $h_v^{(t)}$ is the hidden state of node v at time step t, and e_{vu} represents the edge features between nodes v and u. The readout phase computes a feature vector for the entire graph:

$$\hat{y} = \text{READOUT}(\{h_v^{(T)}, \forall v \in G\}), \tag{13}$$

where T is the total number of time steps.

Both spectral and spatial GNN algorithms have been applied to the problem of power flow. These methods enable modeling complex dependencies and interactions within the power grid, allowing for efficient and scalable solutions.

Integrating GNNs into power flow represents a significant advancement in the field. Spectral algorithms offer a mathematically rigorous approach using the Laplacian eigenvalues and eigenvectors, while spatial algorithms provide a more intuitive and locally focused method. Both approaches have unique advantages and can be applied complementary to achieve optimal performance in power flow optimization tasks.

2.3 Power Flow Problem

Power flow analysis is crucial for understanding how electrical power is distributed and utilized across a network. It involves determining the voltages at different nodes, the power generated and consumed, and the power flows through the transmission lines. This subsection provides a detailed mathematical formulation of the power flow problem and the techniques used to solve it.

In power flow analysis, the primary variables are:

- Magnitude and phase of voltages at each node (bus).
- Active (P) and reactive (Q) power injections at each node.

These variables are interrelated through the network's physical laws and constraints.

The power flow problem is defined by a set of nonlinear equations derived from Kirchhoff's laws. For each bus i in a network with N buses, the active power P_i and reactive power Q_i injected at the bus can be expressed as:

$$P_i = V_i \sum_{j=1}^{N} V_j (G_{ij} \cos \theta_{ij} + B_{ij} \sin \theta_{ij}), \tag{14}$$

$$Q_i = V_i \sum_{j=1}^{N} V_j (G_{ij} \sin \theta_{ij} - B_{ij} \cos \theta_{ij}), \tag{15}$$

where:

- V_i and V_j are the voltage magnitudes at buses i and j.
- $\theta_{ij} = \theta_i - \theta_j$ is the phase angle difference between buses i and j.
- G_{ij} and B_{ij} are the conductance and susceptance of the line connecting buses i and j.

The buses in an electrical power system are categorized into three types based on the known and unknown variables, as shown in Table 1.

Table 1. Bus Types and their known and unknown Variables

Bus Type	known Variables	unknown Variables
PQ Bus	P_i and Q_i	V_i and θ_i
PV Bus	P_i and V_i	Q_i and θ_i
Slack Bus	V_i and θ_i	P_i and Q_i

The Newton-Raphson method [4,20] generally seeks to minimize the difference shown in (16) and (17), where the original result is mapped based on determining voltages and their angles.

$$\Delta P_i = P_i^{\text{specified}} - P_i^{\text{calculated}} \tag{16}$$

$$\Delta Q_i = Q_i^{\text{specified}} - Q_i^{\text{calculated}} \tag{17}$$

Re-writing the (16), (17), the results is:

$$\Delta P_i = P_i^{\text{specified}} - V_i \sum_{j=1}^{N} V_j (G_{ij} \cos \theta_{ij} + B_{ij} \sin \theta_{ij}) \tag{18}$$

$$\Delta Q_i = Q_i^{\text{specified}} - V_i \sum_{j=1}^{N} V_j (G_{ij} \sin \theta_{ij} - B_{ij} \cos \theta_{ij}) \tag{19}$$

3 Methodology

For implementing GNN algorithms we use the PyTorch Geometric (PyG) framework [2,7] an extension of PyTorch designed specifically for processing graph-structured data. PyG includes a variety of datasets and predefined models that facilitate experimentation and comparison with the state of the art in GNNs. In particular, PyTorch Geometric Temporal (PyG Temporal) [3,19] is employed, which extends PyG for developing and researching GNNs with a temporal dimension. This extension is crucial for handling dynamic graphs where the relationships between nodes and their features can change over time.

Due to the lack of open data on electrical systems, synthetic data was created using test cases as benchmark datasets.

To create synthetic data, the methodology involves generating perturbations to the load values in test cases using PandaPower [1,22]. This process includes generating random perturbations in active and reactive power and using the Newton-Raphson method to calculate the voltage magnitude (V_i) and angle (δ_i) at each node. Specifically, the perturbations ΔP_i and ΔQ_i are drawn from a uniform distribution in the range $[-0.5, 0.5]$. These perturbations are applied to the active power P_i and reactive power Q_i at node i, resulting in new values P'_i and Q'_i. The goal is to create operationally feasible power flow scenarios that can be included in the dataset for experiments. The steps are describe in Algorithm 1.

Algorithm 1: Synthetic Data Generation and Power Flow Calculation using Newton-Raphson

---
Data: Initialize a test case in PandaPower
for *each node i in the test case* **do**
    Generate random perturbations $\Delta P_i \sim \mathcal{U}(-0.5, 0.5)$ and $\Delta Q_i \sim \mathcal{U}(-0.5, 0.5)$;
    Apply the perturbations: $P'_i = P_i + \Delta P_i \times P_i \quad Q'_i = Q_i + \Delta Q_i \times Q_i$;
    Use the Newton-Raphson method to update V_i and θ_i;
    if *no convergence* **then**
        | Discard this case and continue with the next;
    end
    Store the final values P_i, Q_i, V_i, θ_i;
end

Once the synthetic data set has been created, we proceed to use these test cases with the synthetic data to test different GNN algorithms with a fixed structure (that is, not optimized) and evaluate their performance. This is shown in Fig. 1 at the schematic level.

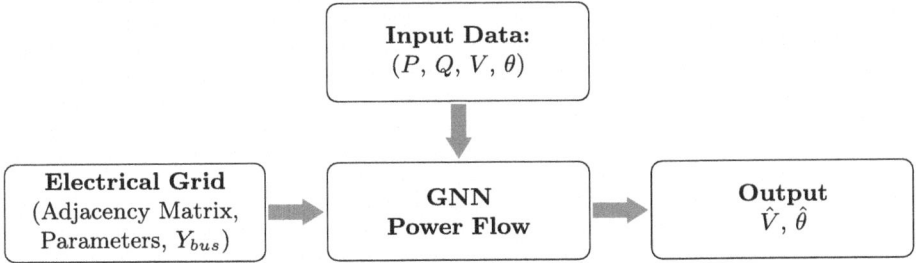

Fig. 1. Flowchart Experiment of GNN Power Flow.

4 Experiments

Based on the methodology used for the evaluation, the development of the experiments will be structured as follows:

- First, two test cases indicated in Figs. 2, 3 are used, one with 14 buses and another with 30 buses
- Regarding the test cases, the 14-bus test case is used only with a generator and PQ buses. In the case of the 30 buses, different types of buses are used
- Both datasets consisting of 2000 independent power flow observations were used. The dataset was partitioned using the holdout method, with 64% of the data allocated for training, 16% for validation, and 20% for testing.
- Figure 4 shows a structure of the type of network used in the general GNNs Block (GCN, GraphSAGE, Chebnet, and GAT) used together with 2 fully connected layers and the loss function indicated in (20)

The loss function utilized in this experiment is shown in (20). A mean squared error (MSE) loss function based on the sum of V and θ was used for simplicity. Although authors [5,16] suggest that physics-based loss functions, which integrate network parameters such as impedance, yield better performance, we have opted for a simpler and easily implemented function.

$$\mathcal{L}(\hat{V}, \hat{\theta}, V, \theta) = \frac{1}{N} \sum_{i=1}^{N} \left((\hat{V}_i - V_i)^2 + (\hat{\theta}_i - \theta_i)^2 \right) \tag{20}$$

The performance of the models will be evaluated using metrics such as the Mean Absolute Percentage Error (MAPE), Root Mean Squared Error (RMSE), Coefficient of Determination (R^2), and Mean Absolute Error (MAE). The formulas for these metrics are as follows:

$$MAPE = \frac{1}{n} \sum_{i=1}^{n} \left| \frac{y_i - \hat{y}_i}{y_i} \right| \times 100 \tag{21}$$

$$RMSE = \sqrt{\frac{1}{n} \sum_{i=1}^{n} (y_i - \hat{y}_i)^2} \tag{22}$$

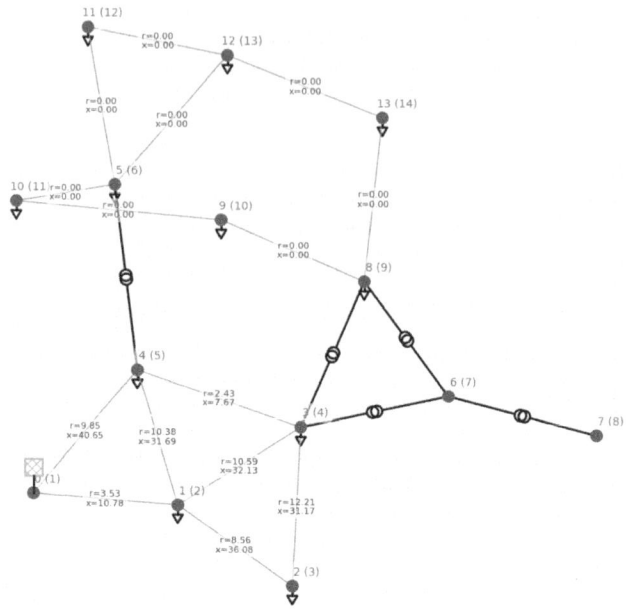

Fig. 2. Test case case14() available in PandaPower are used to measure GNN algorithm performance

$$R^2 = 1 - \frac{\sum_{i=1}^{n}(y_i - \hat{y}_i)^2}{\sum_{i=1}^{n}(y_i - \bar{y})^2} \tag{23}$$

$$\text{MAE} = \frac{1}{n}\sum_{i=1}^{n}|y_i - \hat{y}_i| \tag{24}$$

where y_i represents the actual values, $\hat{y}_i$ represents the predicted values, and n is the number of observations.

All experiments were conducted using Google Colab, which provided a flexible and efficient platform for running the models and performing the computations necessary for this study. The use of Google Colab allowed for easy access to GPU resources, facilitating the processing and analysis of large datasets and the execution of complex GNN architectures.

4.1 Results

Table 2 and Table 3 show the experiments' results for both the 14 and 30 buses networks based on the metrics defined to evaluate their performance.

Figure 5 shows the different training loglosses curves of the algorithms implemented in the experiment.

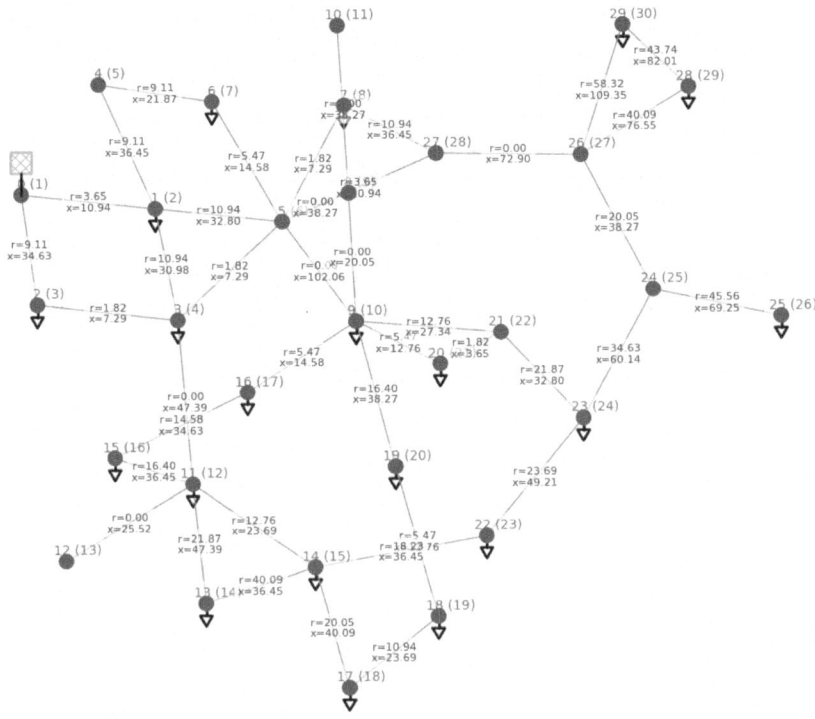

Fig. 3. Test case case30() available in PandaPower are used to measure GNN algorithm performance

Table 2. Performance on test case 14 - PandaPower

Model	# Parameters	Taxonomy	Train Time	MAPE	R^2	RMSE	MAE
GCN	2.308	Hybrid	12 min	3,2%	0.99	0.82	0.31
ChebNet	6.120	Spectral	23 min	4,85%	0.96	1.66	0.68
GraphSAGE	2.768	Spatial	10 min	0,79%	0.99	0.11	0.05
GAT	2.768	Spatial	16 min	0,80%	0.99	0.12	0.06

5 Discussion

Based on the data presented in Table 2 and Table 3 for test cases 14 and 30 buses using PandaPower, it is evident that GraphSAGE consistently delivers the best performance across various metrics, followed closely by GAT. The superiority of GraphSAGE is demonstrated by its minimal MAPE and MAE, as well as its low RMSE and high R^2 values, which indicate strong predictive power and minimal error rates. Specifically, GraphSAGE achieves a remarkably low MAPE of 0.79% and 0.53% for test cases 14 and 30, respectively, which is significantly

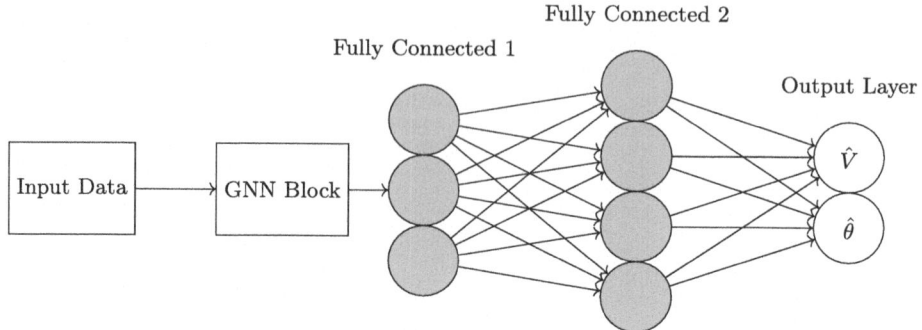

Fig. 4. Network architecture used for all GNN blocks in the experiment.

Table 3. Perfomance on test case 30 - PandaPower

Model	# Parameters	Taxonomy	Train Time	MAPE	R^2	RMSE	MAE
GCN	4.900	Hybrid	13 min	3,25%	0.99	0.86	0.38
ChebNet	23.048	Spectral	27 min	4,04%	0.96	1.58	0.66
GraphSAGE	5.872	Spatial	12 min	0,53%	0.99	0.084	0.04
GAT	5.872	Spatial	17 min	0,60%	0.99	0.1	0.05

lower than that of other models. This is complemented by its high R^2 values of 0.99, showcasing its robustness in predictive accuracy.

Graph Attention Networks (GAT) also perform well, with MAPE values close to those of GraphSAGE, at 0.80% and 0.60% for the respective test cases. GAT's RMSE and MAE values are similarly competitive, reinforcing its efficacy in handling power flow problems. The spatial taxonomy of both models highlights their ability to effectively leverage local information, which is crucial in power system applications where local node characteristics often dominate the problem space. Notably, in the second experiment, GraphSAGE significantly outperforms GAT, further demonstrating its superior capacity in predictive tasks.

In contrast, spectral methods like ChebNet, despite their theoretical sophistication, tend to underperform in these experiments. ChebNet's MAPE values of 4.85% and 4.04% for the two test cases are substantially higher, indicating a less accurate predictive capability. Furthermore, the higher RMSE values suggest ChebNet is less efficient at minimizing prediction errors. This discrepancy can be attributed to the fundamental operational differences between spatial and spectral methods. Spatial methods like GraphSAGE and GAT operate directly on nodes, thereby maximizing local information and making them more suited for power flow problems where local interactions are paramount. On the other hand, spectral methods operate on the global graph structure, which can dilute the influence of local node-specific information and result in higher computational costs.

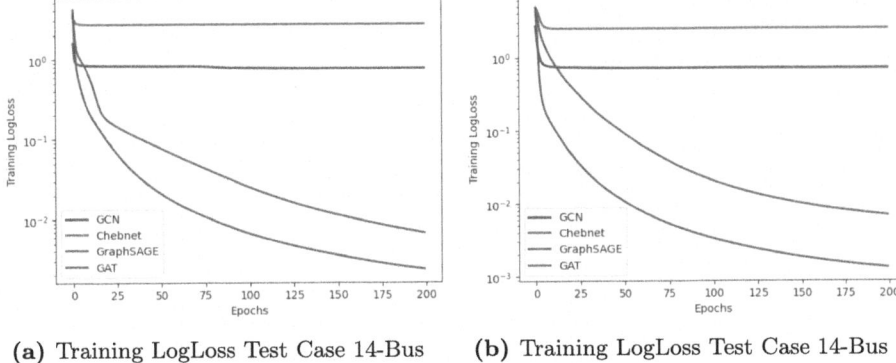

(a) Training LogLoss Test Case 14-Bus **(b)** Training LogLoss Test Case 14-Bus

Fig. 5. Behaviour of Training LogLoss of the GNN algorithms.

Training time also serves as a critical factor in assessing model efficiency. Spatial methods, as shown, are less resource-intensive, with GraphSAGE requiring only 10–12 minutes of training time, while GAT takes 16–17 minutes. ChebNet demands 23-27 min, reflecting its higher computational overhead. Although hybrid in taxonomy, the GCN similarly suffers from longer training times and suboptimal performance metrics, as seen with its relatively higher MAPE and RMSE values. GCN's tendency to overfit, indicated by its performance metrics, further complicates its applicability in practical scenarios.

The experimental results underscore the suitability of spatial methods for power flow problems. Their ability to operate directly on nodes ensures that local information is prioritized, leading to more accurate and computationally efficient models. This is a crucial advantage in power systems, where timely and precise predictions are essential for effective management and operation. While valuable in certain contexts, spectral methods are less ideal for these specific applications due to their global approach and higher computational demands.

6 Conclusion

This paper presented that spatial methods, particularly GraphSAGE and GAT, exhibit superior performance for power flow optimization in electrical networks compared to ChebNet and GCN. These methods effectively leverage local node information, leading to lower prediction errors and improved computational efficiency. While these results are promising, it is important to note that the evaluation was conducted solely on a synthetic dataset. To fully assess the robustness and scalability of these models, future work must scale the approach and test it on real and more complex power systems. Additionally, enhancing scalability and real-time adaptability remains a critical area for further exploration, ensuring that GNN-based power flow optimization can be applied to larger and more dynamic electrical networks.

Acknowledgments. G.A.R. thanks ANID FONDECYT 1230315 and ANID PIA/BASAL FB0002. P.A.E. thanks Enel Distribucion Chile for the support provided in the development of this research.

References

1. pandapower. https://www.pandapower.org/
2. PyG documentation - pytorch_geometric documentation. https://pytorch-geometric.readthedocs.io/en/latest/
3. PyTorch geometric temporal documentation - PyTorch geometric temporal documentation. https://pytorch-geometric-temporal.readthedocs.io/en/latest/
4. Tinney: Power flow solution by newton's method - google académico. https://scholar.google.com/scholar_lookup?title=Power%20flow%20solution%20by%20newtons%20method&publication_year=1967&author=William%C2%A0F.%20Tinney&author=Clifford%C2%A0E.%20Hart
5. Böttcher, L., et al.: Solving AC power flow with graph neural networks under realistic constraints. In: 2023 IEEE Belgrade PowerTech, pp. 1–7. https://doi.org/10.1109/PowerTech55446.2023.10202246. http://arxiv.org/abs/2204.07000
6. Donon, B., Donnot, B., Guyon, I., Marot, A.: Graph neural solver for power systems. In: 2019 International Joint Conference on Neural Networks (IJCNN), pp. 1–8. IEEE. https://doi.org/10.1109/IJCNN.2019.8851855. https://ieeexplore.ieee.org/document/8851855/
7. Fey, M., Lenssen, J.E.: Fast graph representation learning with PyTorch geometric. http://arxiv.org/abs/1903.02428
8. Gilmer, J., Schoenholz, S.S., Riley, P.F., Vinyals, O., Dahl, G.E.: Neural message passing for quantum chemistry. https://doi.org/10.48550/ARXIV.1704.01212, https://arxiv.org/abs/1704.01212, publisher: arXiv Version Number: 2
9. Hamilton, W., Ying, Z., Leskovec, J.: Inductive representation learning on large graphs. In: Guyon, I., et al. (eds.) Advances in Neural Information Processing Systems, vol. 30. Curran Associates, Inc
10. Hammond, D.K., Vandergheynst, P., Gribonval, R.: Wavelets on graphs via spectral graph theory **30**(2), 129–150. https://doi.org/10.1016/j.acha.2010.04.005. https://linkinghub.elsevier.com/retrieve/pii/S1063520310000552
11. Hansen, J.B., Anfinsen, S.N., Bianchi, F.M.: Power flow balancing with decentralized graph neural networks **38**(3), 2423–2433. https://doi.org/10.1109/TPWRS.2022.3195301. https://ieeexplore.ieee.org/document/9847037/
12. Hernandez, C., Sánchez Huertas, W., Gómez, V.: Optimal power flow through artificial intelligence techniques **25**(69), 150–170. https://doi.org/10.14483/22487638.18245. https://revistas.udistrital.edu.co/index.php/Tecnura/article/view/18245
13. Kipf, T.N., Welling, M.: Semi-supervised classification with graph convolutional networks. http://arxiv.org/abs/1609.02907
14. Liao, W., Bak-Jensen, B., Radhakrishna Pillai, J., Wang, Y., Wang, Y.: A review of graph neural networks and their applications in power systems **10**(2), 345–36. https://doi.org/10.35833/MPCE.2021.000058. https://ieeexplore.ieee.org/document/9520300/
15. Liu, S., Wu, C., Zhu, H.: Topology-aware graph neural networks for learning feasible and adaptive AC-OPF solutions **38**(6), 5660–5670. https://doi.org/10.1109/TPWRS.2022.3230555. https://ieeexplore.ieee.org/document/9992121/

16. Lopez-Garcia, T.B., Domínguez-Navarro, J.A.: Power flow analysis via typed graph neural networks **117**, 105567. https://doi.org/10.1016/j.engappai.2022.105567 https://linkinghub.elsevier.com/retrieve/pii/S0952197622005577
17. Marković, M., Bossart, M., Hodge, B.M.: Machine learning for modern power distribution systems: Progress and perspectives **15**(3), 032301. https://doi.org/10.1063/5.0147592. https://pubs.aip.org/jrse/article/15/3/032301/2900695/Machine-learning-for-modern-power-distribution
18. Owerko, D., Gama, F., Ribeiro, A.: Optimal power flow using graph neural networks. In: ICASSP 2020 - 2020 IEEE International Conference on Acoustics, Speech and Signal Processing (ICASSP), pp. 5930–5934. IEEE. https://doi.org/10.1109/ICASSP40776.2020.9053140. https://ieeexplore.ieee.org/document/9053140/
19. Rozemberczki, B., et al.: PyTorch geometric temporal: Spatiotemporal signal processing with neural machine learning models. http://arxiv.org/abs/2104.07788
20. Sereeter, B., Vuik, C., Witteveen, C.: On a comparison of newton-raphson solvers for power flow problems **360**, 157–169. https://doi.org/10.1016/j.cam.2019.04.007. https://www.sciencedirect.com/science/article/pii/S0377042719301876
21. Stock, S., Babazadeh, D., Becker, C.: Applications of artificial intelligence in distribution power system operation **9**, 150098–150119. https://doi.org/10.1109/ACCESS.2021.3125102. https://ieeexplore.ieee.org/document/9599712/
22. Thurner, L., et al.: Pandapower-an open-source python tool for convenient modeling, analysis, and optimization of electric power systems **33**(6), 6510–652. https://doi.org/10.1109/TPWRS.2018.2829021. https://ieeexplore.ieee.org/document/8344496/
23. Tuo, M., Li, X., Zhao, T.: Graph neural network-based power flow model
24. Veličković, P., Cucurull, G., Casanova, A., Romero, A., Liò, P., Bengio, Y.: Graph attention networks. https://doi.org/10.48550/ARXIV.1710.10903. https://arxiv.org/abs/1710.10903, publisher: arXiv Version Number: 3
25. Wu, Z., Pan, S., Chen, F., Long, G., Zhang, C., Yu, P.S.: A comprehensive survey on graph neural networks **32**(1), 4–24. https://doi.org/10.1109/TNNLS.2020.2978386. http://arxiv.org/abs/1901.00596
26. Yaniv, A., Kumar, P., Beck, Y.: Towards adoption of GNNs for power flow applications in distribution systems **216**, 109005. https://doi.org/10.1016/j.epsr.2022.109005. https://linkinghub.elsevier.com/retrieve/pii/S0378779622010549
27. Zhou, J., et al.: Graph neural networks: a review of methods and applications **1**, 57–8. https://doi.org/10.1016/j.aiopen.2021.01.001. https://linkinghub.elsevier.com/retrieve/pii/S2666651021000012

An Effective Artificial Intelligence Pipeline for Automatic Manatee Count Using Their Tonal Vocalizations

Fabricio Quirós-Corella[1]([✉]), Priscilla Cubero-Pardo[1], Athena Rycyk[2], Beth Brady[3,4], César Castro-Azofeifa[1], Sebastián Mora-Ramírez[1], and Juan Pablo Ureña-Madrigal[1]

[1] National High Technology Center, San José, Costa Rica
fquiros@cenat.ac.cr, pcubero@conare.ac.cr,
{s.mora,juurena}@estudiantec.cr
[2] New College of Florida, Florida, USA
arycyk@ncf.edu
[3] Save the Manatee Club, Florida, USA
bbrady@savethemanatee.org
[4] Mote Marine Laboratory, Florida, USA

Abstract. Despite their vulnerable status, manatee conservation efforts are hindered because of limited scientific data due to the data collection challenges. Bio-acoustic studies utilizing advanced computational analysis of long-term recordings offer a promising approach for determining the manatee's presence and abundance. The following manuscript describes an artificial intelligence pipeline that effectively implements automatic manatee identification and counting. The first step is fine-tuning a pre-trained deep neuronal network using transfer learning to detect manatee sounds with a binary accuracy of 96%. In the second phase, the pipeline implements an unsupervised learning method to group acoustic features of the detected sounds with a clustering score of 72%. The article addresses the implications and the outcomes of testing this proof of concept under experimental conditions with passive acoustics monitoring data on the Caribbean coast of Costa Rica and Panama.

Keywords: Artificial intelligence · bio-acoustics · clustering · deep learning · machine learning · music information retrieval · passive acoustics monitoring · transfer learning · unsupervised learning

1 Introduction

Anthropogenic activities have caused *Sirenian* species to be listed as threatened around the world with regional functional extinction [2,12,13,17,19], such as with dugongs (*Dugong dugon*) in China [15] or the extinct Steller's sea cow (*Hydrodamalis gigas*) [28]. The Antillean manatee (*Trichechus manatus manatus*), a subspecies of the West Indian manatee, is a rarely-observed resident of the

Caribbean coast [19] in Costa Rica and has been the national symbol of marine fauna since 2014 [1]. Nevertheless, the conservation efforts towards this species in the region are not enough due to insufficient ecological knowledge [19]. Marine mammals emit underwater sounds for navigation, foraging, and communication [4,29]. A passive acoustics monitoring (PAM) provides insights about species distribution, abundance, and behavior by deploying one or more hydrophones to record marine mammal sounds [4,29]. It normally produces a long-term database (DB) that traditionally requires manual annotation by visual scanning or audible identification for bio-acoustics analysis that can be labor-intensive [11]. To address this, we propose an automated artificial intelligence (AI) solution to detect and count manatees based on their vocalizations within PAM data. This approach aims to provide vital information to researchers and authorities to support manatee conservation efforts.

1.1 Background

Automated bio-acoustic methods are mainly based on signal processing, pattern recognition, and statistical algorithms to analyze field recordings [4,29]. A fundamental step of data-driven algorithms is the feature extraction (FE) to generate audio signal profiles throughout relevant attributes [4,29]. This means a new data structure with observations of acoustic features like relative amplitude, energy statistics, root-mean-square (RMS) values, fundamental frequency (F_0), frequency range, bandwidth (BW) values, Shannon entropy, and more [3,4]. Alternatively, the time-domain signal can be transformed into other representations: wavelet coefficients via discrete wavelet transform (DWT), spectrograms with the fast Fourier transform (FFT), the short-time Fourier transform (STFT), or the Mel-frequency cepstral coefficient (MFCC) calculation, among others [3,4,29]. Many machine learning (ML) models implement single and multiple types of classification for cetaceans' vocalizations [3,4,29]. The classification errors are analyzed with cross-entropy that considers correct and wrong estimations for each class by determining every true positive (TP) (i.e., an accurate prediction) and false positive (FP) which is a missed detection. In contrast, a true negative (TN) is a correct detection of the non-positive class and its incorrect prediction is a false negative (FN). Evaluation metrics like accuracy, precision, and recall measure the classification performance instead. Accuracy reflects overall correctness, while precision and recall focus on TPs and capturing all positive cases, respectively. F1-score balances both metrics. Another common metric is the area under the curve (AUC) of the receiver operating characteristic (ROC) function to compare the TP rate against the FP rate. Regarding ML models for bio-acoustic tasks, the support vector machine (SVM) analyzes MFCCs to execute the binary classification of Humpback whale sounds in song and non-song [29]. Unsupervised learning models like the Gaussian mixture model (GMM) capture the statistical variation of spectral attributes for categorizing sounds of dolphin species and toothed whales [3,4]. In studies of *Sirenian* species, an automatic manatee count method (AMCM) employed a three-step process to detect and count Antillean manatees [8]. After high-pass filtering, kmeans clustering analysis (KMCA) classified signal windows based on the autocorrelation strength of the undecimated discrete wavelet transform (UDWT) coefficients for noise cancellation [8]. It then applied the inverse UDWT to get the denoised signal followed by call detection (f1-score of 90%) using a similarity score with manatee call harmonics [8]. Finally, the expectation-maximization (EM) algorithm with GMMs clustered manatee calls using acoustic patterns that resulted in a count of 4 individuals from 54 detections [8].

1.2 Related Work

Most of ML solutions for bio-acoustic tasks struggle with large, time-varying datasets [29]. Former AMCM showed limitations in denoising and detecting manatee sounds in long-term recordings. Advancements in high-performance computing (HPC) enable the use of AI algorithms for bio-acoustic tasks, potentially improving accuracy and efficiency, but requiring large amounts of labeled data [3]. Recent contributions consider deep learning (DL) architectures as a deep neuronal network (DNN) to identify and categorize cetaceans' sounds that might be adapted to other marine mammals [4,29]. A supervised DL method performs the identification of beluga whales by classifying their whistles and moans with 4 convolutional neuronal network (CNN) models [30]. Using spectrograms from the FFT calculation, authors trained AlexNet from scratch and built the remaining 3 networks upon pre-trained weights of the residual neuronal network (ResNet), VGG-16, and DenseNet with transfer learning, getting a precision of 96.57% and recall of 92.26% on the evaluation set [30]. Transfer learning involves two key phases: a feature extractor and fine-tuning. In the first step, the pre-trained network's learned features are transformed into new, relevant features. Fine-tuning then jointly trains both the pre-trained model and a new classifier, adapting the features to the specific dataset. A contribution implements transfer learning with two pre-trained AlexNet models using the non-uniform FFT spectrograms as color images to perform sound multi-classification with 97.42% overall accuracy of killer whales, long-finned pilot whales, and harp seals [16]. Another transfer learning approach uses a pre-trained GoogLeNet to perform the binary classification (95.90% accuracy) of the African manatee (*Trichechus senegalensis*) by analyzing color spectral images without individual counting [21]. To detect Antillean manatee in western Caribbean Panama, another effort trained a custom CNN pyramidal architecture considering STFT images to implement a binary classifier with a f1-score of 91% and 91.90% for the AUC-ROC metric [20]. Another transfer learning solution proposed a two-step CNN architecture for manatee sound classification, first distinguishing true from false sounds and then categorizing manatee calls into 5 classes [25]. The authors employed EfficientNet as the feature extractor, pre-training the model on a dataset from two German zoos and fine-tuning it on a different dataset from 3 additional zoos [25]. They employed data augmentation techniques to address data imbalance, such as stretching, compressing, adding noise, and muting portions of audio segments.

Their approach utilized the multi-channel capabilities of DL models, transforming audio into 4 time-frequency representations into a single image, achieving a 91.15% binary accuracy for all DBs. For population estimation, the authors employed a robust clustering method based on MFCC coefficients, F_0 statistics, and harmonic frequency bands. Dimensionality reduction using t-SNE was followed by HDBSCAN clustering. While the method achieved a 45.45% deviation from the actual count for known populations, its effectiveness for unknown populations remains uncertain [25].

2 Methodology

2.1 Training and Validation Dataset

We utilized 4 labeled DBs of PAM recordings to collect them in a custom dataset for training and validation. The first dataset consisted of audio samples from digital acoustic recording tag (DTAG) devices deployed on wild Florida manatees (*Trichechus manatus latirostris*) in southwest Florida [22]. This DB considered 917 samples labeled as true and 955 false samples stored as WAV files at a sampling rate of 64 kHz. The second DB consisted of 5 recordings at 48 kHz of approximately 60min each from isolated captive manatees. The third dataset contained audio recordings from a system of land-based acoustic recording stations in Sarasota Bay, FL [22]. This had 72 manatee calls and 322 false vocalizations at 44.1 kHz. The fourth DB, named Bocas del Toro database (BCDB), gathered recordings at 96 kHz collected in the Panamanian Caribbean with 980 audios labeled as true vocalizations and 2051 false calls [8].

2.2 Testing Dataset

The BCDB provided a portion for experimental testing to validate our AMCM using 10 labeled field recordings with variable duration at 96 kHz [8]. From September to December 2021 and from January to May 2022, we conducted a PAM strategy on the Caribbean coast of Costa Rica to collect underwater field recordings by installing hydrophones in three protected areas: Tortuguero National Park (TNP), Barra del Colorado Wildlife Refuge (BCR), and Gandoca Lagoon at Gandoca-Manzanillo Wildlife Refuge (GMR). Five SoundTrap ST600 STD underwater sound recorders (Ocean Instruments, New Zealand) were deployed across the protected areas in a one-month PAM study. These recorders are designed for long-term deployments. Following the initial recordings, the hydrophones were rotated between different points of interest to get 10–20min recordings at 96 kHz for capturing ultrasonic manatee vocalizations [8]. We built a relational DB with structured query language (SQL) to manage PAM metadata and facilitate post-analysis of AMCM results with these long-term recordings. This DB in SQL stores information on protected areas, hydrophone coordinates, deployment times, and retrieval dates for each monitoring point.

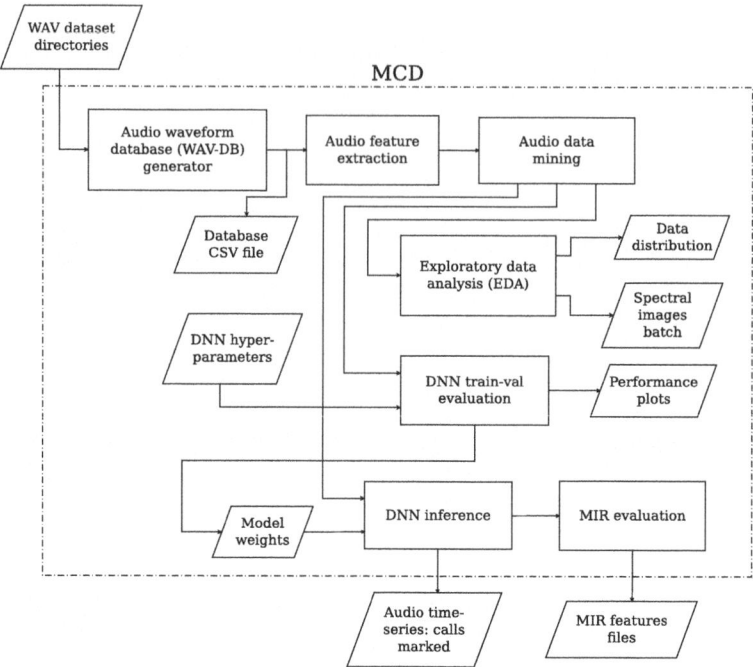

Fig. 1. General diagram that summarizes the MCD component for supervised binary classification and MIR evaluation of manatee calls in audio recordings.

2.3 Manatee Call Detector

Figure 1 illustrates the manatee call detector (MCD) of the proposed AMCM. The WAV-DB generator iterates through input directories containing labeled samples. Then, it creates a comma-separated value (CSV) file containing data-label pairs to perform an initial exploratory data analysis (EDA). The WAV-FE module processes each sample from the CSV file. Here, the 1-dimensional (1D) audio signals are transformed into 2-dimensional (2D) spectrograms. This stage employs audio processing techniques like as onset detection to identify potential manatee calls within the audio by tracking energy peaks [5]. It also considers audio enhancement to reduce noise and isolate harmonic components characteristic of marine mammal calls [9,10,14,23,24]. The WAV-FE incorporates resampling, windowing, and zero-padding to generate audio segments with consistent dimensions. Following this, the spectral magnitude of the audio signals is computed to generate spectrograms. A color channel is then added, resulting in a spectral image. Finally, the generated spectral images are stacked to form a large array of the complete dataset. The data mining (Fig. 1) implements class encoding, converting target labels into a numerical format. Subsequently, vectorization transforms the array of spectrograms into a large data vector. Data scaling ensures a consistent range of values across all features in the dataset. This preprocessed data is then split into training, validation, and testing sets for

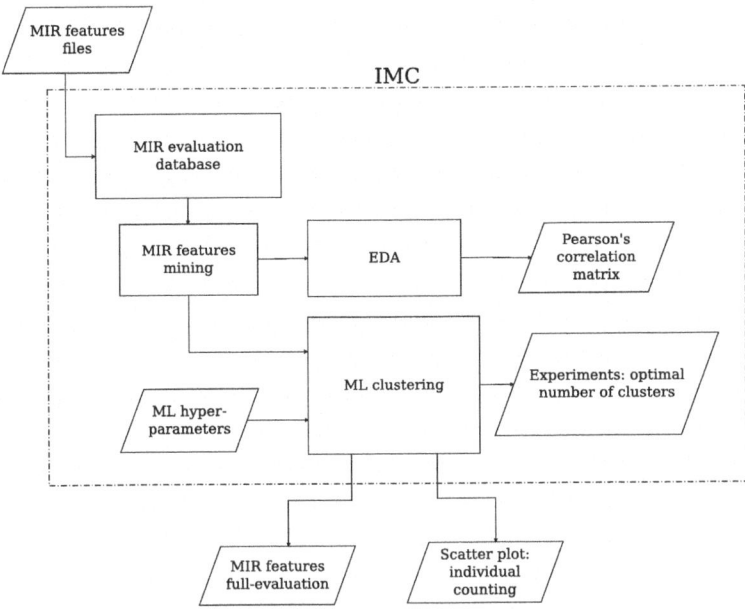

Fig. 2. Diagram of IMC module for manatee individual counting using acoustic features of the vocalizations detected based on unsupervised learning methods.

feeding the DNN architecture. To understand the data characteristics after its mining, the EDA retrieves visualizations that depict batches of spectrograms. The MCD constructs and compiles the model with its hyperparameters defined arbitrarily. The DNN undergoes supervised training with validation, and subsequently stores the trained weights for the binary classifier. To evaluate its performance, the weights are loaded and the testing data is passed through the trained model. During inference (Fig. 1), the MCD iterates through an input directory with experimental recordings. An adaptation of the WAV-FE produces fixed-size spectrograms by generating pre-defined audio segments of the input recording. The loaded model predicts the class between true or false manatee calls by processing a data vector containing the stacked spectrograms. The inference results are then used to generate an audio transcription and a time-domain visualization with the identified calls marked. Finally, the MCD performs the MIR evaluation, generating a numerical transcription of the manatee call estimations.

2.4 Individual Manatee Counter

The individual manatee counter (IMC) (Fig. 2) initiates the data preprocessing phase for the MIR evaluation files. This stage establishes a structured dataset that catalogs all identified manatee vocalizations and their associated acoustic characteristics derived from various field recordings. The features encompass two primary categories: temporal and spectral. In total, the MIR evaluation considers

11 attributes. 3 of them represent numerical values within the temporal domain, including the signal's F_0 in Hz, estimated with the probabilistic YIN method [18]. The mean energy level expressed in dB and the zero-crossing rate (ZCR) is defined as the number of times the signal crosses a zero value per second. The remaining attributes correspond to the spectral domain and include mean values of the spectrogram's RMS magnitude, spectral flatness (dB), also dimensionless features like the spectrogram's kurtosis and skewness, and the spectral centroid, roll-off, contrast, and bandwidth (BW)—all measured in Hz. The respective data mining not only involves scaling but dimensionality reduction, like principal components analysis (PCA) or non-negative matrix factorization (NMF), to enhance clustering performance. Also, invalid values handling implies removing observations with not-a-number (NaN) values for any descriptor, since invalid MIR features imply incorrect manatee calls. The EDA in Fig. 2 calculates the Pearson's correlation matrix to assess potential redundancy between MIR descriptors. Finally, the clustering module employs a ML model to analyze the MIR dataset. Our approach uses an automatic pre-specification of the anticipated number of clusters or expected individuals. The IMC fine-tunes the model by retraining it with the k-most relevant attributes of the MIR dataset. We determined these features based on the predictions made and the acoustic data. Figure 2 includes an EDA component to visualize the manatee counting with a bivariate analysis that plots the distribution of the estimations. This visualization depicts the observations with their prediction labels written into the complete transcription of the MIR evaluation for a set of recordings.

3 Experimental Results

Table 1 details the hardware and software used (programming language, libraries). Focusing on the execution time of critical stages of the MCD, we considered the V100 NVIDIA-GPUs suited for DL algorithms. The WAV-FE stage took 17 min and 37 s to process our personalized DB and 18 min with 15 s for model training and validation. Manatee inference on 10 variable-length field recordings took 15 min and 18 s. To train the supervised MCD model, we built from the 4 DBs available a custom dataset of 7344 audio samples (5324

Table 1. Hardware and software components used in each system allocation to run the complete AMCM pipeline.

allocations	CPUs	memory	GPUs	OS	software
Kabré	2 Intel Xeon Silver 4214R CPU@2.40 GHz	31 GiB	1 T V100-PCIE-32 GB	Linux-64bits	Python 3.10, librosa 0.10.1, CUDA 11.8, cuDNN 8.4, TensorFlow 2.14.
JUWELS-BOOSTER	2 AMD EPYC 7402 24-Core	514 GiB	4 NVIDIA A100-SXM4-40GB	Linux-64bits	Python 3.10, librosa 0.9.2, CUDA 11.7, cuDNN 8.6, TensorFlow 2.11.
Laptop	1 11th-Gen Intel Core i5-11300H@3.10 GHz	16 GiB	1 NVIDIA GeForce GTX 1650	Linux-64bits	Python 3.11, librosa 0.10.1, CUDA 12.1, cuDNN 8.9, TensorFlow 2.14.

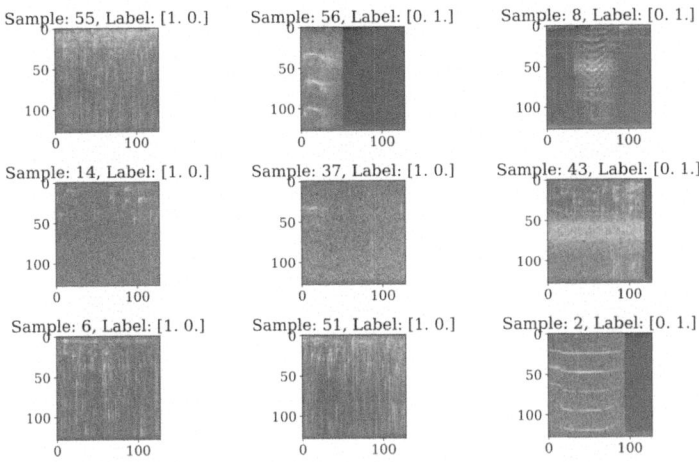

Fig. 3. Training data batch and the binary label for each image, where the x-axis corresponds to its height, and the y-axis to its width, both in number of pixels.

Table 2. Confusion matrix for the MCD model to identify the presence of the manatee in an evaluation set of labeled audio samples.

		Predicted	
		false vocalization	true vocalization
Actual	false vocalization	1568 (0.98)	30 (0.02)
	true vocalization	54 (0.09)	552 (0.91)

Table 3. Prediction metrics as a binary classification report for true and false manatee calls using pre-trained weights of large-scale DL model.

class	precision	recall	f1-score
false vocalization	0.97	0.98	0.97
true vocalization	0.95	0.91	0.93

false vocalizations, 2020 true vocalizations). The main configurable options of the WAV-FE that influence detection's performance correspond to the sampling rate, FFT parameters, image dimensions (i.e., height, width, and depth), and data resolution. The experimental setup consisted of resampling at 32 kHz, along with call pre-detection and zero padding steps to create signal windows of 0.512 s. Then, it converted the signal segments into spectrograms using successive FFTs of 254×128 size at 32-bit floating point. Figure 3 depicts a data batch of color images (128×128×3) with the binary labels as a one-hot scheme: [1. 0.] for false vocalization and [0. 1.] for true vocalization. Data mining involved scaling and splitting the labeled dataset. Standardization was applied to approximate a Gaussian distribution, a common practice in DL methodologies. Random strat-

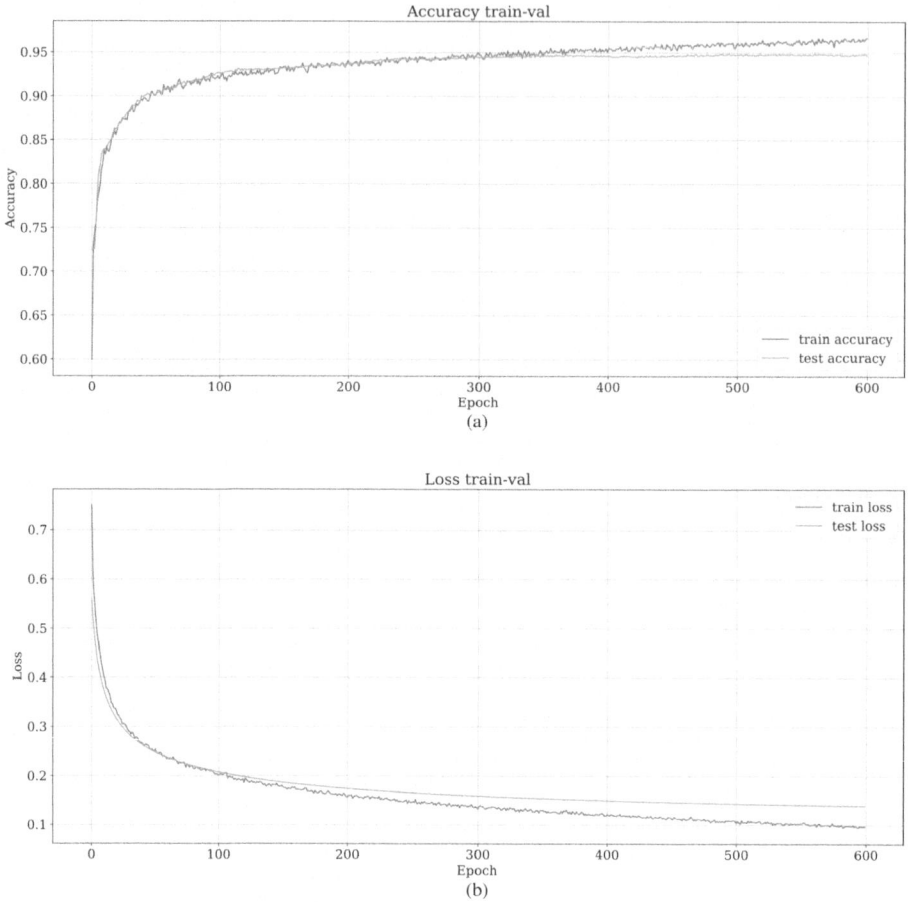

Fig. 4. Training and validation metrics of the DNN model implemented in the MCD module: (a) accuracy plot and (b) loss function.

ified splitting (70%/30%) yielded 3598 training samples (2608 false, 990 true calls), 1542 validation samples (1118 false, 424 true calls), and 2204 evaluation samples (1598 false, 606 true calls). We explored transfer learning with large-scale DL model, motivated by its success with the African manatee, and other cetaceans [16,21,25,30]. Its key configurable options were determined by trail-and-error including hyper-parameters, such as the number of frozen layers, the top classifier architecture, the number of neurons, and fine-tuning arguments (i.e., dropout, batch size, epochs) alongside the optimizer's learning rate and momentum. Due to the promising result of beluga sound classification using pre-trained weights of large-scale DL models [30], we present a transfer learning assembly that consists of a pre-trained VGG-16 with ImageNet [26]. This base model has an input shape of 128×128×3 that serves as the feature extractor, followed by fine-tuning a multi-layer perceptron (MLP) classifier. A data flatten-

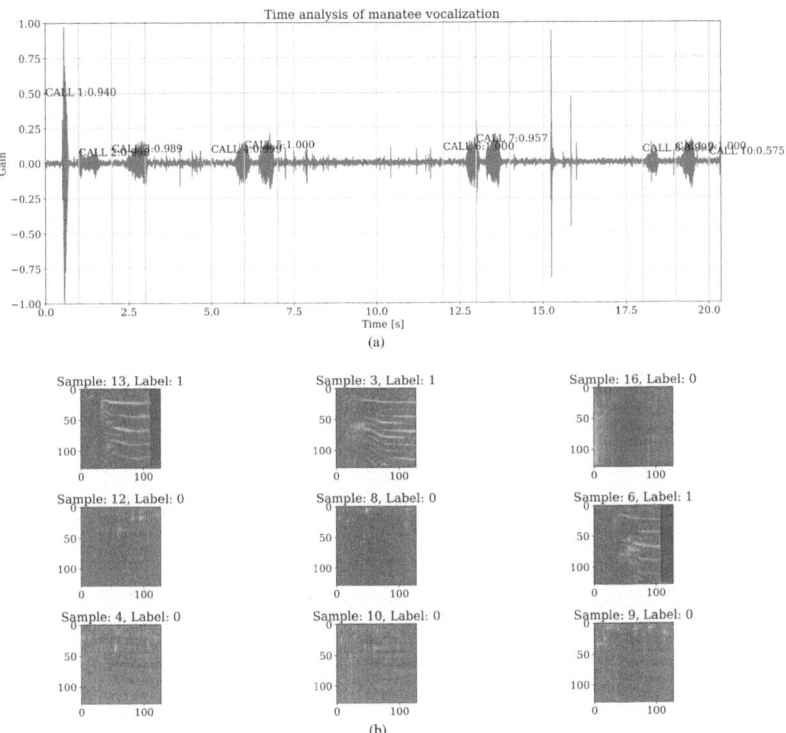

Fig. 5. Inference results of the MCD with a labeled testing sample: (a) time domain plot with the calls detected and (b) random spectral images with its resulting label.

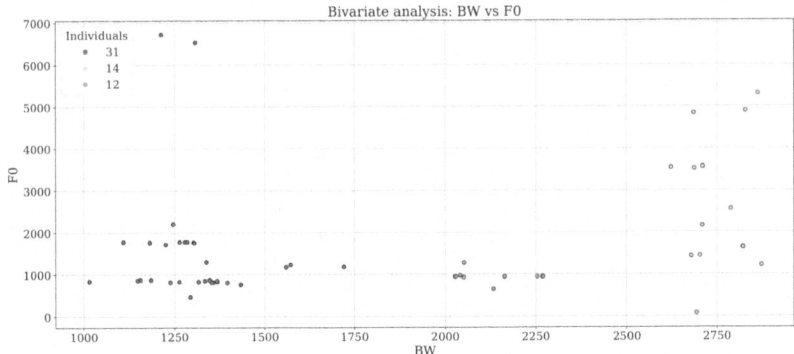

Fig. 6. Bivariate analysis of the results of counting manatees in an experimental set of recordings using the IMC module.

ing layer vectorizes pre-trained features, feeding into an MLP with 256 neurons, a 0.6 dropout layer to prevent overfitting, and a final MLP with an output size of 2 units for binary classification. During training, the model compiler uses the stochastic gradient descent (SGD) as an optimizer with a momentum of 0.55 and a learning rate 0.0001 to balance convergence speed and stability. We fine-tune the frozen base model (except the last 2 layers) with batch size 64 to speed up CNN operations. Figure 4 shows training and validation metrics after 600 epochs for optimal fine-tuning of the learned features to the specific spectral images. The evaluation results of the loaded model were an overall accuracy of 0.962, a binary cross-entropy of 0.098, and a AUC-ROC score of 0.946, as the capability to distinguish between false and true manatee calls. According to the binary confusion matrix (Table 2) along with the classification report of Table 3, our model produces 1568 TNs and 30 FNs from 1598 false vocalizations, and it generates 552 TPs and 54 FPs from 606 true vocalizations. To validate the MCD inference, we analyzed an experimental split from the BCDB that consists of 10 WAV audio recordings, labeled with the number of expected calls. Figure 5.a corresponds to the time-domain visualization of a testing sample that indicates 9 TP and 1 FP calls, where the FP has a low prediction probability (0.575). Figure 5.b depicts spectrograms with their label estimations. The system generates 10 MIR transcriptions with the corresponding calls detected for the experimental set. We tuned IMC parameters with trial-and-error, such as the data scaler, the unsupervised ML models (e.g., GMMs, HDBSCAN, and k-means (KM)) used, and the descriptors selection or the dimensionality reduction with principal components analysis (PCA) or non-negative matrix factorization (NMF). The IMC created a data structure from MIR transcriptions with 68 observations and 11 MIR attributes. Under this unlabeled data, 11 samples have invalid values (NaNs) for the F_0 attribute, associated with FP candidates (i.e., non-manatee vocalization or underwater noise). The MIR dataset (57×11) was then vectorized and normalized. Feature selection based on Gini importance [7] from the MIR evaluation identified the top-3 relevant spectral attributes: BW, spectral centroid, and spectral roll-off that exhibited high positive correlations (0.92 and 0.98). Due to the reduced number of observations in the MIR dataset and its low dimensionality (57×3), we used the KM model that implements the EM algorithm to count the individuals based on acoustic information. Our KMCA facilitated the inertia evaluation (i.e., the sum-of-squares distance within clusters) to calculate the optimal number of groups. The clustering strategy identified automatically 3 individuals in the MIR dataset (Fig. 6). The silhouette coefficient supported the clustering performance (0.722). Spectral BW and F_0 analysis revealed distinct clusters that may correspond to individual manatees. The first manatee (red, 31 calls) produced vocalizations with a mean F_0 of 1531.67 Hz and BW between 1–1.75 kHz. The second one (yellow, 14 calls) had a mean F_0 of 2.7 kHz and BW between 2.5–3 kHz. Finally, the third manatee (green, 12 calls) displayed a mean F_0 of 949.35 Hz and BW between 2–2.25 kHz.

4 Discussion

We leveraged supercomputing infrastructures with GPU units to address limitations in denoising and call detection of the original AMCM [8] on diverse audio samples and long-term recordings. Kabré provided a suitable execution environment with wide storage for training/validation data and PAM-DBs. The MCD modules required a total time of 35 min and 51.87 s for training and evaluation, and inference exceeded 20 min on an experimental dataset of varying length. These findings highlighted performance constraints in the inference stage compared to other AMCM components. The custom labeled DB exhibited variations in duration and sampling rate for audio samples. To achieve consistency within the windowing analysis, we downsampled all recordings. The training batch (Fig. 3) captured the spectral content of manatee calls within an approximated duration of 100–400 ms [6,22]. Despite our weakly labeled DB, the onset detection step improved the evaluation metrics and the inference results of the MCD model. Pre-processing the 1D audio signal in both time and spectral domains before 2D-conversion did not enhance the MCD performance and notably hindered WAV-FE step. Therefore, unaltered spectrograms (Fig. 3) were used for data batch generation. A CNN architecture initially suffered from overfitting due to a limited labeled DB. Transfer learning was implemented to address this issue, resulting in an acceptable fit between training and validation metrics (Fig. 4). Although slight overfitting persisted, it did not significantly affect model performance during evaluation. This was confirmed by achieving a binary accuracy of 96% with an f1-score of 93% for TPs to get a well-balanced between precision and recall. Our metrics surpassed the former AMCM (f1-score of 90% for true vocalizations) [8] and outperformed the end-to-end transfer learning for binary classification of captive Antillean manatees (accuracy slightly above 91%) [25], demonstrating the effectiveness of the pre-trained weights from very large-scale CNN architectures [26]. We preserved data imbalance to increase noise sensitivity in our model by training with more TNs than TPs improving TN rates without increasing FNs. Regarding the IMC behavior, F_0 with NaN values are related to non-manatee sounds causing clustering errors. Figure 5 suggested potential FPs from the F_0, where the expected manatee vocalization range is around 2 kHz–4 kHz [6]. The corresponding ML model that showed the best clustering performance was the KM model compared with GMMs and hierarchical clustering. We identified strong redundancy among MIR attributes using Pearson's coefficients that selected a more highly correlated subset for KMCA. This reduced set improved clustering metrics compared to using all 11 MIR descriptors. The feature selection with Gini importance had better clustering metrics than 2D features decomposition like PCA or NMF. While Fig. 6 shows clustering results below 80% accuracy, distinct call groups emerged within 3 BWs with varying mean F_0s. A manatee with a high-pitched F_0 (2.7 kHz) might be related to a female [27], or a calf since manatees with smaller body sizes produce calls at higher frequencies [6]. And lower-pitched F_0s (950 Hz, 1 kHz–1.5 kHz) could imply larger body size manatee vocalizing since it produces calls with lower frequencies [6]. The overall results of the proposed AMCM (57 calls, 3 manatees) are similar to the original solution (54 calls, 4 manatees), considering the same set of 10 experimental recordings from the BCDB [8]. Our PAM data from Costa Rica's Caribbean provided long-term data for evaluating AMCM. However, the MCD faced challenges due to underwater noise from navigation and potentially suboptimal PAM site placement. Ideally, hydrophone positioning near food sources or further offshore could improve call reception. Additionally, the training set generated from the 4 DBs available focus on clean tonal calls limited effectiveness in real-world scenarios with less harmonic sounds [27]. Bio-acoustic cross-validation revealed constraints with true calls masked by chewing sounds or vibrations. While MCD achieved acceptable identification for confirmed calls (>75% probability), FPs remained an issue. The IMC module addressed this by treating observations with NaNs, enabling valid acoustic abundance and population size estimations.

4.1 Conclusions and Future Work

This study considers large-scale transfer learning and clustering techniques to present a data-driven pipeline for manatee identification and counting in audio recordings. It combines DL for binary call detection and ML clustering to predict which calls are from the same individual. Addressing the functionality and performance limitations found, we implemented efficient AI tools for noise reduction and call processing of the previous AMCM. HPC enabled training, evaluation, and long-term call inference experiments. While complex underwater conditions limited MCD results in some recordings, the method's ability to count individuals based on BWs remains unaffected. Robust post-analysis of the AMCM results in PAM samples can increase manatee's ecological knowledge. This is valuable information for the manatee conservation and their ecosystems across the Caribbean coast of Costa Rica. Future efforts include continuing post-analysis of AMCM using Costa Rica's PAM metadata under our experimental DB in SQL. To enhance model throughput, increasing the number of labeled samples with human-annotated data from the Caribbean coast would be beneficial. Exploring data augmentation layers and synthetic data with target labels for transfer learning while mitigating overfitting, and optimizing PAM deployment for better signal quality. These advancements will enhance the pipeline's performance and contribute to manatee conservation efforts on the Caribbean coast of Costa Rica.

Acknowledgments. This research was supported by the National Geographic Society (NGS-84535T-21) and the AI4Earth program (69005a29-9390-4178-b6a2-4e4b68b470c6). We thank the Advanced Computing Laboratory at CeNAT for access to the Kabré infrastructure. We thank the ZooTampa project and the creators of the HaikuMarine system (David Mann and Austin Anderson) for sharing data. We also thank the SBLN and the manatee DTAG project for sharing the data collected by Florida Fish and Wildlife Conservation Commission staff and partners. Fieldwork was facilitated by the administrations of TNP, BCR, and GMR, along with local collaborators who volunteered their time. We are grateful to Dr. Stefan Kesselheim and the Jülich Supercomputing Center (JSC) team at FSZJ for their guidance in DL matters and granting access to JUWELS-BOOSTER supercomputer.

References

1. Declaratorio del manatí Antillano (*Trichechus manatus manatus*) como símbolo nacional de la fauna marina de Costa Rica (2014). http://www.pgrweb.go.cr/scij/Busqueda/Normativa/Normas/nrm_texto_completo.aspx
2. Human activity devastating marine species from mammals to corals - IUCN Red List (2023). https://www.iucn.org/press-release/202212/human-activity-devastating-marine-species-mammals-corals-iucn-red-list
3. Bianco, M.J., et al.: Machine learning in acoustics: theory and applications. J. Acoust. Soc. Am. **146**(5), 3590–3628 (2019)
4. Bittle, M., Duncan, A.: A review of current marine mammal detection and classification algorithms for use in automated passive acoustic monitoring. In: Proceedings of Acoustics, vol. 2013. Citeseer (2013)
5. Böck, S., Widmer, G.: Maximum filter vibrato suppression for onset detection. In: Proceedings of the 16th International Conference on Digital Audio Effects (DAFx), Maynooth, September 2013, vol. 7, p. 4 (2013)
6. Brady, B.: Manatee calf call contour and acoustic structure varies by species and body size. Sci. Rep. **12**(1), 19597 (2022)
7. Breiman, L.: Random forests. Mach. Learn. **45**, 5–32 (2001)
8. Castro, J.M., Rivera, M., Camacho, A.: Automatic manatee count using passive acoustics. In: Proceedings of Meetings on Acoustics 169ASA, vol. 23, p. 010001. Acoustical Society of America (2015)
9. Driedger, J., Müller, M., Disch, S.: Extending harmonic-percussive separation of audio signals. In: ISMIR, pp. 611–616 (2014)
10. Fitzgerald, D.: Harmonic/percussive separation using median filtering (2010)
11. Fleishman, E., et al.: Ecological inferences about marine mammals from passive acoustic data. Biol. Rev. **98**(5), 1633–1647 (2023)
12. Freitas, K.: Detecção de zoonoses em carnes de caça comercializadas na região do médio rio solimões–coari-am (2023)
13. Keith Diagne, L.: Trichechus senegalensis. The IUCN Red List of Threatened Species, pp. 2015–4 (2015)
14. Lim, J.S.: Two-dimensional Signal and Image Processing. Englewood Cliffs (1990)
15. Lin, M., et al.: Functional extinction of dugongs in china. Royal Society Open Science **9**(8), 211994 (2022)
16. Lu, T., Han, B., Yu, F.: Detection and classification of marine mammal sounds using AlexNet with transfer learning. Eco. Inform. **62**, 101277 (2021)
17. Marsh, H.: *Dugong dugon* (amended version of 2015 assessment). The IUCN Red List of Threatened Species 2019, pp. e–T6909A160756767 (2019)
18. Mauch, M., Dixon, S.: pYIN: a fundamental frequency estimator using probabilistic threshold distributions. In: 2014 IEEE International Conference on Acoustics, Speech and Signal Processing (ICASSP), pp. 659–663. IEEE (2014)
19. May-Collado, L.: Marine mammals. In: Wehrtmann, I.S., Cortés, J. (eds.) Marine Biodiversity of Costa Rica, Central America, pp. 479–495. Springer, Dordrecht (2009). https://doi.org/10.1007/978-1-4020-8278-8_45
20. Merchan, F., Guerra, A., Poveda, H., Guzmán, H.M., Sanchez-Galan, J.E.: Bioacoustic classification of Antillean manatee vocalization spectrograms using deep convolutional neural networks. Appl. Sci. **10**(9), 3286 (2020)
21. Rycyk, A., Bolaji, D.A., Factheu, C., Kamla Takoukam, A.: Using transfer learning with a convolutional neural network to detect African manatee (*Trichechus senegalensis*) vocalizations. JASA Exp. Lett. **2**(12) (2022)

22. Rycyk, A.M., et al.: Manatee behavioral response to boats. Mar. Mamm. Sci. **34**(4), 924–962 (2018)
23. Sainburg, T.: Timsainb/noisereduce: v1.0 (2019). https://doi.org/10.5281/zenodo.3243139
24. Sainburg, T., Thielk, M., Gentner, T.Q.: Finding, visualizing, and quantifying latent structure across diverse animal vocal repertoires. PLoS Comput. Biol. **16**(10), e1008228 (2020)
25. Schneider, S., Von Fersen, L., Dierkes, P.W.: Acoustic estimation of the manatee population and classification of call categories using artificial intelligence. Front. Conserv. Sci. **5**, 1405243 (2024)
26. Simonyan, K., Zisserman, A.: Very deep convolutional networks for large-scale image recognition. arXiv preprint arXiv:1409.1556 (2014)
27. Sousa-Lima, R.S., Paglia, A.P., da Fonseca, G.A.: Gender, age, and identity in the isolation calls of Antillean manatees (Trichechus manatus manatus). Aquat. Mamm. **34**(1), 109–122 (2008)
28. Turvey, S.T., Risley, C.L.: Modelling the extinction of Steller's sea cow. Biol. Let. **2**(1), 94–97 (2006)
29. Usman, A.M., Ogundile, O.O., Versfeld, D.J.: Review of automatic detection and classification techniques for cetacean vocalization. IEEE Access **8**, 105181–105206 (2020)
30. Zhong, M., Castellote, M., Dodhia, R., Lavista Ferres, J., Keogh, M., Brewer, A.: Beluga whale acoustic signal classification using deep learning neural network models. J. Acoust. Soc. Am. **147**(3), 1834–1841 (2020)

Exploring Neural Joint Activity in Spiking Neural Networks for Fraud Detection

Dylan Perdigão[✉][iD], Francisco Antunes[iD], Catarina Silva[iD], and Bernardete Ribeiro[iD]

Department of Informatics Engineering, Centre for Informatics and Systems of the University of Coimbra, University of Coimbra, Coimbra, Portugal
{dgp,fnibau,catarina,bribeiro}@dei.uc.pt

Abstract. Spiking Neural Networks (SNNs), inspired by the real brain's behavior, offer an energy-efficient alternative to traditional artificial neural networks coupled with their neural joint activity, also referred to as population coding. This population coding is replicated in SNNs by attributing more than one neuron to each class in the output layer. This study leverages SNNs for fraud detection through real-world datasets, namely the Bank Account Fraud dataset suite, addressing the fairness and bias issues inherent in conventional machine learning algorithms. Different configurations of time steps and population sizes were compared within a 1D-Convolutional Spiking Neural Network, whose hyperparameters were optimized through a Bayesian optimization process. Our proposed SNN approach with neural joint activity enables the classification of fraudulent opening of bank accounts more accurately and fairly than standard SNNs. The results highlight the potential of SNNs to surpass non-population coding baselines by achieving an average of 47.08% of recall at a business constraint of 5% of false positive rate, offering a robust solution for fraud detection. Moreover, the proposed approach attains comparable results to gradient-boosting machine models while maintaining predictive equality towards sensitive attributes above 90%.

Keywords: Spiking Neural Networks · Population Coding · Fraud Detection · Energy Efficiency · Responsible AI · Fair ML

1 Introduction

Home banking solutions have gained significant popularity in recent years due to technological advancements. However, these solutions have also become excellent targets for fraudsters, resulting in considerable financial losses within the sector. Despite the effectiveness of Artificial Intelligence (AI) algorithms in detecting such frauds, they often suffer from energy inefficiency, primarily due to the algorithm's extensive computational abundance of floating point operations [20]. As a possible solution, the popularity of Spiking Neural Networks was raised to fulfill the high energy demand [17].

Spiking Neural Networks (SNNs) have the particularity to mimic the brain as a complex network of neurons interacting through electrochemical signals and a coordinated population of cells to process information [10]. Inspired by this efficiency, SNNs are more energetically viable than traditional Artificial Neural Networks (ANNs), being remarkably power-efficient when implemented on neuromorphic hardware [6]. Moreover, brain neurons do not work in isolation, which means that an entire area of the brain is activated with synaptic impulses of spiking neurons [12,18]. This coordinated activity underscores the importance of population coding, where groups of neurons collectively represent and process information, offering enhanced robustness, efficiency, and accuracy compared to the simplistic approach of assigning one neuron per target class in Machine Learning (ML) problems.

In this work, we study SNNs coupled with their neural joint activity to improve financial fraud detection within a real-world benchmark problem using the Bank Account Fraud (BAF) dataset suite [13]. The improvement consists in increasing the True Positive Rate (TPR) of fraud detection, constrained with 5% of False Positive Rate (FPR). This threshold enables the model to minimize the impact on non-fraudster clients, making it suitable for deployment in a production environment. In addition, we develop a fair model [3] that does not discriminate against clients by age, income, or employment status. Our proposed SNN aims to overcome these limitations by offering a more equitable and energetically efficient solution while providing comparable performance with unfair Gradient-Boosting Machines (GBM) models and more performant than SNNs without neural joint activity.

This paper is structured as follows. Section 2 presents the functioning of SNNs. Then, Sect. 3 overviews the main works done for fraud detection, namely with the BAF dataset. Sections 4 and 5 describe our approach and the experimental setup, whose results are discussed in Sect. 6. Finally, Sect. 7 concludes this work and presents future directions.

2 Background

In SNNs, the exchanges of chemical ions between neuronal cells enable the production of electrical impulses that occur at time t when the potential U surpasses the threshold θ of the membrane, which has a decay rate β. Modeled as an electrical circuit the Leaky Integrate-and-Fire (LIF) neuron [11,15] consists in a capacitor C in parallel with a resistor R driven by a current I [9], described by the differential equation:

$$\tau_m \frac{dU}{dt} = -U(t) + RI(t) \qquad (1)$$

where $\tau_m = RC$ is the membrane time constant of the neuron. This system can be approximated using the forward Euler method for compatibility with sequential networks. In the simplified model, the input current combines the weighted sum of inputs WX of the neural network, with the potential represented as:

$$U[t] = \underbrace{\beta U[t-1]}_{\text{decay}} + \underbrace{WX[t]}_{\text{input}} - \underbrace{S_{\text{out}}[t-1]\theta}_{\text{reset}} \qquad (2)$$

The raw input is converted into a matrix X of spikes, where the superposition of spikes on each layer of the SNN creates the current I at time t. This current is translated into the potential of the membrane U, generating spikes for the next layer. The reset is defined by subtracting the threshold via the Heaviside step function of the last recorded spike, as shown in Fig. 1.

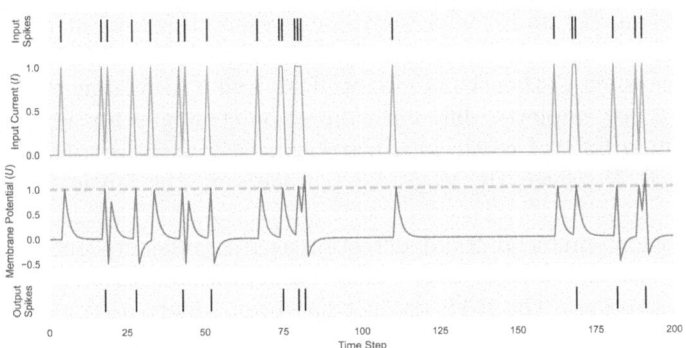

Fig. 1. Simulation of a LIF neuron, modeling the input and output spikes, the input current I, and the membrane potential U.

This modeling of the neuron's membrane allows for classification tasks by counting the number of spikes in the final layer of the ANN, with the predicted class having the highest number of spikes. With population coding, assigning multiple neurons of the last layer population P to a specific class is possible, mimicking the real behavior of the brain's neurons [10].

3 Related Work

Fraudulent activities have become a significant concern for financial institutions. Several works have been proposed to improve fraud detection in this sector, namely within the recent BAF dataset developed by Jesus et al. [13], which has been used to improve fraud detection with highly imbalanced data. Moreover, this dataset, which will be used in our work, is biased toward groups of people, being a benchmark to assess the performance and fairness of the models. As a baseline, the dataset was primarily assessed using Light Gradient-Boosting Machine (LightGBM) algorithm [14], with performance varying between 25% and 75% of recall at 5% of FPR, and fairness between 30% and 75% of predictive equality [21], depending on the dataset variant. Later, they improved the LightGBM algorithm by introducing Fair Gradient-Boosting Machine (FairGBM) [5], a fairer gradient-boosting model that uses Lagrangian-constrained optimization to ensure fairness. However, Fair Gradient-Boosting Machine (FairGBM) was only tested on datasets such as the Account Opening Fraud (AOF) and other American Community Survey (ACS)-based datasets [8].

Other works have used strategies to improve fraud detection performance on the BAF dataset. For instance, Uwaoma [25] analyzed multiple algorithms, their strengths and limitations, and the fairness of the models. He provided insights into several practical techniques for improving fraud detection with the BAF dataset. As a result, he fine-tuned different models as the logistic regression performed with 44% of recall, 11% with a decision tree, 48% with a random forest, and 47% with LightGBM. Regarding fairness, the fine-tuned model achieves a predictive equality above 95% for those models. Luzio et al. [16] also used the BAF dataset with the business restriction and improved the performance of ML models, such Catboost [23], LightGBM, and a Multi-Layered Perceptron (MLP) [22]. They employed different calibration strategies for increased performance, reaching 52% of recall with Catboost, 54% with LightGBM, and 49% with the MLP. However, the work did not measure the fairness of the tested algorithms.

In conclusion, financial fraud detection is a significant concern for financial institutions, and several works have been proposed to address this issue. Despite of its recency, the BAF dataset has been widely used as a benchmark to improve fraud detection performance, but only a few attempted to address fairness in their algorithms. Further research is still needed to enhance fairness while ensuring the maximization of the recall at the 5% of FPR, the business constraint for such critical application of models. Regarding SNNs, there is a gap in using these networks for fraud detection, namely with the BAF dataset. Nevertheless, some work explored SNNs in other fraud detection problems, such as website phishing detection [2], or anomaly detection for critical infrastructures [7], where population coding is employed.

4 Proposed Approach

Our proposed approach involves a two-stage process to optimize and evaluate the 1D-Convolutional Spiking Neural Network (CSNN) model of Fig. 2 using Bayesian optimization[1]. The model is composed of two convolutional layers, with 32, 128, and 256 neurons, respectively, and a kernel size of 2. The fully connected layer is flattened, resulting in 768 neurons connected to P outputs, the population of our SNN. Consequently, we assign $P/2$ neurons to each data class (fraud and non-fraud) as we study the neural joint activity.

4.1 Hyperparameter Optimization

In the first stage, the hyperparameters of the CSNN model were optimized with six different configurations, varying the number of steps ($S \in \{20, 50\}$) and the population size of the last layer of the network ($P \in \{2, 20, 200\}$). The multi-objective Bayesian optimization was performed with a Tree-structured Parzen

---

[1] Experimental work available on Github at https://github.com/DylanPerdigao/Neural-Joint-Activity-in-SNNs, and archived on Zenodo at https://doi.org/10.5281/zenodo.13546088.

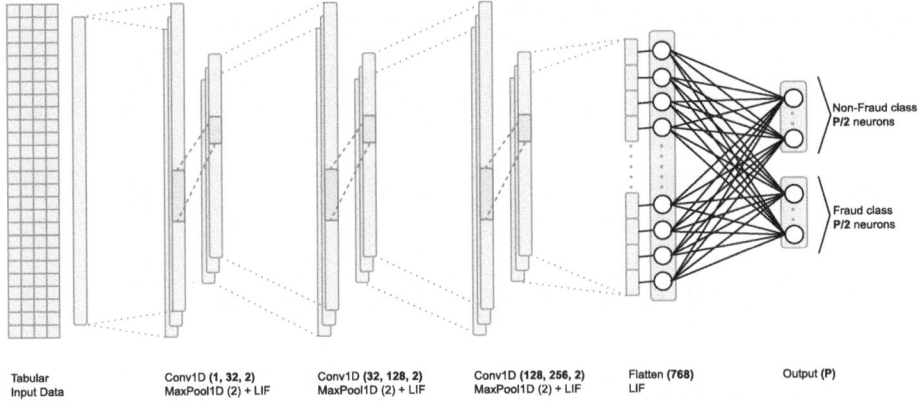

Fig. 2. 1D-Convolutional Spiking Neural Network architecture used for the experiments with a population P on the Fully Connected layer.

Estimator (TPE) sampler, enabling the identification of the optimal hyperparameter values of the 1D-CSNN through the Optuna[2] Python library [1]. The objectives consisted of maximizing the recall and minimizing the FPR. This approach allowed us to efficiently navigate the hyperparameter space of Table 1 and identify the best recall at a specific point of the Receiver Operating Characteristic (ROC) curve, which is 5% of FPR business constraint. The TPE sampler was chosen to explore the hyperparameter space as the SNNs are extremely sensitive to their hyperparameters, requiring fewer runs to converge to the objective. The best combination for each configuration (steps/population) was identified over 1,000 trials (runs) by fixing the batch size at 1024 and the number of epochs at 5. The specific optimized hyperparameters optimized include the decay ($\beta_1, \beta_2, \beta_3, \beta_4$) and threshold ($\theta_1, \theta_2, \theta_3, \theta_4$) of the neuron's membrane, the slope (σ) of the sigmoid function of the surrogate gradient. We use a surrogate gradient to approximate the backpropagation derivative since SNNs are not differentiable [19]. Finally, the Adam optimizer parameters, such as the weight of the minority class (ω), betas (β_A, β_B), and learning rate (λ), are also optimized. The objective of this optimization is to find the set of hyperparameters that maximizes the performance of the CSNN model at a particular point on the ROC curve (5% of FPR).

4.2 Best Configurations Evaluation

In the second stage, we evaluate the CSNN model using the best hyperparameters obtained from the Bayesian optimization for each of the six configurations. With these optimized hyperparameters, we run the CSNN model 100 times on each dataset variant. This extensive evaluation allows us to compare the performance and robustness of the configurations across the different variants of the

---

[2] Optuna: https://optuna.readthedocs.io.

Table 1. Optimization search space

Hyperparameters	Search Space	Distribution
Decays (β_L)	0.1–0.95	log.
Slope (σ)	10–50	log.
Thresholds (θ_L)	0.1–1.0	log
Adam weight (ω)	0.95–1.0	log.
Adam betas (β_k)	0.97–0.99	uniform
Adam learning rate (λ)	10^{-6}–10^{-3}	log.

BAF dataset suite, ensuring that the model is optimized for a specific setting and performs well across varied data conditions. By following this approach, we aim to thoroughly explore the hyperparameter space and ensure that the resulting model parameters are robust and generalizable across different data scenarios. The process combines the precision of Bayesian optimization with extensive empirical validation, providing a comprehensive assessment of the model's performance.

5 Experimental Setup

The experimental setup to test our approach consisted of a server with an Intel(R) Xeon(R) Silver 4310 CPU @ 2.10 GHz, an NVIDIA RTX A6000 GPU, and 252 GB of RAM. The spiking neurons were implemented through the snnTorch[3] Python library [10] were assessed using the different BAF dataset variants.

5.1 Benchmark Dataset

The Bank Account Fraud[4] suite of datasets [13] is a fairness-oriented benchmark comprising six variants of real-world online bank account opening applications. The collected data is used to generate synthetic data through a Generative Adversarial Network (GAN), assuring the clients' anonymity. This collection comprises 30 features and eight months of data (1,000,000 instances) divided into six months of training (90%) and two months for testing (10%). The dataset variants present different types of biases and challenges for training models. All dataset variants comprise three sensitive attributes, such as the client's age, income, and employment status, whose fraud prevalence is presented in detail in Fig. 3. For this work, the groups are divided as follows:

- *Age*: Older people whose ages are above or equal to 50 years old and younger people are under;

---

[3] snnTorch: https://snntorch.readthedocs.io.
[4] Bank Account Fraud suite of datasets: https://www.kaggle.com/datasets/sgpjesus/bank-account-fraud-dataset-neurips-2022.

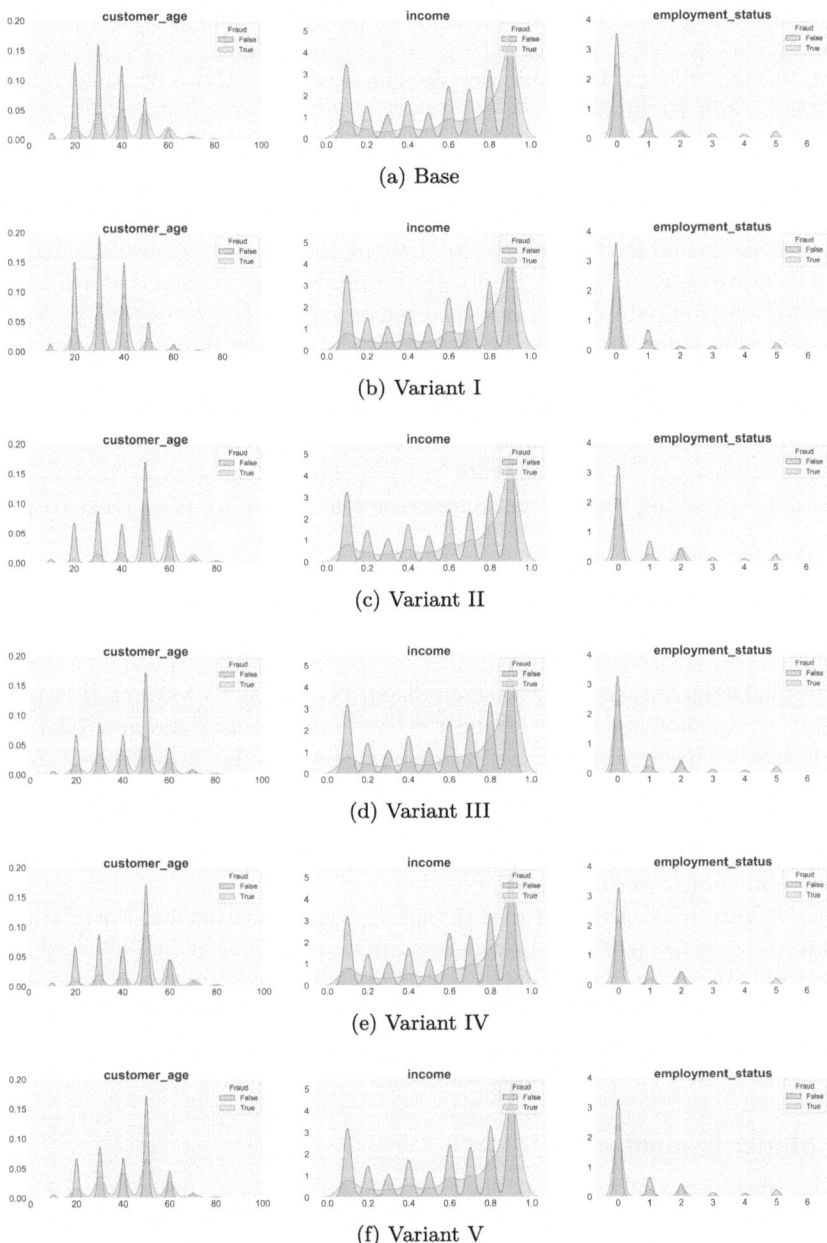

Fig. 3. Distribution of instances per sensitive attribute for each dataset variant. In red are represented fraudulent instances, and in blue are represented non-fraud instances. (Color figure online)

- *Income*: Rich people whose value is above or equal to 0.5 and poor people are under;
- *Employment Status*: Unstable people whose value is above or equal to 3 and stable people are under.

Those divisions enable us to compute the fairness of the models between groups with the FPR ratios between groups, also known as the predictive equality [4].

In Variant I, the dataset features a group size disparity where the minority group size is reduced from 20% to 10% of the dataset. However, the fraud prevalence rates are equal at 1.1% for both groups. This means that while models trained on this dataset will not face issues related to group-wise prevalence imbalance, they must be robust to the smaller size of the minority group, which may be underrepresented.

Variant II introduces a more pronounced prevalence disparity than Variant I and the base dataset. In this variant, the majority group has a fraud rate of 0.4%, while the minority group's fraud rate is 1.9%. The group sizes in Variant II are equal at 50%, making this dataset a stress test for handling prevalence disparity bias.

In Variant III, the dataset is characterized by separability disparity, where the majority group benefits from increased separability with two additional synthetic features. Both groups have equal sizes (50%) and equal fraud prevalence rates (1.1%).

Variant IV introduces a prevalence disparity similar to Variant II, with the majority group showing a 0.3% fraud rate and the minority group a 1.7% fraud rate. However, when adjusted for temporal aspects, the prevalence rates are equal at 1.5%. This setup mimics real-world scenarios where a biased training data collection process can impact model performance over time. Accounting for such temporal biases is essential for models trained on this variant to ensure realistic and robust performance.

Lastly, Variant V also features changes in data bias patterns over time. In this variant, group sizes and prevalence rates are balanced at 50% and 1.1%, respectively. However, the majority group benefits from increased separability (with the two additional features) in the first six months, which becomes equal afterward. Models must adapt to these changing bias patterns, addressing different types of bias in the training and testing periods.

5.2 Model Evaluation

After the Bayesian optimization, the best configurations of models are chosen to reproduce the experiments with 100 trials to evaluate the models more precisely. As an evaluation metric, we measure the performance with the recall (known as the TPR), which is closer to the 5% threshold of the FPR on the ROC curve:

$$TPR_{@5FPR} = \max_t TPR(t), \text{ where } \forall t,\ FPR(t) \leq 0.05 \qquad (3)$$

where the FPR and the TPR are computed as follow:

$$FPR = \frac{FP}{FP + TN} \qquad (4)$$

$$TPR = \frac{TP}{TP + FN} \quad (5)$$

In addition, we compute the fairness [4,21] of the models as the Predictive Equality (PE), which is the FPR ratio between two groups $(g1, g2)$ of a sensitive attribute $attr \in \{customer\_age, income, employment\_status\}$:

$$PE_{attr} = \begin{cases} \frac{FPR_{g1}}{FPR_{g2}}, & \text{if } FPR_{g2} \geq FPR_{g1} \\ \frac{FPR_{g2}}{FPR_{g1}}, & \text{otherwise} \end{cases} \quad (6)$$

This enables us to analyze, with Aequitas[5] Python library [24], the disparities of fraud between the customer's age, income, and employment status. Higher value implies fairness towards the sensitive attribute. Finally, the trade-off T between the performance and fairness of a sensitive attribute is computed as follows:

$$T_{attr} = \alpha \cdot TPR_{@5FPR} + (1 - \alpha) \cdot PE_{attr} \quad (7)$$

This metric uses $\alpha \in [0,1]$ to measure the trade-off for all importance given to the performance towards fairness. $\alpha = 0.5$ implies equal importance between performance and fairness.

6 Results and Discussion

The first experimentation consisted of the Bayesian optimization of the six configurations of steps $S \in \{20, 50\}$ and population $P \in \{2, 20, 200\}$, which resulted in the Pareto's fronts of Fig. 4. Empirically, for the Base dataset variant, we observe that using 20 steps and a population of 2 performs worse than configurations with more steps and a more significant population of neurons. The Pareto's front crosses the 5% FPR at 40% of recall while the other configurations cross near 50%. Consequently, the increasing number of steps and population of the neural activity helps the CSNN improve its recall performance. The best obtained parameters at this threshold are presented in Table 2 for each configuration.

The second experiment used the best hyperparameters for each configuration to measure the performance and fairness, resulting in Fig. 5. Using a population of 2, we observe a more significant standard deviation between points, whose best result is 34.96% of recall for Variant V with 20 steps. For a population of 20, this disparity in terms of recall is reduced, achieving at his best 43.15% of recall with Variant II and 50 steps. With a population of 200, the best overall result is 47.08% of recall with Variant II and 50 steps. Consequently, we observe that the number the CSNN achieves its best performance with a more significant population, while the number of steps contributes to reducing the disparities between points. Regarding fairness globally, the results are promising, with values above 90% for all sensitive attributes and indicating that the CSNN is adequate for this problem of bank account fraud. Table 3 presents performance and fairness

---

[5] Aequitas: https://dssg.github.io/aequitas/.

Table 2. Best hyperparameters resulting from the Bayesian optimization process

Hyperparameters		20 Steps			50 Steps		
		P = 2	P = 20	P = 200	P = 2	P = 20	P = 200
Decays	β_1	0.8502	0.9406	0.2624	0.8876	0.7504	0.7228
	β_2	0.8034	0.3749	0.7039	0.9375	0.8773	0.2994
	β_3	0.7913	0.1530	0.2622	0.8347	0.2536	0.5142
	β_4	0.7979	0.9497	0.8862	0.8439	0.9376	0.9603
Slope	σ	43	13	12	12	12	14
Thresholds	θ_1	0.4633	0.3603	0.1516	0.3369	0.3261	0.5725
	θ_2	0.2299	0.1057	0.4408	0.1028	0.8492	0.5919
	θ_3	0.6039	0.6745	0.1275	0.8687	0.4361	0.8743
	θ_4	0.2337	0.4338	0.5481	0.4018	0.8395	0.8419
Adam Weight	ω	0.9755	0.9604	0.9698	0.9707	0.9743	0.9762
Adam Betas	β_A	0.9830	0.9730	0.9790	0.9760	0.9780	0.9700
	β_B	0.9800	0.9850	0.9890	0.9710	0.9740	0.9740
Adam Learning Rate	λ	$0.2426e^{-3}$	$0.4680e^{-3}$	$0.2950e^{-3}$	$0.1807e^{-3}$	$0.7998e^{-3}$	$0.9235e^{-3}$

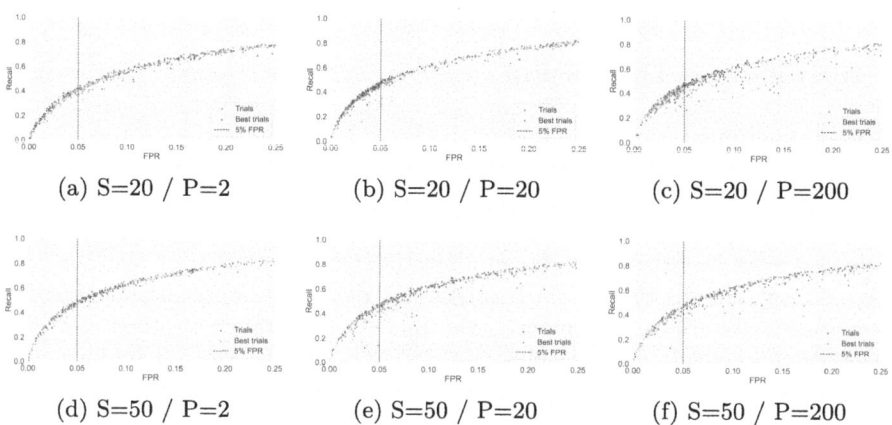

(a) S=20 / P=2   (b) S=20 / P=20   (c) S=20 / P=200

(d) S=50 / P=2   (e) S=50 / P=20   (f) S=50 / P=200

Fig. 4. Pareto's fronts of each configuration (steps S and population P) for the Base dataset variant with 5 epochs and 1024 of batch size.

results for each configuration and dataset. Moreover, compared with the LightGBM algorithm, recall varies between 32% and 74%, resulting in classification instability depending on the dataset variant. These higher results are also due to the optimization's ability to approximate the 5% FPR threshold closely. However, this algorithm is highly concerned by unfairness with predictive equality varying between 12% and 77% for age, 39% and 52% for income, and 11% and 18% for employment status.

By integrating all variants, we observe that the trade-off per configuration ranges from 42% to 99%. This indicates that the $[S = 20/P = 200]$ configuration

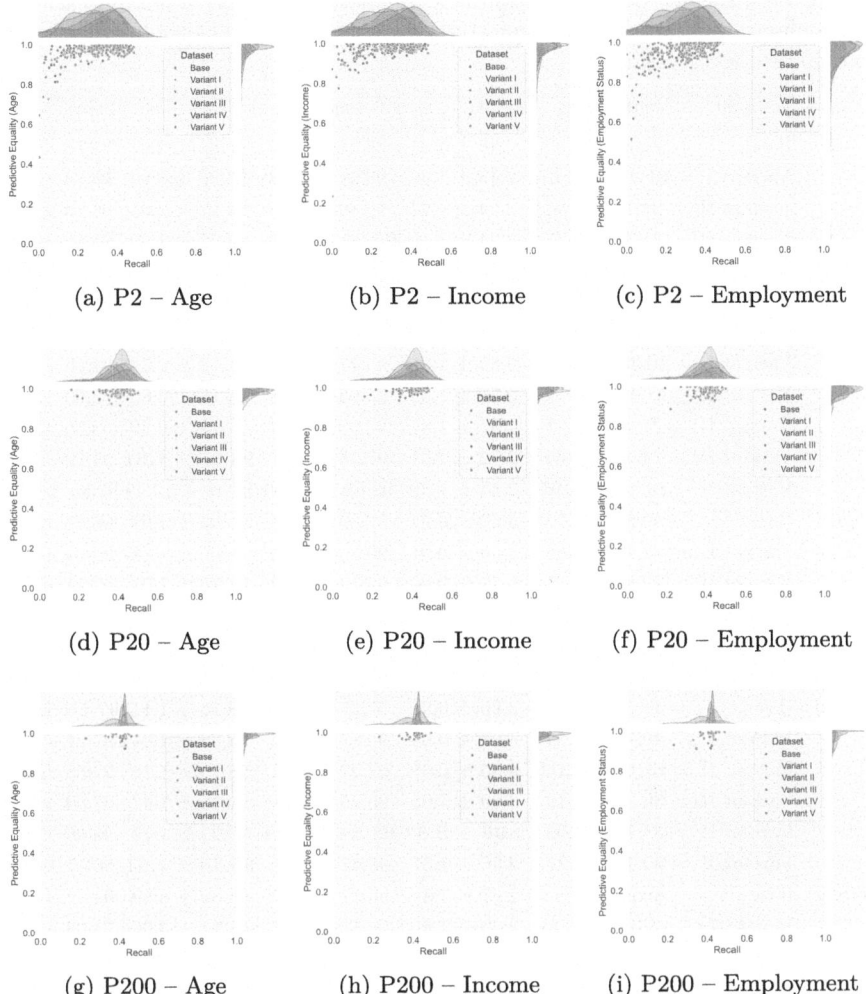

Fig. 5. Performance (recall) and fairness (predictive equality) for each population $P \in \{2, 20, 200\}$ and sensitive attributes for a time step $S = 50$.

achieves its lowest performance in terms of recall (42%) and its highest fairness with 99% for income and 98% for both age and employment status. A graphical representation of these trade-offs per sensitive attributes is illustrated in Fig. 6. Moreover, compared with LightGBM, the trade-offs have a negative slope, meaning that our SNN approach prioritizes fairness rather than performance.

In sum, our SNN approach demonstrates higher performance with more significant time steps and a population of neurons. In addition, the approach enables a more constant classification performance towards different biased datasets while ensuring fairness towards sensitive attributes in opposition to LightGBM.

Table 3. Performance (recall) and fairness (predictive equality) results of each optimized configuration (population P and time steps S) of the 1D-Convolutional Spiking Neural Networks

Pop.	Steps	Dataset	TPR$_{@5FPR}$	FPR$_{@5FPR}$	PE$_{age}$	PE$_{income}$	PE$_{employ.}$
P2	S20	Base	30.30 ± 11.45	2.48 ± 1.41	96.58 ± 2.67	96.81 ± 2.97	**96.78 ± 2.89**
		Variant I	31.95 ± 11.67	2.93 ± 1.44	96.21 ± 3.43	97.09 ± 3.84	95.84 ± 3.31
		Variant II	34.78 ± 8.98	2.94 ± 1.27	97.89 ± 1.56	97.27 ± 2.01	96.13 ± 3.45
		Variant III	31.73 ± 8.33	3.24 ± 1.19	**98.05 ± 1.45**	97.77 ± 1.90	95.66 ± 3.30
		Variant IV	25.26 ± 11.02	2.73 ± 1.61	97.83 ± 2.47	97.95 ± 2.52	96.62 ± 2.66
		Variant V	**34.96 ± 10.08**	**3.46 ± 1.32**	97.70 ± 2.12	97.88 ± 2.20	95.40 ± 8.55
	S50	Base	33.08 ± 13.14	2.84 ± 1.60	95.63 ± 4.28	96.71 ± 3.06	94.30 ± 5.41
		Variant I	30.33 ± 11.19	2.59 ± 1.39	95.77 ± 4.52	97.13 ± 2.53	94.66 ± 5.92
		Variant II	28.91 ± 14.28	2.20 ± 1.53	95.65 ± 5.38	96.15 ± 3.60	92.90 ± 9.26
		Variant III	27.97 ± 9.77	2.57 ± 1.47	97.90 ± 1.51	97.15 ± 2.39	95.59 ± 3.81
		Variant IV	24.40 ± 9.64	2.32 ± 1.31	97.38 ± 2.27	**98.07 ± 1.61**	94.79 ± 4.52
		Variant V	28.84 ± 11.97	2.55 ± 1.39	93.68 ± 16.36	93.18 ± 17.49	92.95 ± 14.75
P20	S20	Base	40.68 ± 5.78	3.60 ± 0.94	97.80 ± 1.94	98.05 ± 1.57	96.47 ± 2.64
		Variant I	38.41 ± 5.68	3.56 ± 0.91	96.75 ± 2.55	98.06 ± 1.58	96.94 ± 2.00
		Variant II	42.86 ± 5.63	3.77 ± 0.90	98.16 ± 1.33	97.73 ± 1.71	97.11 ± 2.10
		Variant III	36.06 ± 4.93	3.60 ± 0.84	98.45 ± 1.14	98.12 ± 1.23	96.83 ± 2.12
		Variant IV	32.99 ± 5.11	3.58 ± 0.92	97.77 ± 1.84	98.19 ± 1.51	97.06 ± 1.82
		Variant V	36.41 ± 6.70	3.40 ± 1.00	98.03 ± 1.53	97.94 ± 1.49	96.47 ± 2.48
	S50	Base	41.57 ± 6.15	3.90 ± 0.98	97.27 ± 1.40	98.15 ± 1.11	**97.55 ± 1.89**
		Variant I	40.71 ± 3.94	4.01 ± 0.64	96.87 ± 2.43	98.33 ± 1.04	97.27 ± 2.13
		Variant II	**43.15 ± 5.02**	4.03 ± 0.67	97.97 ± 1.11	98.25 ± 1.18	97.38 ± 1.70
		Variant III	36.98 ± 6.19	3.90 ± 1.05	98.25 ± 1.48	97.36 ± 1.43	97.50 ± 1.86
		Variant IV	34.79 ± 4.99	**4.06 ± 0.89**	**98.54 ± 1.24**	98.24 ± 1.73	96.66 ± 3.13
		Variant V	39.48 ± 7.27	3.96 ± 1.07	98.47 ± 1.16	**98.38 ± 1.61**	96.17 ± 2.05
P200	S20	Base	29.65 ± 11.11	2.67 ± 1.34	96.17 ± 3.60	96.59 ± 6.59	94.71 ± 5.99
		Variant I	24.04 ± 11.45	2.29 ± 1.45	95.52 ± 4.46	96.11 ± 3.96	94.09 ± 5.73
		Variant II	30.16 ± 10.61	2.78 ± 1.39	97.76 ± 2.11	97.42 ± 2.31	94.71 ± 4.64
		Variant III	25.03 ± 8.92	2.62 ± 1.29	97.85 ± 1.75	97.09 ± 2.71	95.48 ± 3.95
		Variant IV	25.43 ± 7.69	2.95 ± 1.27	97.56 ± 1.51	97.87 ± 1.67	96.60 ± 2.72
		Variant V	24.97 ± 11.45	2.55 ± 1.50	97.37 ± 2.50	96.87 ± 3.90	95.72 ± 4.38
	S50	Base	42.79 ± 0.59	4.07 ± 0.15	96.54 ± 2.98	98.02 ± 1.07	96.25 ± 3.47
		Variant I	40.21 ± 4.87	3.95 ± 0.98	96.61 ± 2.38	98.34 ± 0.99	**97.35 ± 1.41**
		Variant II	**47.08 ± 2.85**	4.32 ± 0.53	98.97 ± 0.34	98.67 ± 1.39	99.48 ± 0.19
		Variant III	41.83 ± 0.78	**4.85 ± 0.11**	97.76 ± 1.16	98.93 ± 0.87	96.98 ± 3.21
		Variant IV	35.54 ± 5.71	4.20 ± 1.05	98.68 ± 1.02	97.23 ± 1.32	95.58 ± 3.11
		Variant V	43.10 ± 2.83	4.58 ± 0.53	**99.31 ± 0.56**	**98.94 ± 0.65**	97.80 ± 1.59
LightGBM		Base	51.76 ± 3.11	4.99 ± 0.01	33.52 ± 1.75	39.93 ± 3.24	14.07 ± 4.75
		Variant I	49.40 ± 2.94	4.99 ± 0.01	77.54 ± 14.09	41.10 ± 4.65	11.20 ± 4.97
		Variant II	51.86 ± 5.16	4.96 ± 0.37	12.39 ± 3.54	44.64 ± 3.44	11.36 ± 5.70
		Variant III	73.56 ± 2.29	4.99 ± 0.01	18.81 ± 3.33	52.71 ± 2.94	18.38 ± 6.20
		Variant IV	39.63 ± 4.26	4.94 ± 0.48	13.00 ± 2.83	44.71 ± 2.79	13.06 ± 3.87
		Variant V	32.39 ± 3.73	4.94 ± 0.44	20.91 ± 4.25	50.80 ± 3.41	17.57 ± 6.77

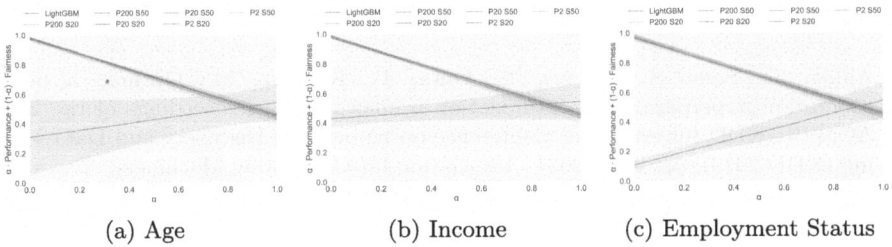

Fig. 6. Trade-offs between performance and attribute's fairness.

7 Conclusion and Future Work

In this work, we explored different configurations of 1D-CSNN models that use the joint activity of neurons, or population coding, for financial fraud detection. As a result, our best configuration achieved a remarkable 47% recall at the business constraint of 5% of FPR, performing better than non-population coding models. Moreover, our approach is comparable in performance to LightGBM, with our model being more consistent in classification and fairer with more than 90% of predictive equality towards sensitive attributes of the BAF dataset such as the customer's age, the income, and the employment status. This is particularly important in fraud detection to avoid discriminatory practices based on individual characteristics not predictive of fraudulent behavior. In addition, within the future evolution of neuromorphic hardware, SNN architectures are advantaged with their energetic efficiency, contributing to a more sustainable AI in comparison to traditional ANNs.

In future work, we aim to enhance SNNs by incorporating a causal inference approach. This will bring us closer to mimicking the brain's natural behavior, allowing us to identify causal relationships between spikes generated by neural activity. Moreover, this approach will help make these black-box models more explainable and transparent, providing more precise insights into their decision-making processes. This improvement will also ensure compliance with regulations and boost acceptance among the general public.

Acknowledgments. This research was supported by the Portuguese Recovery and Resilience Plan (PRR) through project C645008882-00000055, Center for Responsible AI, and within the scope of the project CISUC (UID/CEC/00326/2020).

References

1. Akiba, T., Sano, S., Yanase, T., Ohta, T., Koyama, M.: Optuna: a next-generation hyperparameter optimization framework. In: Proceedings of the 25th ACM SIGKDD International Conference on Knowledge Discovery and Data Mining (KDD 2019), pp. 2623–2631. Association for Computing Machinery, New York (2019). https://doi.org/10.1145/3292500.3330701
2. Arya, A.S., Ravi, V., Tejasviram, V., Sengupta, N., Kasabov, N.: Cyber fraud detection using evolving spiking neural network. In: 2016 11th International Conference on Industrial and Information Systems (ICIIS), pp. 263–268 (2016). https://doi.org/10.1109/ICIINFS.2016.8262948
3. Barocas, S., Hardt, M., Narayanan, A.: Fairness and Machine Learning: Limitations and Opportunities. MIT Press (2023)
4. Corbett-Davies, S., Pierson, E., Feller, A., Goel, S., Huq, A.: Algorithmic decision making and the cost of fairness. In: Proceedings of the 23rd ACM SIGKDD International Conference on Knowledge Discovery and Data Mining (KDD 2017), pp. 797–806. Association for Computing Machinery, New York (2017). https://doi.org/10.1145/3097983.3098095
5. Cruz, A.F., Belém, C., Jesus, S., Bravo, J., Saleiro, P., Bizarro, P.: FairGBM: gradient boosting with fairness constraints. In: The Eleventh International Conference on Learning Representations (2023). https://doi.org/10.48550/arXiv.2209.07850
6. Davies, M., et al.: Loihi: a neuromorphic manycore processor with on-chip learning. IEEE Micro **38**(1), 82–99 (2018). https://doi.org/10.1109/MM.2018.112130359
7. Demertzis, K., Iliadis, L., Bougoudis, I.: Gryphon: a semi-supervised anomaly detection system based on one-class evolving spiking neural network. Neural Comput. Appl. **32**(9), 4303–4314 (2020). https://doi.org/10.1007/s00521-019-04363-x
8. Ding, F., Hardt, M., Miller, J., Schmidt, L.: Retiring adult: new datasets for fair machine learning. In: Advances in Neural Information Processing Systems, vol. 34, pp. 6478–6490. Curran Associates, Inc. (2021). https://doi.org/10.48550/arXiv.2108.04884
9. Dutta, S., Kumar, V., Shukla, A., Mohapatra, N.R., Ganguly, U.: Leaky integrate and fire neuron by charge-discharge dynamics in floating-body MOSFET. Sci. Rep. **7**(1), 8257 (2017). https://doi.org/10.1038/s41598-017-07418-y
10. Eshraghian, J.K., et al.: Training spiking neural networks using lessons from deep learning. Proc. IEEE **111**(9), 1016–1054 (2023). https://doi.org/10.1109/JPROC.2023.3308088
11. Gerstner, W., Kistler, W.M.: Spiking Neuron Models: Single Neurons, Populations, Plasticity. Cambridge University Press, Cambridge; New York (2002)
12. Gerstner, W., Kistler, W.M., Naud, R., Paninski, L.: Neuronal Dynamics: From Single Neurons to Networks and Models of Cognition, UK edn. Cambridge University Press (2014)
13. Jesus, S., et al.: Turning the Tables: biased, imbalanced, dynamic tabular datasets for ML evaluation. In: 36th Conference on Neural Information Processing Systems Datasets and Benchmark Track (2022). https://doi.org/10.48550/arXiv.2211.13358
14. Ke, G., et al.: LightGBM: a highly efficient gradient boosting decision tree. In: Advances in Neural Information Processing Systems, vol. 30. Curran Associates, Inc. (2017)
15. Lapicque, L.: Recherches quantitatives sur l'excitation électrique des nerfs traitée comme une polarisation. Journal de physiologie et de pathologie générale, pp. 1–16 (1907)

16. Luzio, E., Ponti, M.A., Arevalo, C.R., Argerich, L.: Decoupling decision-making in fraud prevention through classifier calibration for business logic action (2024). https://doi.org/10.48550/arXiv.2401.05240
17. Maass, W.: Networks of spiking neurons: the third generation of neural network models. Neural Netw. **10**(9), 1659–1671 (1997). https://doi.org/10.1016/S0893-6080(97)00011-7
18. Maass, W., Bishop, C.M.: Pulsed Neural Networks, 1st edn. Bradford Books, Cambridge (1998)
19. Neftci, E.O., Mostafa, H., Zenke, F.: Surrogate gradient learning in spiking neural networks (2019). https://doi.org/10.48550/arXiv.1901.09948
20. Paugam-Moisy, H.: Spiking Neuron Networks: A Survey. Tech. Rep. 06-11, IDIAP Research Report (2006)
21. Pombal, J., Cruz, A.F., Bravo, J., Saleiro, P., Figueiredo, M.A.T., Bizarro, P.: Understanding unfairness in fraud detection through model and data bias interactions (2022). https://doi.org/10.48550/arXiv.2207.06273
22. Popescu, M.C., Balas, V.E., Perescu-Popescu, L., Mastorakis, N.: Multilayer perceptron and neural networks. WSEAS Trans. Circuits Syst. **8**(7), 579–588 (2009)
23. Prokhorenkova, L., Gusev, G., Vorobev, A., Dorogush, A.V., Gulin, A.: CatBoost: unbiased boosting with categorical features (2019). https://doi.org/10.48550/arXiv.1706.09516
24. Saleiro, P.,et al.: Aequitas: a bias and fairness audit toolkit (2018). https://doi.org/10.48550/arXiv.1811.05577
25. Uwaoma, C.: Detecting Bank Account Opening Fraud Using Machine Learning. Master's thesis, Dublin Business School (2024). https://hdl.handle.net/10788/4517

Rethinking the Quality of Synthetic Palm Vein Images from Spectral Analysis

Colton Clarke[1,4], Edwin H. Salazar-Jurado[2,4],
and Ruber Hernández-García[3,4(✉)]

[1] School of Financial Mathematics and Economics, University of Victoria, Victoria, Canada
[2] Facultad de Ciencias Básicas, Universidad Católica del Maule, Maule, Chile
[3] Departamento de Computación e Industrias, Facultad de Ciencias de la Ingeniería, Universidad Católica del Maule, Maule, Chile
rhernandez@ucm.cl
[4] Laboratorio of Technological Research on Pattern Recognition, Universidad Católica del Maule, Maule, Chile

Abstract. Palm vein-based biometric identification offers a higher level of security than traditional methods such as fingerprinting, iris, or facial recognition. One of its main advantages lies in using internal body features, which makes it highly secure and less susceptible to external changes. However, its large-scale application is limited by the need for large-scale public databases. In this context, synthetic palm vein image databases partially address this challenge, as there will always be a difference between synthetic and real. To mitigate these gaps, we propose to evaluate the differences using a spectral perspective and present techniques to fit the magnitude spectrum and power spectral distribution. We evaluated the similarity of the resulting synthetic images to the real images from the most representative state-of-the-art palm vein databases. The proposed approaches help to reduce the difference between synthetic and real images from the CASIA database, improving the accuracy in the representation of synthetic palm veins for the evaluation of biometric recognition algorithms.

Keywords: Biometrics · Palm vein recognition · Spectral analysis · Synthetic datasets

1 Introduction

Nowadays, biometric recognition is essential to ensure user security while still being widely accepted. In this context, palm vein biometrics stands out for its contactless acquisition capability, accuracy and privacy. Despite its advantages, the creation of research databases faces challenges regarding time, security, and cost, which restricts its large-scale application. Existing databases have limitations in terms of the number of individuals included, which hampers the generalization of models trained with convolutional neural networks (CNNs) [17]. To approach these limitations, the use of synthetic image databases for large-scale recognition tasks has been explored [12,13].

Current models for generating synthetic images of palm veins for large-scale applications fall into two groups: automatic generation through the use of generative adversarial networks (GANs) and generation by fusing the vascular pattern, based on biological optimization, with the palm print, generated using GANs [12,13]. The quality of the generated images depends on the ability to be coherently integrated into real images. In this sense, [14] uses qualitative and quantitative metrics to evaluate the similarity between real and generated images. Although these metrics yield good results, spectral analysis research has revealed defects in the generated images that are not detectable in the spatial domain [2]. This issue underscores the importance of spectral analysis as an effective technique for discerning between authentic and artificially generated images, providing a promising route for addressing the challenges associated with identifying generated content in various application contexts [4,19]. Then, some questions arise, which we try to answer in the present study: do all real palm vein databases have similar spectral distributions? do the synthetically generated palm vein images match the spectral distribution of the real databases? can we adjust the synthetic vein images by altering the magnitude spectrum and spectral power distribution features?

In this context, this paper evaluates the quality of synthetic images of palm veins from the spectral domain in relation to the spectrum of the most referenced real databases. This analysis also allows testing two techniques to adjust the magnitude spectrum and the spectral power distribution characteristics [2], improving the similarity of the synthetic images with the real ones. The first method uses the difference between the average spectral domain of real and synthetic databases to correct the magnitude spectrum of the generated images. The second method adjusts the spectral power distribution characteristics from a power correction dictionary constructed from real databases. Within the framework of this evaluation, the synthetic databases Synthetic-sPVDB [13] and NS-PVDB [12] are experimentally evaluated in relation to the real databases CASIA [5], POLYU [18], PUT [6], and VERA [15]. These comparisons allow us to understand how the generated images behave compared to the real ones in terms of spectral characteristics. In this sense, the present study focuses on evaluating the similarity between synthetic and real images of palm veins from a spectral perspective, with the aim of reducing the existing gap, highlighting the following contributions:

- Spectral evaluation of similarity between real and synthetic palm vein images of palm veins to compare images and evaluate differences.
- Reduction of the gap between real and synthetic palm vein images based on adjusted magnitude spectrum and power spectral distribution, reducing the discrepancy with real images from the CASIA database.
- Analysis of spectra between real databases to evaluate differences in magnitude and phase in the spectra of the databases due to the lack of standardization of the acquisition devices, suggesting the possibility of preprocessing

the images in the spectrum to standardize image quality without resorting to the standardization of acquisition devices.

The remaining sections of the paper are organized as follows. Section 2 summarizes the related works of the state-of-the-art. Section 3 presents the proposed methodology to perform our research. Section 4 analyzed the obtained experimental results. Finally, Sect. 5 gives the conclusions and future works.

2 Related Works

In the field of palm vein biometry, synthetic image generation and analysis have been poorly addressed [14]. In general, generative adversarial networks (GANs) are resorted to capture the complexity of palm texture. Different approaches vary in how they simulate the vascular structure and integrate it with the fingerprint [12,13]. In this regard, techniques that allow distinguishing images generated by GANs play a key role, especially when seeking to produce realistic images. Similarity analysis has not considered GAN-based techniques in the context of synthetic palm veins. The metrics usually employed are structural similarity index SSIM, gray level concurrence matrix GLCM, Frechet Inception Distance (FID), Bessel K-forms, UMAP projections, precision, and recall. The metrics are not very robust (they do not take into account deep features of the images) and may miss details generated by GANs, which may affect the models trained with these synthetic data.

Techniques for detecting real and GAN-generated images can abstract deep image features. These techniques can be divided into two main categories: spatial domain and frequency domain. The spatial domain includes methods such as the steganalysis method based on photo response non-uniformity (PRNU) patterns [10], using saturation signals [11], and implementation of neural network-based detectors [9,16]. On the other hand, in the frequency domain, spectrum detectors trained on GAN-generated images, power spectral distribution analysis, and detection of deficiencies in the reproduction of high-frequency modes are employed [3,4,19]. These detectors require fewer parameters than spatial domain-based detectors and have demonstrated superior performance [2]. Indeed, they have the ability to detect artifacts in the spectral domain that pass undetected in the spatial domain.

In this regard, work has been proposed to adjust the generated images to achieve greater similarity to the real ones. Particularly, in [1] the suppression of high-frequency defects in generative networks is studied. They demonstrated that altering the upsampling process for the generator's final layer can prevent abnormal power distributions. However, most GANs still have difficulty mitigating spectral abnormalities because the kind and number of the up-sampling modules must be manually and carefully adjusted. On the other hand, in [2] three algorithms are proposed and evaluated to improve the images generated in the spectral domain. Their proposal showed promising and effective results by altering the magnitude spectrum and spectral power distribution features and reducing the accuracy of fake detection.

Considering the above context, this paper presents a new perspective for the analysis of synthetic palm vein images proposed in [12,13], using the techniques proposed in [2] and adjusting the spectral domain to minimize the presence of artifacts.

3 Methodology

This section details the methodology used for the present study. The Fig. 1 summarizes the workflow carried out to adjust the spectral domain of the synthetically generated palm vein images. Our approach is based on the use of four of the most referenced real databases in the literature [5,6,15,18] and the only two existing synthetic databases to our knowledge [12,13], described in Sect. 3.1. Taking each of the real datasets as a reference, we analyze how the spectral distribution of the synthetically generated images matches the real ones; such transformation is detailed in Sect. 3.2. From the spectral characteristics obtained, the two methods implemented to reduce the differences detected in the spectral domain between real and synthetic images are described in Sect. 3.3.

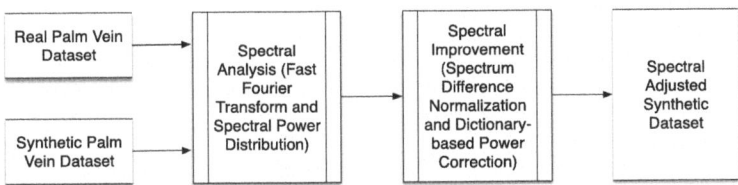

Fig. 1. Overview of the proposed methodology.

3.1 Palm Vein Datasets

Two sets of synthetic datasets, Synthetic-sPVDB [13] and NS-PVDB [12], were compared against four real databases: CASIA [5], PUT [6], PolyU [18], and VERA [15]. According to our knowledge, Synthetic-sPVDB [13] and NS-PVDB [12] are currently the only synthetic datasets available in the state-of-the-art. Synthetic-sPVDB was generated using a StyleGAN2-based model [8], trained with 850 nm band images from the CASIA dataset [5]. On the other hand, NS-PVDB [12] employed an adaptive dynamics model to recreate the hand vein plexus, and the palm print was also generated using StyleGAN2 [8], trained with visible spectrum band images from the CASIA dataset [5].

The chosen real databases were selected for their relevance in research, ensuring the quality and validity of the study results. The most relevant information is summarized in Table 1. These databases are characterized by a limited number of subjects and the use of various acquisition devices for image capture.

To achieve visual consistency across all database images, we conducted a preprocessing process in two main stages: first, a region of interest (ROI) was selected based on the approach described in [7], defining a square area around the midpoint between the index and little fingers; then, the images were resized to 128×128 pixels and converted to grayscale. This method not only optimizes computational efficiency but also preserves essential features for biometric recognition, complemented by the application of a normal filter to enhance quality and highlight relevant details.

Table 1. Overview of Palm Vein Images used for Evaluation. Samples per Subject are Represented as Hands × Acquisition Sessions × Samples.

Dataset	Acquisition Technology	Subjects × Samples	Total Images	Source
CASIA [5]	Multispectral Reflection (460 nm, 630 nm, 700 nm, 850 nm, 940 nm, VIS)	100 × (2 × 2 × 3)	7,200	Real
PolyU [18]	Multispectral Reflection (470 nm, 525 nm, 660 nm, 880 nm)	250 × (2 × 2 × 6)	6,000	Real
PUT [6]	Reflection (880 nm)	50 × (3 × 2 × 4)	1,200	Real
VERA [15]	Reflection (940 nm)	110 × (2 × 2 × 5)	2,200	Real
Synthetic-sPVDB [13]	Synthetic images	20,000 × (1 × 1 × 6)	120,000	Synthetic
NS-PVDB [12]	Synthetic images	16,000 × (1 × 1 × 6)	96,000	Synthetic

3.2 Spectral Analysis

To evaluate the spectral domain of synthetic palm vein images, we used the approach proposed in [2] to estimate the differences in the spectral distributions of real images and generated images. Firstly, we transform the images to obtain the spectral power distribution (SPD) of each spectrum by applying the Fast Fourier Transform (FFT). We compute FFT from Discrete Fourier Transform (DFT), resulting in the magnitude and phase of the image. Since the DFT is a sampled Fourier Transform, it only contains a collection of samples big enough to adequately characterize the spatial domain image. The number of frequencies and pixels in the spatial domain image are equal, meaning that the size of the images in both domains is the same. The two-dimensional DFT for a square image of size $N \times N$ is computed as follows:

$$\mathfrak{F}(k,l) = \sum_{m=0}^{N-1} \sum_{n=0}^{N-1} f(m,n) e^{-2\pi i (\frac{km}{N} + \frac{ln}{N})} \quad (1)$$

where $f(m,n)$ is the image in the spatial domain, and the exponential term is the basis function corresponding to each point $\mathfrak{F}(k,l)$ in the Fourier space.

On the contrary, the Fourier image can be re-transformed to the spatial domain by computing the two-dimensional Inverse Discrete Fourier Transform of $\mathfrak{F}(k,l)$, which is given by:

$$f(m,n) = \frac{1}{N^2} \sum_{k=0}^{N-1} \sum_{l=0}^{N-1} \mathfrak{F}(k,l) e^{2\pi i (\frac{km}{N} + \frac{ln}{N})} \qquad (2)$$

In terms of complexity, we need an order of N^2 operations to compute the two-dimensional DFT for a square image of size $N \times N$ (N products for each of the N components). However, it can be computed by the Fast Fourier Transform, which decreases the number of required operations down to $N \log N$.

The FFT results in a complex-valued image, one representing the real component and another for the imaginary component. Alternatively, it can be shown as magnitude spectrum and phase (frequency distribution). The magnitude spectrum gives information on the amplitudes of different frequencies in the image, capturing most of the spatial structure information. The phase spectrum gives information on the phase shifts of different frequencies. However, for accurate reconstruction of the original image after manipulations in the frequency domain (e.g., filtering), both the magnitude and phase are required.

From the magnitude spectrum (S), we obtain the spectral power distribution, which gives us information on the wavelength and its power in the magnitude spectrum. In Implementation 1, we show the Python code of the algorithm proposed in [2], which computes the Spectral Power Distribution from the magnitude spectrum of any given image of size $M \times M$. This algorithm does not alter the images but is rather just a crucial visualization tool to analyze the differences between real and synthetic databases. Besides, it is a key step in the Implementation 3 to create the Power Distribution Dictionary.

Algorithm 1. Compute the Spectral Power Distribution

```
1: def COMPUTE_SPD_FEATURE(S):
       ▷ Takes a spectrum S and returns its SPD feature.
2:     M = len(S)
3:     SPD = np.zeros(M)
4:     center = 0.5 * M
5:     for i in range(M):
6:         for j in range(M):
7:             index = int(np.sqrt((i - center) ** 2
                       + (j - center) ** 2))
8:             if index < M:
9:                 SPD[index] += S[i][j]
10:    SPD /= SPD[0]
11:    return SPD
```

Algorithm 2. Compute the Spectrum Difference Normalization

1: def SPECTRUM_DIFFERENCE_NORM(S, R, G):
     ▷ Takes three parameters, a spectrum S, the average real spectra R, and the average synthetic spectra G. Computes the difference between the two spectra $Diff$ and returns the normalized spectrum S_norm.
2:     Diff = G - R
3:     S_norm = S - Diff
4:     return S_norm

Algorithm 3. Compute the Power Dictionary-based Correction

1: def CONSTRUCT_POWER_DICTIONARY(I):
     ▷ Takes a set of real images I and computes the SPD features. Returns the power distribution dictionary D as a list of SPD features.
2:     D = []
3:     for i in I:
4:         fft = DFT2(i)
5:         fft_shifted = np.fft.fftshift(fft)
6:         spectrum = np.abs(fft_shifted)
7:         SPD_feature = compute_spd_feature(spectrum)
8:         D.append(SPD_feature)
9:     return D

10: def DICTIONARY_BASED_CORRECTION(S, D):
     ▷ Takes two parameters, a spectrum S and the power distribution dictionary D. Returns the adjusted spectrum S_adj after the power dictionary corrections.
11:    M = len(S)
12:    SPD_feature = compute_spd_feature(S)
13:    k = SPD_feature.shape[0]

14:    SPD_diff = []
15:    for d in D:
16:        SPD_diff.append(sum((SPD_feature[j] -
                        d[j])**2 for j in range(k//4)))

17:    center = 0.5 * M
18:    Dr = D[np.argmin(SPD_diff)]
19:    S_adj = np.zeros((M,M))
20:    for i in range(M):
21:        for j in range(M):
22:            index = int(np.sqrt((i - center) ** 2
                        + (j - center) ** 2))
23:            S_adj[i,j] = SPD_feature[index]
                        and S[i,j] * Dr[index] /
                        SPD_feature[index] or 0
24:    return S_adj

3.3 Spectral Improvement

For spectral improvement, we utilize the algorithms proposed in [2], whose Python versions are shown in Implementation 2 and Implementation 3. The first, called the Spectrum Difference Normalization or Method 1, takes the average real image spectrum and the average synthetic spectrum to compute the difference $Diff$ between both. Thus, the correction is applied by subtracting $S - Diff$ from each synthetic spectrum S. We take each pair of real and synthetic databases and process them to compute their average spectrums. Then, for each magnitude spectrum corresponding to an image of the evaluated synthetic dataset, we call the function to obtain the normalized spectrum S_norm. Finally, the normalized magnitude spectrum is combined with the original phase to reconstruct the corrected image by using the IFT.

Implementation 3, named Power Dictionary-based Correction or Method 2, uses Implementation 1 to create a power distribution dictionary D from the spectral power distribution (SPD) features of a real dataset. For this purpose, we implement a function to transform a real image dataset into a power distribution dictionary as a list of their SPD features. With D as input, Method 2 computes the SPD feature from a given synthetic spectrum S and searches for the most similar power distribution Dr in the dictionary. Then, the adjusted spectrum S_adj is calculated by multiplying each spectrum value by the proportion of Dr/SPD. Likewise, as in Method 2, the image is reconstructed from the adjusted magnitude spectrum and the original phase to obtain a corrected image.

4 Experimental Results

To evaluate the quality of synthetic palm vein images in the spectral domain, we used databases reported in previous studies, both real and synthetic. The synthetic databases employed are Synthetic-sPVDB [13] and NS-PVDB [12], while the real databases are CASIA [5], PUT [6], PolyU [18], and VERA [15]. To ensure a balance in the number of images, we randomly sampled 500 images from each dataset. Subsequently, we applied spectral transformations to the synthetic images using the algorithms described in Algorithm 2 and Algorithm 3.

Figure 2 shows the effects of the Spectrum Difference Normalization (center column) and Power Dictionary-based Correction (right column) processes in relation to the SPD and magnitude spectrum of the Synthetic-sPVDB dataset compared to the four real databases studied. The spectral distributions exhibit two characteristic forms: a clearly defined peak at a wavelength around 6 for Synthetic-sPVDB, PolyU, and VERA, and a form that tends to be constant in the range of 25 to 65 in the wavelength values for CASIA and PUT, which differentiates them from the Synthetic-sPVDB spectrum.

The proposed methods for improving the quality of synthetic images proved to be highly sensitive to small variations in magnitude or phase, as observed in the spectra of CASIA and PUT. Although they have similar spectra, the enhancement methods are effective only in CASIA. Despite this, the spectral similarity of both real databases suggests that if the enhancement method works for one of them, it will also approximate the spectrum of the other. Regarding the similarity of the spectral distribution of Synthetic-sPVDB with PolyU and VERA, the enhancement method did not have the expected effect; instead of narrowing the gap between synthetic and real images, it widened. In this case, the spectral similarity can be used to standardize the spectra of all real databases. This is possible since the enhancement methods can be viewed as an application that transforms the spectra of Synthetic-sPVDB, PolyU, and VERA to the spectrum of CASIA, where the methods was effective.

Regarding the NS-PVDB data set, its spectral distribution is similar to CASIA and PUT but differs from VERA and PolyU, as shown in Fig. 3. The methods to improve the similarity of synthetic images are effective with the CASIA database, as in the previous case. Unlike Synthetic-sPVDB, where we were unable to improve the spectral similarity with VERA and PolyU, applying the improvement method to the NS-PVDB dataset produces a spectral distribution comparable to VERA and PolyU. In general, we observe that the enhancement method works consistently with CASIA and in cases where the spectral distribution of the real databases differs from the synthetic ones.

The consistent effectiveness of the methods with CASIA suggests that we can standardize the spectra from different real databases. This is particularly relevant because palm vein imaging devices are not standardized [17], causing images from different databases to reflect the specific characteristics of the devices used. This variability negatively affects the performance of recognition algorithms. Therefore, standardizing the spectra could improve the robustness and precision of these algorithms, promoting greater uniformity and reliability in biometric recognition.

From the obtained results, we can point out the following insights:

- Although the real images used in the various experiments were scaled to the same size and gray level to achieve visual uniformity across all databases, the spectral distributions differed. This variation likely stems from differences in acquisition devices and NIR illumination spectra, as discussed in Sect. 3.1.
- In cases where the spectral distributions were similar before applying the adjustments, the results were contrary to expectations after applying both correction methods, except in the CASIA database.
- The spectral distributions of the databases can be standardized using the spectrum of the CASIA dataset as a reference. Synthetic data, as well as the PolyU and VERA databases, have similar spectra and can be adjusted to the CASIA spectrum using the proposed methods. On the other hand, the PUT database already has a spectral distribution similar to CASIA, so no additional adjustments are needed.
- The best results were obtained for the CASIA database across both synthetic datasets and with both correction methods. This success is attributed to the StyleGAN training, which used the CASIA database as a reference for generating synthetic images.
- When comparing the outcomes of both correction methods, there is no clear overall winner. However, Method 1 yielded lower average spectral difference

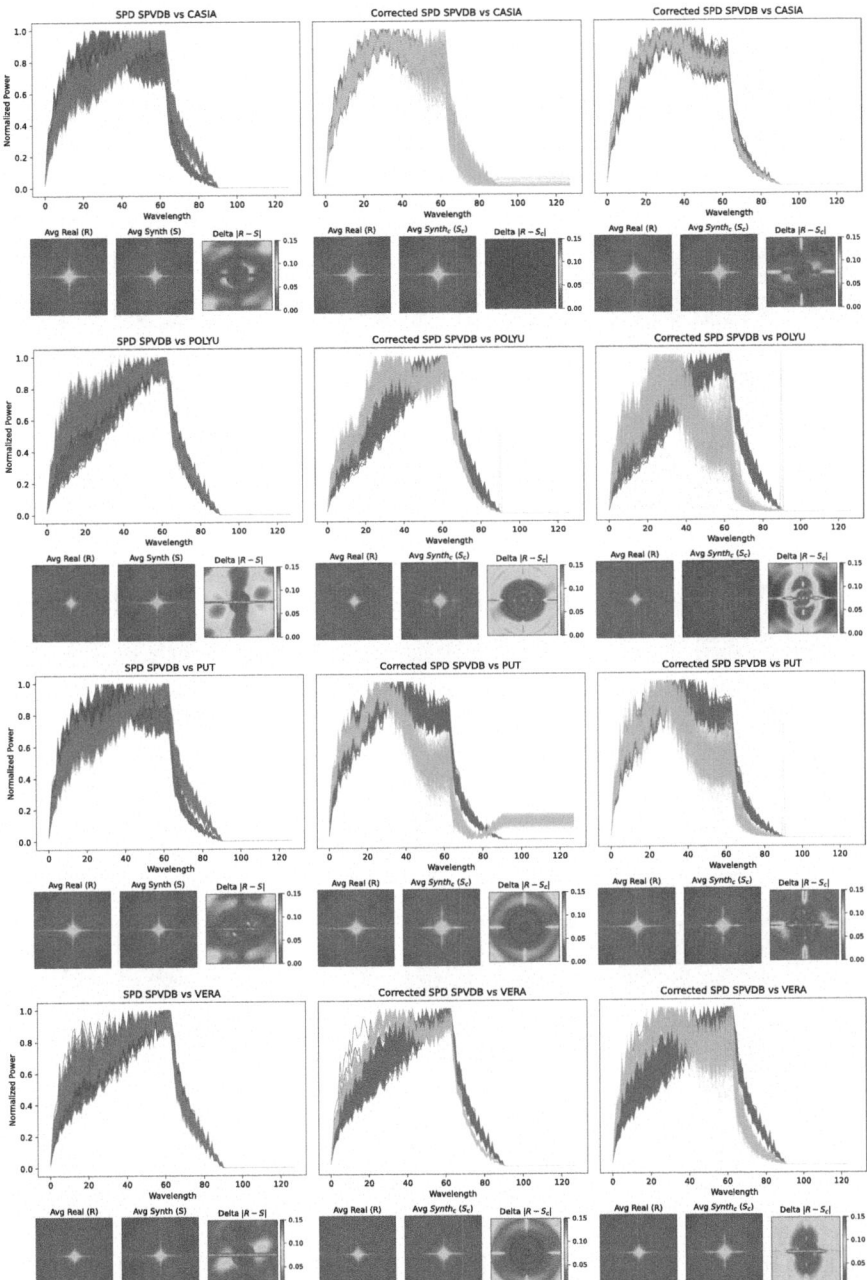

Fig. 2. Comparison results from the Synthetic-sPVDB dataset versus the four real databases studied. Each row shows in the top the SPD plots and the average spectrums along with their differences in three columns: before any improvement (left), after Method 1 (center), and after Method 2 (right). In all SPD plots, the blue representation corresponds to the real database. (Color figure online)

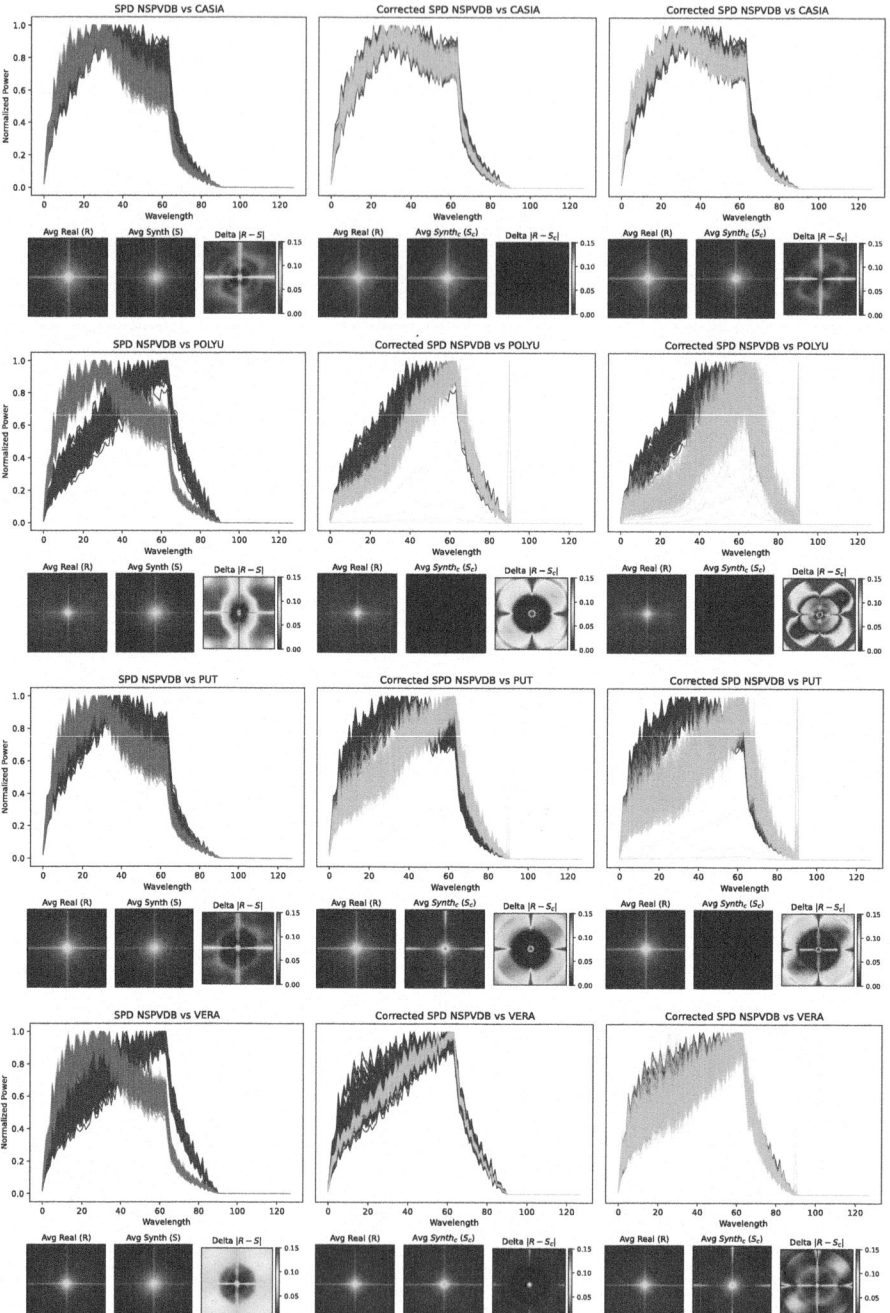

Fig. 3. Comparison results from the NS-PVDB dataset versus the four real databases studied. Each row shows in the top the SPD plots and the average spectrums along with their differences in three columns: before any improvement (left), after Method 1 (center), and after Method 2 (right). In all SPD plots, the blue representation corresponds to the real database. (Color figure online)

values post-application, indicating a slight edge in effectiveness for reducing the spectral gap.

5 Conclusions

In this article, we presented a methodology for improving synthetic palm vein images by adjusting their magnitude spectrum and spectral power distributions to better match real-world images. This approach is founded on the need for computer-generated images to reduce the differences in the frequency domain and avoid spectral artifacts that do not appear in the spatial domain. The obtained results show the difference between real databases and synthetic datasets, highlighting the need to take into account the frequency domain in the generation of synthetic palm vein images and their quality evaluation.

The analyzed data determined that two types of spectral distributions distinguished by their shape characterize the palm vein databases. We found that applying the proposed methods to reduce the gap between the spectra of synthetic and real images was more effective when using the CASIA database. In other real databases, if the spectrum resembled that of the synthetic database, the gap increased; however, if it differed, the spectrum of the synthetic database tended to adopt the shape of the spectral distribution of the real database.

In future work, we propose to develop a spectrum-based GAN model that considers differences in spatial and frequency domains within the discriminator module or loss function. Another important consideration is the types of spectral distributions found in real databases identified in this study, as using other databases may reveal different spectra. Furthermore, we should evaluate the possibility of transforming one spectrum into another using the techniques developed in this work. This is particularly relevant due to the lack of standardization in palm vein capture devices, which causes data to reflect specific characteristics of each device. Standardizing spectra could improve the robustness and accuracy of these algorithms, promoting greater uniformity and reliability in biometric recognition.

Acknowledgments. This work was partially funded by Agencia Nacional de Investigación y Desarrollo (ANID), Ministerio de Ciencia, Tecnología, Conocimiento e Innovación, Gobierno de Chile, Research Project ANID FONDECYT Iniciación en Investigación 2022, Grant number No. 11220693. The authors also thank the research project ANID Subdirección de Investigación Aplicada/Concurso IDeA I+D ID23i10242 and the Laboratory of Technological Research in Pattern Recognition. Portions of the research in this paper used the CASIA-MS-PalmprintV1 collected by the Chinese Academy of Sciences' Institute of Automation (CASIA). Portions of the research in this paper used the VERA-Palmvein Corpus made available by the Idiap Research Institute, Martigny, Switzerland.

Disclosure of Interests. All authors contributed to the study conception and design. Material preparation, software, and visualization were performed by Colton Clarke and Ruber Hernández-García. Conceptualization, formal analysis, and development were carried out by Edwin H. Salazar-Jurado and Ruber Hernández-García. Methodology, supervision, and funding acquisition were conducted by Ruber Hernández-García.

References

1. Chandrasegaran, K., Tran, N.T., Cheung, N.M.: A closer look at fourier spectrum discrepancies for cnn-generated images detection. In: Proceedings of the IEEE/CVF Conference on Computer Vision and Pattern Recognition, pp. 7200–7209 (2021)
2. Dong, C., Kumar, A., Liu, E.: Think twice before detecting gan-generated fake images from their spectral domain imprints. In: 2022 IEEE/CVF Conference on Computer Vision and Pattern Recognition (CVPR), pp. 7855–7864 (2022). https://doi.org/10.1109/CVPR52688.2022.00771
3. Durall, R., Keuper, M., Keuper, J.: Watch your up-convolution: Cnn based generative deep neural networks are failing to reproduce spectral distributions. In: Proceedings of the IEEE/CVF Conference on Computer Vision and Pattern Recognition, pp. 7890–7899 (2020)
4. Frank, J., Eisenhofer, T., Schönherr, L., Fischer, A., Kolossa, D., Holz, T.: Leveraging frequency analysis for deep fake image recognition. In: International Conference on Machine Learning, pp. 3247–3258. PMLR (2020)
5. Hao, Y., Sun, Z., Tan, T., Ren, C.: Multispectral palm image fusion for accurate contact-free palmprint recognition. In: 2008 15th IEEE International Conference on Image Processing, pp. 281–284. IEEE (2008)
6. Kabaciński, R., Kowalski, M.: Vein pattern database and benchmark results. Electron. Lett. **47**(20), 1127–1128 (2011)
7. Kang, W., Wu, Q.: Contactless palm vein recognition using a mutual foreground-based local binary pattern. IEEE Trans. Inf. Forensics Secur. **9**(11), 1974–1985 (2014)
8. Karras, T., Laine, S., Aittala, M., Hellsten, J., Lehtinen, J., Aila, T.: Analyzing and Improving the Image Quality of StyleGAN. In: Proceedings of the IEEE/CVF Conference on Computer Vision and Pattern Recognition, pp. 8110–8119 (2020)
9. Marra, F., Gragnaniello, D., Cozzolino, D., Verdoliva, L.: Detection of gan-generated fake images over social networks. In: 2018 IEEE Conference on Multimedia Information Processing and Retrieval (MIPR), pp. 384–389 (2018). https://doi.org/10.1109/MIPR.2018.00084
10. Marra, F., Gragnaniello, D., Verdoliva, L., Poggi, G.: Do gans leave artificial fingerprints? In: 2019 IEEE Conference on Multimedia Information Processing and Retrieval (MIPR), pp. 506–511 (2019). https://doi.org/10.1109/MIPR.2019.00103
11. McCloskey, S., Albright, M.: Detecting gan-generated imagery using saturation cues. In: 2019 IEEE International Conference on Image Processing (ICIP), pp. 4584–4588 (2019). https://doi.org/10.1109/ICIP.2019.8803661
12. Salazar, E., Hernández-García, R., Barrientos, R.J., Vilches, K., Mora, M., Vásquez, A.: Automatic generation of synthetic palm vein images: a nature-based approach. In: 11th International Conference of Pattern Recognition Systems (ICPRS 2021), pp. 38–43(5). Institution of Engineering and Technology (2021)
13. Salazar, E., Hernández-García, R., Barrientos, R.J., Vilches, K., Mora, M., Vásquez, A.: Generating style-based palm vein synthetic images for the creation of large-scale datasets. In: 11th International Conference of Pattern Recognition Systems (ICPRS 2021), pp. 182–187(5). Institution of Engineering and Technology (2021)

14. Salazar-Jurado, E.H., Hernández-García, R., Vilches-Ponce, K., Barrientos, R.J., Mora, M., Jaswal, G.: Towards the generation of synthetic images of palm vein patterns: a review. Inf. Fusion **89**, 66–90 (2023)
15. Tome, P., Marcel, S.: Palm vein database and experimental framework for reproducible research. In: 2015 International Conference of the Biometrics Special Interest Group (BIOSIG), pp. 1–7. IEEE (2015)
16. Wang, S.Y., Wang, O., Zhang, R., Owens, A., Efros, A.A.: Cnn-generated images are surprisingly easy to spot... for now. In: Proceedings of the IEEE/CVF Conference on Computer Vision and Pattern Recognition, pp. 8695–8704 (2020)
17. Wu, W., Elliott, S.J., Lin, S., Sun, S., Tang, Y.: Review of palm vein recognition. IET Biometrics **9**(1), 1–10 (2019)
18. Zhang, D., Guo, Z., Lu, G., Zhang, L., Zuo, W.: An online system of multispectral palmprint verification. IEEE Trans. Instrum. Meas. **59**(2), 480–490 (2009)
19. Zhang, X., Karaman, S., Chang, S.F.: Detecting and simulating artifacts in gan fake images. In: 2019 IEEE International Workshop on Information Forensics and Security (WIFS), pp. 1–6. IEEE (2019)

An Uncertainty-Driven ScaledYOLOv4 for Open-Pit Mining Helmet Detection

Roger Calle[1] and Eduardo Aguilar[1,2]

[1] Department of Computing and Systems Engineering,
Catholic University of the North, Antofagasta, Chile
eduardo.aguilar@ucn.cl

[2] Departament de Matematiques i Informatica, Universitat de Barcelona, Barcelona, Spain

Abstract. Failure to use proper personal protective equipment can lead to fatal consequences or complex accidents. Binary helmet detection is a very active research topic that involves detecting whether a worker is wearing a safety helmet or not. Despite the great interest, almost all of the work focuses on the analysis of images acquired in construction environments. To support mining safety monitoring, we propose a new dataset collected from mining environments, consisting of 1,510 images labeled for helmet and head object detection. Furthermore, we propose an uncertainty-driven model aiming to improve the performance by reducing the false positive rate. The effectiveness of the proposed method was demonstrated on the proposed dataset, and also by a performance analysis between cross-datasets strategy. The proposed method provides 95.3% in terms of mAP and 91.4% in terms of F1, which is a promising performance for binary helmet detection in real solutions applied to mining environments. This work contributes to better control of the use of safety helmets to reduce the complexity of head injuries related to work-related accidents, thus protecting workers and guaranteeing the continuity of the operational process.

Keywords: Binary helmet detection · ScaledYOLOv4 · Ensemble learning · Uncertainty quantification · Mining industry

1 Introduction

Identifying and managing critical controls will prevent a serious or fatal accident from occurring, or mitigate the consequences at most if an undesirable event occurs [10]. According to the International Labor Organization, head accidents are quite common in industrialized regions or countries, and they fluctuate between 3% and 6% of all injuries [2]. To mitigate the severity of an accident and ensure the continuity of the production process, the industries need to increase their capacity to monitor and control risk behaviors, such as the non-use of safety helmets. Therefore, it is important to have additional tools that facilitate the analysis of compliance with safety protocols, such as automatic detection of whether or not a worker is wearing a safety helmet. We will call this research problem as binary helmet detection.

Fig. 1. Complexity of detecting helmets in images belonging to the proposed dataset where they appear (a) small, (b) in poor illumination, (c) in various weather conditions or (d) with air pollution.

Focusing on the mining industry, there is an 89% increase in research between 2011 and 2022 [1], expanding the development of artificial intelligence technology [5,17,25]. However, regardless of the implementation of these technologies, their application in the field of mining safety has encompassed a more delayed development [11]. Unlike this industry, in others such as construction and energy, several methods based on Deep Learning (DL) have been successfully applied to occupational safety, where object detection becomes relevant to monitor the worker [20], mobile equipment [23] or Personal Protective Equipment (PPE) [22].

The success of these methods depends not only on the model design but also on the availability of a large and diverse dataset [15]. In this regard, there are several datasets with images acquired in the construction and energy environment to address the problem of binary helmet detection [18,22,24,26], but only a few images of mining sites are available. A dataset of images acquired in mine site environments would benefit the mining industry by serving as the basis for the development of DL-based object detection models applicable to this industry.

The mining industry contains different sites than other industries and the collected data are often poorly illuminated, with the presence of air pollution, or acquired from far away (see Fig. 1), making it difficult to learn a model. The main difference in the data collected concerning the construction industry (see Fig. 2) is that the latter visualizes buildings, beam structures, columns or foundations with mobile equipment from this industry, whereas mining presents various aspects associated with polluted environments, mobile equipment (mining truck, bulldozer, rotary shovel) and weather conditions (e.g., sunny, cloudy). While categories such as weather conditions (cloudy or sunny) may be shared with construction environments. Pollution and the presence of mobile equipment from the mining industry are not found in construction, so the imaging environment is completely different. Moreover, in open-pit mining, since data acquisition is performed by conducting its operations in an open environment, it is exposed to various random noises that could severely affect model performance. To improve model robustness in this type of data, in addition to a dataset with images that represent the variability of conditions that can occur in the mining environment, a DL method that quantifies uncertainty is desirable to provide better handling on the certainty of the predictions. In particular, when there is a distribution shift, that is, in those data that could differ from the data used during training [7]. So far there is evidence in the literature related to object detection, but not in helmet detection, of adopting uncertainty-based methods when analyzing highly challenging data [9,12]. The uncertainty has been quantified for the bounding box coordinates and classes, separately [4] and also together [8,16]. In most of them, a probabilistic approach is considered [12,14,16], providing good results but increasing the complexity of the model design [4,8] and the computational resources necessary during training or inference [14,16].

Fig. 2. A sample of images to illustrate the difference between images acquired in the construction (first row) and mining (second row) industry.

Motivated by previous works and considering the challenges posed by images acquired in mining environments, we propose an uncertainty-based approach to perform binary helmet detection. Specifically, we selected the ScaledYOLOv4 [21] method as the baseline due to its superior performance in this problem demonstrated on two public datasets compared to several DL methods and also its usability for real-time detection [3]. Next, we endow ScaledYOLOv4 with the ability to quantify uncertainty using Deep Ensemble [13], a simple and effective deterministic approach. Finally, inspired by [4], we designed a confidence score that considers estimated uncertainty to improve decision making at inference time. Regarding the latter, unlike the confidence score proposed by [4], our proposal includes not only bounding box uncertainty, but also class uncertainty.

Our main contributions are as follows: 1) We make public the first dataset to perform the binary helmet detection with images acquired in mining environments, where complex scenarios such as pollution, occlusion, abrupt illumination changes, among other factors are present; 2) We adapt the ScaledYOLOv4 object detection method by allowing it to quantify the uncertainty of the predicted class and bounding box coordinates. With the estimated uncertainty, we propose a new confidence score used to refine object detection; 3) We demonstrate the benefits of the proposed approach by improving the reduction in false detection rate compared to the baseline model on the proposed challenging dataset.

2 Mining Safety Helmet Dataset

There are some datasets for helmet detection [18,22,24,26], however, most of them were acquired in buildings under construction, which differs from mining environments. In the absence of a public dataset for helmet detection with images taken in mining environments, in this article, we propose a new Mining Safety Helmet Dataset (MSHD), which will be publicly available for research or academic purposes. Next, we describe the data acquisition, cleaning, and annotation process (see Fig. 3), while also providing some statistical details of MSHD.

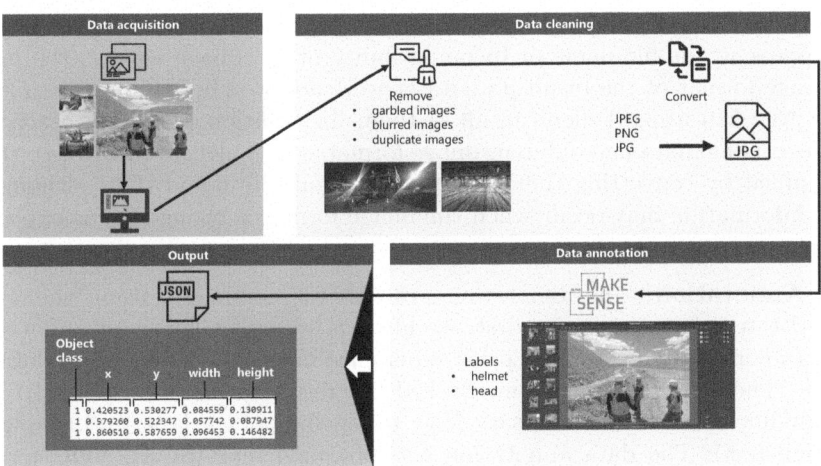

Fig. 3. Graphical examples of the activities involved in the process of creating the proposed dataset.

Data Acquisition: Image acquisition was carried out during 3 months in mining sites considering 5 h each week to collect a minimum of 1,000 images. It should be noted that the images are collected exclusively in open-pit mining environments, as the mining company where the data acquisition was performed conducts its operations in an open environment. The data were collected from the point of view of a security professional, manager, or mining executive visiting or conducting a safety inspection in the mining sector. This was done sporadically every Thursday at different times of the day, mainly from mobile device cameras and video recordings from the site in the mine areas. All data were obtained during normal worker operations, and no restrictions were taken into account to acquire data as close to reality as possible. After each month, the security specialist provided the data to the research team. In the specific case of video recording data, the research team extracts the frame in which a group of workers with or without hard hats are present, taking care to avoid overlapping frames. All the images have been captured in mining activities, where they are present in mining or open pit blasting, heavy machinery maintenance, cathode production, hoisting pulley storage, safety training activities, sanitary inspections, gas measurement, hot work (welding), pipeline maintenance, electrical maintenance and drilling activities. As a result, 371, 1,072 and 145 images were obtained from March to May 2022 (each month respectively), obtaining a total of 1,588 images, exceeding the minimum expected.

Data Cleaning: At the end of data collection, the data were manually reviewed to process unwanted data. In particular, data with garbled images, blurred images, and duplicate images are removed. After manual review, a total of 78 images were removed and the remaining 1,510 images were used for the annotation process. To avoid conflicts in reading the images due to special symbols that may appear in the file name or incompatibility of the image format, the name and image format of the resulting dataset were adapted by naming the images from 1 to 1,510 using 4 digits in all of them. In addition, taking into account that the original data had different image formats (JPG, JPEG and PNG), these were unified by converting them to JPG. It should be noted that, although a loss of information may occur when the image format is changed, it is expected that the model will be able to cope with it.

Data Annotation: The annotation process was conducted using the *Make Sense* [19] tool in two phases: first, we hired a bounding box data annotation expert (a person with extensive prior experience in this work) for a total of 90 h of work. Then the research team checked the resulting annotations and made some modifications where necessary (e.g., misspelling of the label and misplaced bounding box). The data annotation was obtained in PASCAL VOC format from this tool. Then, the annotations were automatically converted to YOLO format with a developed Python code. In this way, the annotations were compatible with training the models used in the experimentation. For the annotation, two types of objects have been considered: head, which corresponds to the head of the worker without a helmet; and helmet, which corresponds to the head of the worker with a helmet. The importance of covering the helmet with the face is that otherwise, the model could detect a helmet placed on a location other than the head, which is not appropriate from a safety point of view. Therefore, helmets that are not worn are not considered valid instances for the helmet class.

Data Statistics: The proposed MSHD contains 1,510 images acquired in a mining environment with sizes ranging from 640×352 to 12000×9000 pixels, with a total number of annotations of 7,644 distributed in 6,660 for helmet and 984 for head. The data were acquired from a variety of mining activities, environmental conditions and acquisition conditions, ensuring a broad representation for application to the problem at hand. For experimentation purposes, MSHD was randomly divided into training and test sets. The training set contains 80% of the images, with 5,434 instances for the helmet object and 775 for the head object. The remaining 20% of the images were used for testing, with 1,226 instances for the helmet object and 209 for the head object. Both the number of images with the object instances are distributed in a stratified manner with approximately the same proportion.

3 Methodology

3.1 Baseline Architecture

ScaledYOLOv4 is an object detector developed using the same framework as YOLOv5. This method redesigns YOLOv4 and provides an efficient architecture (in terms of accuracy and inference time) taking into consideration the size of the input image. Several optimizations are proposed over YOLOv4 to provide a better balance between accuracy and response time, such as: a) modification of the CSPDarknet53 backbone with the incorporation of residual layers; b) in the neck of the network, CSP technology is applied over PAN and a SPP module is incorporated in the central position, reducing parameters and improving accuracy; and c) the multi-head strategy proposed in YOLOv3 is maintained, which consists in the generation of features maps of different dimensions for the detection of object instances at multiple scales. ScaledYOLOv4 offers small preconfigured models (tiny variant), to be used in low-end devices, and large ones (to be used in mid-range or high-end devices). In our experimentation we have considered the P5 variant of ScaledYOLOv4-large because it is possible to use it on mid-range devices and offers real-time performance [3].

3.2 Uncertainty Quantification

Uncertainty in DL can be quantified by single deterministic methods, Bayesian methods, ensemble methods and test time augmentation [7]. The simplest form, but one of the most effective to quantify the uncertainty, is employing ensemble learning. One of the first methods published in this group was Deep Ensemble [13], which serves as the baseline for the proposed approach. The method is inspired by MC-dropout [6] uncertainty estimation. The latter can be interpreted as a Bayesian inference approach in which a Bernoulli distribution is placed over the neural weights [6]. In MC-dropout, it can be observed that multiple randomly set fine networks, with shared weights, are used at inference time to samples from the predictive distribution. Based on this observation, in [13], the performance of the non-Bayesian Deep Ensemble method for quantifying uncertainty in image recognition was evaluated. They showed that using an ensemble built from the same architecture trained with random initialization of

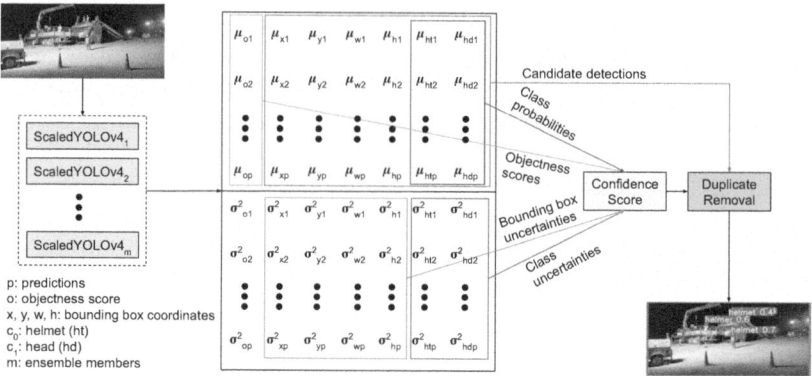

Fig. 4. Pipeline of the proposed Uncertainty-driven ScaledYOLOv4-based technique for helmet detection.

weights can effectively quantify uncertainty and improve model performance. In addition, different numbers of members were evaluated and it was concluded that 5 members is a good balance between resources and quality of uncertainty.

Consider the Deep Ensemble method for the image classification problem, with input data d, M ensemble members and the model output $f_m^W(d)$ with weights W belonging to the m-th member. Then the prediction (p) and the epistemic uncertainty ($\mathcal{H}_1$) is formally calculated as follows:

$$p(d) = \frac{1}{M} \sum_{m=1}^{M} f_m^W(d) \qquad (1)$$

and

$$\mathcal{H}_1(d) = -\sum_{i=0}^{C} p(d)_i \times log(p(d)_i) \qquad (2)$$

Therefore, in the Deep Ensemble method, the prediction is obtained by averaging the prediction of each member belonging to the ensemble ($p(d)$), and the uncertainty is calculated based on the Shannon entropy ($\mathcal{H}_1$).

3.3 Uncertainty-Driven ScaledYOLOv4 for Helmet Detection

The proposed UDScaledYOLOv4 provides object detection and uncertainty related to both bounding box coordinates and predicted classes. The uncertainty is quantified based on the Deep Ensemble approach, computed from several independently trained ScaledYOLOv4 models with random initialization of the weights. As observed in Fig. 4, during inference time, the P predictions generated for each ensemble member are used to compute the mean prediction (μ_p) and variance (σ_p^2), concerning the objectness score (o), bounding box coordinates (x, y, h and w) and class score for each one (e.g., helmet ht and head

hd). Instead of the Shannon entropy used for image classification in the Deep Ensemble approach, for object detection, the variance ($\mathcal{H}_2$) is seen as a better measure to quantify the uncertainty of both the bounding box coordinates and the predicted class. As discussed in [4] the original versions of YOLO cannot quantify uncertainty. The same authors demonstrate that by incorporating location uncertainty into the detection process it is possible to reduce false positives (FP) by filtering out highly uncertain bounding boxes. Although YOLO provides a class score, it cannot be used as a measure of uncertainty because it is calculated from data known to the model, but says nothing about the behavior of the model with unknown data. We argue that, in real solutions, the objective data can be drastically different from those used in model training, since it is not possible to capture the wide variety of scenarios that occur in an open environment and, therefore, we must be aware of the uncertainty for both the bounding box coordinates and the predicted class to provide robust solutions and prevent wrong detection. Therefore, we propose a new criterion (Cr) to compute the confidence score on detection which considers the mean i-th class probability (cs_i), the mean objectness score (o), bounding box uncertainty ($\mathcal{H}_{bbox}$) and class uncertainty ($\mathcal{H}_2(cs_i)$). Note that Cr acts at the time of inference as a post-processing method to improve decision-making and therefore does not interfere during the training stage. Let's consider four individual lists: x', y', w' and h'; containing M elements each one corresponding to each value of M bounding box coordinates (x, y, w, h) predicted by the ensemble members, then, the Cr for the i-th class is formally defined as:

$$Cr_i = cs_i \times o \times (1 - \sqrt{\mathcal{H}_{bbox}}) \times (1 - \sqrt{\mathcal{H}_2(cs_i)}) \qquad (3)$$

where

$$\mathcal{H}_{bbox} = \frac{1}{4}(\mathcal{H}_2(\frac{x'}{max(x')}) + \mathcal{H}_2(\frac{y'}{max(y')}) + \mathcal{H}_2(\frac{w'}{max(w')}) + \mathcal{H}_2(\frac{h'}{max(h')})) \qquad (4)$$

,

$$\mathcal{H}_2(z) = \sigma^2(z) = \frac{1}{M-1} \sum_{m=0}^{M-1} (z_m - \mu(z))^2 \qquad (5)$$

and

$$\mu(z) = \frac{1}{M} \sum_{m=0}^{M-1} z_m \qquad (6)$$

As can be seen in the Eq. 3, when there is no uncertainty in the bounding box coordinate and class, the criterion Cr_i is the same as the original one used in the YOLO family models. When some of them are completely uncertain, Cr_i is 0, avoiding providing this prediction. Otherwise, Cr_i tends to be smaller depending on the degree of uncertainty computed. Take into account that after calculating the proposed criterion, candidate detections that have a Cr_i above a fixed confidence threshold are selected. These detections are then filtered by the NMS method, using a predefined IoU threshold, to remove duplicates and provide the final detections. The proposed Cr is expected to avoid providing erroneous predictions when the data differ from what the model learned during training.

4 Experimental Setup

The variant P5 of ScaledYOLOv4 was selected because of the performance achieved in object detection, its use in real-time and the capacity of the available hardware. We keep the same architecture of this network and only modify the last layer to be able to detect two different types of objects (helmet and head). In addition, we considered 896 × 896 pixels as the input size of the network.

Before starting the training, ScaledYOLOv4 automatically calculates the 12 anchors (4 for each scale), using a genetic algorithm, to fix the ratio to the target dataset. The network is then pre-trained on COCO datasets for better initialization of the weights and retrained for 300 epochs. Optimization is performed by Stochastic Gradient Descent (SGD) with a Nesterov momentum of 0.937, weight decay of 5e-4, and batch size equal to 4. As for learning rate (lr) scheduling, warm-up is performed for 3 epochs using an lr of 1e-5, increasing linearly to 1e-2. After that, the lr decreases at the end of each training epoch by a cosine decay. All values were assigned taking into account the default configuration of the ScaledYOLOv4 and the computational resources.

The only data preprocessing applied is a scaling of the image pixel intensity to the range 0-1. In addition, several data augmentation techniques are applied at training time to avoid overfitting and improve generalization ability. In particular, color adjustment in HSV space (with a maximum adjustment of ± 0.015, ± 0.7 and ± 0.4 for h, s and v respectively), random horizontal and vertical shift across the entire image, scaling (with a minimum of 0.2 and a maximum of 0.8), random horizontal flips (with a probability of 0.5) and Mixup [27] (with a probability of 0.5) are used. During testing, no data enhancement is applied.

To perform the final detection, the confidence threshold has been set to 1e-3 and the IoU threshold is determined empirically. On the other hand, we have considered 5 members to form the ensemble, which is the minimum number of members to ensure a good quality uncertainty quantification [13]. All models are trained following the same configuration described above, with the only difference being the random initialization of the weights.

To evaluate the performance of the baseline and the proposed method we used the standard metric used in object detection methods corresponding to mAP@.50. This metric evaluates the Average Precision (AP) for each class (head and helmet) individually and then averages the results obtained. The AP is calculated with the resulting bounding boxes after filtering out those with an Intersection over Union (IoU) of 50% or more. In addition, Precision (P), Recall (R) and $F1_{score}$ are calculated by considering the average performance of the helmet and head classes. These metrics allow us to show the ability of the model to detect instances avoiding miss or duplicate detection.

5 Results

5.1 Helmet Detection Performance

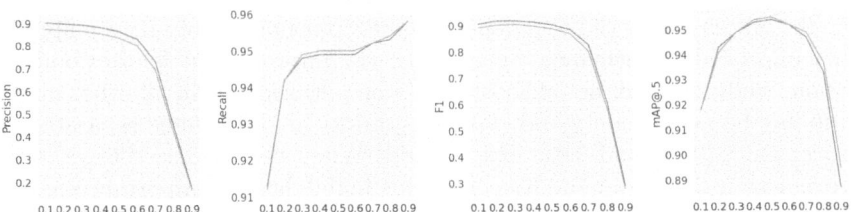

Fig. 5. Plots with precision, recall, F1 score and mAP@.50 obtained with the proposed approach (blue line) and the baseline (red line) varying the IoU threshold. (Color figure online)

ScaledYOLOv4 provides two thresholds (confidence and IoU) that can be set to provide a better balance between true and false positive detection. The confidence threshold was set to a very small value, 1e-3 (default value), and the IoU threshold was changed from 0.1 to 0.9, with increments of 0.1, to check the difference in performance between the baseline model (ensemble model) and the proposed model (uncertainty-driven ensemble model). As can be seen in Fig. 5, UDScaledYOLOv4 provides a marked improvement in performance over the baseline for various thresholds. Although the baseline incorporates the same members, the proposed uncertainty-based confidence allows for the reduction of false positives without significantly impairing correct detections. After analyzing the obtained results, an IoU threshold of 0.4 was selected for the remainder of the discussion of the results.

Table 1. Helmet detection results of ScaledYOLOv4, baseline and proposed method on MSHD dataset. CProb, Obj, BUnc and CUnc represent each term (from left to right) of the Cr defined in Eq. 3.

Method	Obj	CProb	BUnc	CUnc	P	R	F1	mAP@.50
ScaledYOLOv4	✓	✓	-	-	84.6 ± 1.3	93.6 ± 0.6	88.9 ± 0.8	93.7 ± 0.6
Ensemble	✓	✓	-	-	85.3	95.0	89.9	95.4
UDScaledYOLOv4	✓	✓	✓	-	85.9	94.9	90.2	95.4
UDScaledYOLOv4	✓	✓	-	✓	87.6	95.0	91.2	95.4
UDScaledYOLOv4	✓	✓	✓	✓	**88.1**	94.9	**91.4**	95.3

Table 1 shows the performance obtained by ScaledYOLOv4 (mean and standard deviation of the results for five trained models), the baseline model corresponding to an ensemble that averages the predictions provided by all members, and the proposed method that has the same members as the baseline, but considers individually the bounding box uncertainty and the class uncertainty, and also both together to generate the final prediction. As expected, the ensemble model immediately provides an improvement in performance over the single model, in this case by about 1.0% in the F1 and 1.5% in mAP@.50. Using the same ensemble model but changing the confidence score for our uncertainty-driven approach further improved performance. Specifically, the improvement focuses on model precision, while maintaining at most the same performance in all other metrics. If bounding box uncertainty and class uncertainty are considered separately, the improvement is 0.6% and 2.6%, respectively. When both are used together, the improvement is 2.8% over the baseline. This highlights the importance of quantifying uncertainty at the class level, which in some cases allows us to avoid erroneous detection when it is very difficult to distinguish between helmet or head classes.

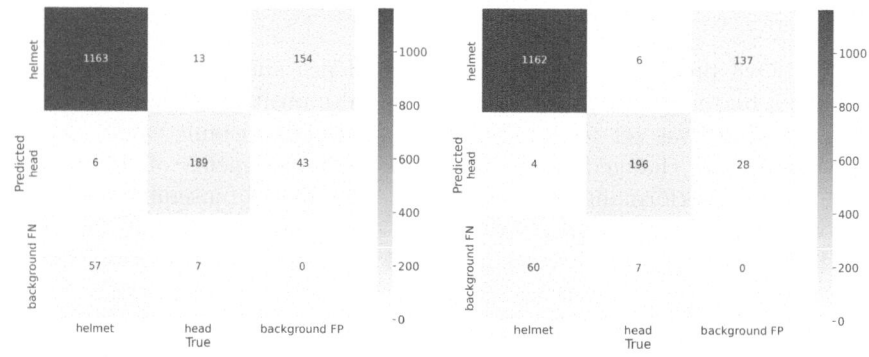

Fig. 6. Confusion matrix corresponding to baseline (left) and UDScaledYOLOv4 (right) performance.

Figure 6 shows the confusion matrix with the results achieved by the baseline and UDScaledYOLOv4 on the MSHD test set when using a confidence threshold of 0.1. Three classes can be distinguished: helmet, head, and background. The last class corresponds to the entire surface of the image where there are no instances for the first two classes. When both confusion matrices are compared, an improvement is evident in the reduction of false positive detection, mainly in the background but also between helmet and head classes. This reaffirms the fact that the proposed approach can rule out unreliable predictions. Regarding the detection of true positives, a small improvement is observed in the proposal, particularly in the minority class when the head is not easy to differentiate from the helmet. As for the false negative, it is observed that UDScaledYOLOv4 has a slightly worse performance than the baseline produced in those predictions with very low confidence that was finally discarded when the confidence was refined with our criterion.

Concerning computational resources, both the baseline and UDScaledYOLOv4 are larger than ScaledYOLOv4, where the increase is proportional to the number of members considered in the ensemble. Despite this, interestingly, the proposed approach can improve the results much more while maintaining the computational resources required by the baseline. However, in critical environments, this cost can be assumed when the increase is related to greater prediction accuracy.

Table 2. Helmet detection results of the baseline and proposed method on the MSHD dataset for different acquisition conditions.

Method	P	R	F1
Poor illumination			
Ensemble (Baseline)	86.30%	85.50%	85.90%
UDScaledYOLOv4	86.80%	84.00%	85.38%
Air Pollution			
Ensemble (Baseline)	87.70%	97.60%	92.39%
UDScaledYOLOv4	89.00%	97.40%	93.01%
Long-distance			
Ensemble (Baseline)	81.90%	93.33%	87.24%
UDScaledYOLOv4	86.20%	93.30%	89.61%
Others			
Ensemble (Baseline)	85.89%	95.65%	90.48%
UDScaledYOLOv4	88.63%	95.65%	91.98%

Table 2 presents the results in terms of Precision, Recall and F1, distributed according to the main acquisition conditions, which are: Poor illumination, Air pollution, Long-distance (small objects) and Others. From the results, it is observed that the proposed method has a better Precision than the baseline. In particular, in long-distance image acquisition, the improvement is more than 4%, highlighting the advantages of our method in reducing false detection in difficult conditions. As for Recall, in almost all cases the improvement provided by the proposed method does not affect this metric. However, we observed a 1.5% decrease in performance compared to baseline in poorly illuminated images. We assume that the reduction in Recall is because in this condition the helmets are not distinguishable and therefore a high uncertainty is associated with this prediction which is finally filtered out with the proposed uncertainty-based criterion. Finally, when both metrics are weighted to calculate F1, it can be seen that the proposed method outperforms in all cases, except for poor lighting. In this case, the difference is very low (around 0.5%).

Table 3. Cross-dataset results of the baseline and the proposed method. The size of the training and test sets is indicated in parentheses below the dataset names.

Dataset	Method	P	R	F1	mAP
MSHD/HHW (1,208)/(1,766)		All			
	Ensemble (Baseline)	75.4%	89.2%	81.7%	89.0%
	UDScaledYOLOv4	81.0%	88.3%	84.5%	89.3%
		Helmet			
	Ensemble (Baseline)	87.9%	87.5%	87.7%	90.9%
	UDScaledYOLOv4	90.4%	86.6%	88.5%	91.0%
		Head			
	Ensemble (Baseline)	62.9%	90.8%	74.3%	87.2%
	UDScaledYOLOv4	71.5%	90.1%	79.7%	87.7%
HHW/MSHD (5,297)/(302)		All			
	Ensemble (Baseline)	72.3%	76.6%	74.4%	72.9%
	UDScaledYOLOv4	74.0%	76.3%	75.1%	73.0%
		Helmet			
	Ensemble (Baseline)	67.8%	78.6%	72.8%	71.6%
	UDScaledYOLOv4	68.6%	78.4%	73.2%	71.3%
		Head			
	Ensemble (Baseline)	76.8%	74.6%	75.7%	74.3%
	UDScaledYOLOv4	79.5%	74.2%	76.8%	74.6%

5.2 Cross-Dataset Evaluation Between HHW and MSHD

The performance of the models was also evaluated using cross-datasets. That is, the models were trained using the training set from one dataset and evaluated using the test set of the other. Table 3 shows the performance obtained with the MSHD/HHW and HHW/MSHD dataset, that is, using the MSHD training set and the HHW test set and vice versa. Overall, performance increases by more than 10% in terms of F1 score and by more than 15% in terms of mAP when we compare the performance obtained in MSHD/HHW concerning HHW/MSHD. It is interesting to note that the improvement is achieved even though the proposed dataset contains a smaller number of images. This suggests that the diversity of the data contained in the proposed dataset, much higher than that of HHW, allows us to obtain good performance even on data acquired in other domains. Specifically, when we compare the results obtained by the proposed method concerning the baseline, we again obtain a significant improvement of the accuracy in both cross-data sets, being higher in MSHD/HHW. When the results are analyzed at the class level, the greatest improvement is achieved in the head class, which suffers from a strong imbalance compared to the helmet class. In the same line, it is observed that using the MSHD/HHW set the proposed method provides better performance for the helmet class and in HHW/MSHD for the head class. This agrees with the results obtained when using both datasets separately suggesting that it is due to the difficulty of the images belonging to the test set in each dataset to detect these classes. Specifically, the results for the method trained and evaluated on HHW in terms of mAP was 99.0% for the helmet and 98.0% for the head, and on MSHD it was 94.7% for the helmet and 96.0% for the head. Finally, it is interesting to note that the performance in terms of mAP of the proposed method trained on MSDH data provides about 10% less when evaluated on HHW compared to MSDH test data. On the other hand, when trained on HHW data it provides more than 20% less on MSDH compared to evaluation on HHW test data. This reinforces the fact that the data variability of the proposed dataset allows for better generalizability of the method.

6 Conclusions

This article proposes a new uncertainty-driven object detection method, based on Deep Ensemble and ScaledYOLOv4, to reduce false detections. In addition, to validate the proposed approach in mining environments, a new dataset called MSHD is proposed for binary helmet detection. From the results, we observe that uncertainty quantification helps us to reduce false positives without significantly affecting the good detection achieved. This matches what is needed in safety-critical applications, such as mining helmet detection, where reducing false positives is as important as detecting all true positives. Specifically, the performance achieved was 91.4% and 95.3% in terms of F1 and mAP, respectively. An additional analysis of the model performance was performed considering different image acquisition conditions, obtaining a high improvement in images taken in the presence of air pollution and also in long-distance images. Furthermore, performing cross-dataset evaluation highlights the complexity of the proposed datasets that allow us to increase generalization even in domains other than the mining industry. Although the proposed UDScaledYOLOv4 shows remarkable results, we have identified some limitations, such as 1) Computational complexity and 2) Lack of data from Underground Mining Environments. Therefore, in future work, we will analyze the integration of different uncertainty quantification methods in an object detection framework to find a method that avoids false detections without requiring an increase in hardware resources and we will also further scale the proposed dataset to cover other mining environments.

Acknowledgments. We acknowledge the support of the management and executive teams for giving us permission to publish the images and use them for non-commercial purposes, and especially to the safety specialists at the Radomiro Tomic mine for capturing and sharing the data from the mining sector. In addition, we are grateful for the support of the Catholic University of the North for the partial funding provided through the project: 202203010030-VRIDT-UCN.

Disclosure of Interests. The authors have no competing interests to declare that are relevant to the content of this article.

References

1. ANANI, A., Risso, N., Nyaaba, W., Tenorio, V.: Application of machine learning in mine safety: a state-of-the-art review. Available at SSRN 4314075 (2022)
2. Balty, I., Mayer, A.: Head protection - ilo encyclopedia of occupational health and safety [internet] (2011). https://www.iloencyclopaedia.org/part-iv-66769/personal-protection-59388/item/689-head-protection
3. Calle Quispe, R.M., Aghaei Gavari, M., Aguilar Torres, E.: Hacia una detección precisa de cascos de seguridad en tiempo real a través de un método basado en el aprendizaje profundo. Ingeniare. Revista chilena de ingeniería **31** (2023)
4. Choi, J., Chun, D., Kim, H., Lee, H.J.: Gaussian yolov3: An accurate and fast object detector using localization uncertainty for autonomous driving. In: Proceedings of the IEEE/CVF Int. Conf. on Computer Vision, pp. 502–511 (2019)
5. De Silva, C.W.: Intelligent control: fuzzy logic applications. CRC press (2018)
6. Gal, Y., Ghahramani, Z.: Dropout as a bayesian approximation: representing model uncertainty in deep learning. In: Int. Conf. on Machine Learning, pp. 1050–1059. PMLR (2016)
7. Gawlikowski, J., et al.: A survey of uncertainty in deep neural networks. Artificial Intelligence Review, pp. 1–77 (2023)
8. Harakeh, A., Smart, M., Waslander, S.L.: Bayesod: a bayesian approach for uncertainty estimation in deep object detectors. In: 2020 IEEE Int. Conf. on Robotics and Automation (ICRA), pp. 87–93. IEEE (2020)
9. He, Y., Zhu, C., Wang, J., Savvides, M., Zhang, X.: Bounding box regression with uncertainty for accurate object detection. In: Proceedings of the IEEE/CVF Conf. on Computer Vision and Pattern Recognition, pp. 2888–2897 (2019)
10. ICMM: Icmm - critical control management: Good practice guide (2015). https://www.icmm.com/en-gb/guidance/health-safety/2015/ccm-good-practice-guide
11. Koopialipoor, M., Jahed Armaghani, D., Hedayat, A., Marto, A., Gordan, B.: Applying various hybrid intelligent systems to evaluate and predict slope stability under static and dynamic conditions. Soft. Comput. **23**, 5913–5929 (2019)
12. Kraus, F., Dietmayer, K.: Uncertainty estimation in one-stage object detection. In: 2019 IEEE Intelligent Transportation Systems Conf. (ITSC), pp. 53–60. IEEE (2019)
13. Lakshminarayanan, B., Pritzel, A., Blundell, C.: Simple and scalable predictive uncertainty estimation using deep ensembles. Advances in neural information processing systems **30** (2017)
14. Miller, D., Dayoub, F., Milford, M., Sünderhauf, N.: Evaluating merging strategies for sampling-based uncertainty techniques in object detection. In: 2019 Int. Conf. on Robotics and Automation (ICRA), pp. 2348–2354. IEEE (2019)
15. Otgonbold, M.E., Gochoo, M., Alnajjar, F., Ali, L., Tan, T.H., Hsieh, J.W., Chen, P.Y.: Shel5k: an extended dataset and benchmarking for safety helmet detection. Sensors **22**(6), 2315 (2022)
16. Peng, L., Wang, H., Li, J.: Uncertainty evaluation of object detection algorithms for autonomous vehicles. Automotive Innov. **4**(3), 241–252 (2021)
17. Ristovski, K., Gupta, C., Harada, K., Tang, H.K.: Dispatch with confidence: integration of machine learning, optimization and simulation for open pit mines. In: Proceedings of the 23rd ACM SIGKDD Int. Conf. on Knowledge Discovery and Data Mining, pp. 1981–1989 (2017)
18. Shen, J., Xiong, X., Li, Y., He, W., Li, P., Zheng, X.: Detecting safety helmet wearing on construction sites with bounding-box regression and deep transfer learning. Comput.-Aided Civil Infrastruct. Eng. **36**(2), 180–196 (2021)

19. Skalski, P.: Make sense (2019). https://www.makesense.ai/
20. Son, H., Choi, H., Seong, H., Kim, C.: Detection of construction workers under varying poses and changing background in image sequences via very deep residual networks. Autom. Constr. **99**, 27–38 (2019)
21. Wang, C.Y., Bochkovskiy, A., Liao, H.Y.M.: Scaled-yolov4: scaling cross stage partial network. In: Proceedings of the IEEE/CVF Conference on Computer Vision and Pattern Recognition, pp. 13029–13038 (2021)
22. Wang, Z., Wu, Y., Yang, L., Thirunavukarasu, A., Evison, C., Zhao, Y.: Fast personal protective equipment detection for real construction sites using deep learning approaches. Sensors **21**(10), 3478 (2021)
23. Xiao, B., Kang, S.C.: Development of an image data set of construction machines for deep learning object detection. J. Comput. Civil Eng. **35**(2) (2021)
24. Xiao, B., Lin, Q., Chen, Y.: A vision-based method for automatic tracking of construction machines at nighttime based on deep learning illumination enhancement. Autom. Constr. **127**, 103721 (2021)
25. Zarie, M., Jahedsaravani, A., Massinaei, M.: Flotation froth image classification using convolutional neural networks. Miner. Eng. **155**, 106443 (2020)
26. Zeng, T., Wang, J., Cui, B., Wang, X., Wang, D., Zhang, Y.: The equipment detection and localization of large-scale construction jobsite by far-field construction surveillance video based on improving yolov3 and grey wolf optimizer improving extreme learning machine. Constr. Build. Mater. **291**, 123268 (2021)
27. Zhang, H., Cisse, M., Dauphin, Y.N., Lopez-Paz, D.: mixup: beyond empirical risk minimization. In: International Conference on Learning Representations (2018)

A Generative Algorithm to Compute NanoFingerprints

Francesc Serratosa[✉][iD]

Universitat Rovira i Virgili, Catalonia, Spain
francesc.serratosa@urv.cat
https://webs-deim.urv.cat/francesc.serratosa/

Abstract. NanoFingerprints have been recently presented as a new representation of chemical nanocompounds designed for toxicity prediction. Only one algorithm to compute NanoFingerprints have been reported, which needs the original 3D structure of the nanocompound in XYZ format and it has a computational cost to the power of four with respect to the number of atoms. This paper presents a generative algorithm to compute NanoFingerprints, which does not need the exact nanocompound structure and it has a linear computational cost with respect to the number of atoms. Our practical experiments report a low error while generating NanoFingerprints and thus a huge capacity of properly representing the initial nanocompound. This is a crucial advance since the toxicity of larger nanocompounds can be estimated without the need of generating the huge representation of its 3D structure.

Keywords: graph embedding · graph regression · metal-oxide nanocompound · chemical 3D-structure · NanoFingerprint

1 Introduction

Chemical metal-oxide nanocompounds consist of tens of thousands of atoms, only composed by a metal and oxygen. Their internal sub-structures remain nearly constant. Although several models exist to predict properties like toxicity, they rely on global properties without utilizing three-dimensional structural information. Funded by the European projects NanoInformatix[1] and Sbd4nano[2], the toxicity level of nanocompounds composed of metal-oxide was parameterised on their three-dimensional structure and modelled for the first time in [14]. NanoFingerprintwas the name of the model. It is composed of a vector that each cell counts the appearance number of some local structural combination of atoms connected by bonds.

Supported by AGAUR research group 2021SGR-00111: "ASCLEPIUS: Smart Technology for Smart Healthcare" and the Spanish project NextPandemics (PID2022-138327OB-I00).

[1] https://www.nanoinformatix.eu.
[2] https://www.sbd4nano.eu.

Figure 1 shows a tiny nanocompound composed of a TiO_2 of size 0.6 nm. Moreover it also shows its NanoFingerprint. The first six elements are global properties. After that, the first column is the cell number of the NanoFingerprint, the second column is the local structure that represents and the third column is the number of appearances. Note, the empty cells of the NanoFingerprint are not shown to make the figure much more comprehensible. In Sect. 3, this representation is detailed. As an example, position 8 counts the number of Oxygen that has two bonds and position 171 counts the number of pairs Titanium-Oxygen, which are connected by a bond and that the Titanium is connected to two other Titanium and the Oxygen is connected to another oxygen.

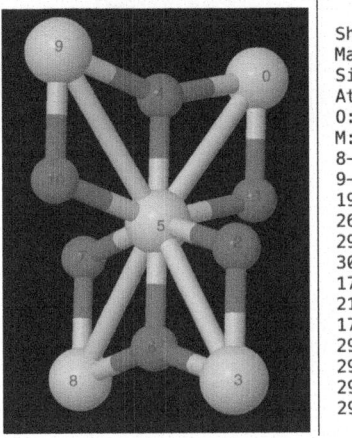

Fig. 1. TiO_2 of size 0.6 nm and its NanoFingerprint. The first column is the number cell of the vector. The empty cells of the NanoFingerprint are not shown.

As a second example, Fig. 2 shows the chemical nanocompound Al_2O_3 that has size 13 Angstroms. This compound only has 30 atoms, 12 of them (in grey) are aluminium and 18 of them (in red) are oxygen. Due to we have its three-dimensional structure[3], we can draw this compound. Moreover, Its NanoFingerprint can be seen in Fig. 3. The maximum number of bonds per atom was set to 10 since no atom has a higher number of bonds.

The aim of this article is to present a fast way to generate NanoFingerprints given some global features of the nanocompound such as size and the type of metal, instead of generating them through the features of an existing nanocompound. In this way, we can predict the toxicity level of a nanocompound before having physically generated it. Thus, reducing the time and computational cost of the research in chemical nanocompounds toxicity.

---

[3] '.XYZ' is the most common file format to describe three-dimensional structures of chemical compounds. It is a very simple and well-know format. Per each atom, it informs of the position (x, y, z) and the type of atom. The information about the bonds is not included and for this reason, it has to be deduced through chemical-physical properties.

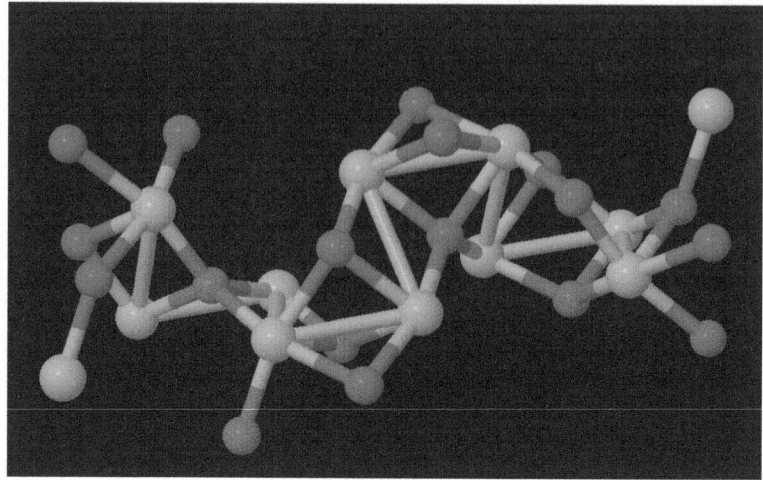

Fig. 2. Chemical nanocompound Al_2O_3 of size 13 Angstroms. Oxygen atoms in red and aluminium atoms in grey. (Color figure online)

```
Section 1
max: 10                     Section4
attr(O): 8                  17841-> M[2,2]_M[2,2]: 1
attr(M): 13                 19183-> M[3,2]_M[3,2]: 1
num(O): 18                  20525-> M[4,2]_M[4,2]: 1
num(M): 12                  21713-> M[5,1]_M[2,2]: 1
Section2                    21724-> M[5,1]_M[3,2]: 1
5-> O[1]: 5                 21834-> M[5,2]_M[2,2]: 1
6-> O[2]: 6                 21845-> M[5,2]_M[3,2]: 1
7-> O[3]: 5                 21856-> M[5,2]_M[4,2]: 2
8-> O[4]: 2                 29732-> O[0,1]_M[5,1]: 4
16-> M[1]: 2                29733-> O[0,1]_M[5,2]: 1
19-> M[4]: 2                29808-> O[0,2]_M[1,0]: 1
20-> M[5]: 2                29821-> O[0,2]_M[2,2]: 1
21-> M[6]: 4                29843-> O[0,2]_M[4,2]: 3
22-> M[7]: 2                29853-> O[0,2]_M[5,1]: 3
                            29854-> O[0,2]_M[5,2]: 4
Section 3                   29929-> O[0,3]_M[1,0]: 1
27-> O[0,1]: 5              29942-> O[0,3]_M[2,2]: 1
28-> O[0,2]: 6              29953-> O[0,3]_M[3,2]: 5
29-> O[0,3]: 5              29964-> O[0,3]_M[4,2]: 3
30-> O[0,4]: 2              29974-> O[0,3]_M[5,1]: 2
158-> M[1,0]: 2             29975-> O[0,3]_M[5,2]: 3
171-> M[2,2]: 2             30063-> O[0,4]_M[2,2]: 2
182-> M[3,2]: 2             30074-> O[0,4]_M[3,2]: 1
193-> M[4,2]: 2             30085-> O[0,4]_M[4,2]: 2
203-> M[5,1]: 2             30095-> O[0,4]_M[5,1]: 1
204-> M[5,2]: 2             30096-> O[0,4]_M[5,2]: 2
```

Fig. 3. NanoFingerprint of Al_2O_3 that has a maximum size of 13 Angstroms shown in Fig. 2. The first element of the NanoFingerprint $max:10$, informs about the maximum number of bonds. Then, the next two elements inform about the nanocompound is composed of oxygen (atomic number: 8) and aluminium (atomic number: 13). Note that only the non-null values have been selected to be shown. Its position appears at the beginning of each line.

An attributed graph is a data structure that represents entities as nodes and their pairwise relationships as edges. Attributed graphs have been used in machine learning and pattern recognition for a long time to represent objects, particularly in representing chemical compounds to predict their toxicity. Their effectiveness stems from their ability to characterise complex structures composed of rich attributes and the application of graph distances, such as the classical Graph Edit Distance. This distance captures dissimilarities between graphs. However, computing graph distances typically incurs a high computational cost, despite using sub-optimal algorithms that return very accurate results. This poses a significant challenge because the computational time required for learning and using the model could be prohibitive, especially when dealing with chemical compounds composed of thousands of atoms and databases containing millions of compounds.

Additionally, graphs with attributes representing the three-dimensional structure of nanocompounds composed of metal-oxide consist of nearly constant sub-structures and only two types of node attributes. This lack of internal variability renders the graph edit distance ineffective, as it fails to discriminate between graphs adequately.

Graph Embedding have been applied to address this issue, Graph Convolutional Networks (GCNs), or Graph Autoencoders to classify or predict chemical compound properties. These embedding methods learn the semantics of attributed graphs and their local structures and convert them into a latent space, represented as a vector of real numbers or a matrix. This latent space allows for the application of classical machine learning methods. It is crucial for these methods that the involved graphs not only have large graph distances but also exhibit rich internal variability, which is composed of several different sub-structures and node attributes. Unfortunately, attributed graphs representing chemical nanocompounds have very low internal variability, as mentioned earlier.

NanoFingerprints [14] were recently presented as a new way to embed in a vector the atomic structure of a nanocompound to later perform regressions for nanocompound toxicity prediction. The computation of this embedding, which has been demonstrated to be very useful, is to the power of four with respect to the number of atoms. ATENA[4] is a web server that allows users to compute NanoFingerprints given a metal oxide nanocompound.

We present a new algorithm for the computation of an approximation of the NanoFingerprints in linear time and without the need of having the whole structure of the nanocompound. The NanoFingerprint is based on the idea of structuring the nanocompound into a graph and latter embedding it into a vector. For this reason, in the next section, we summarise the old graph embedding and graph regression methods, and also we explain the new graph convolutional and graph autoencoder methods. In Sect. 3, we detail the embedding method called NanoFingerprint and in Sect. 4, we first explain the algorithm used to compute the exact NanoFingerprints in ATENA and then we move to

---

[4] https://atena.urv.cat/model/.

our proposal to approximate NanoFingerprints. Section 5 shows the generation of some NanoFingerprints together with their quality. Finally, Sect. 6 concludes this paper.

2 Embedding of Graphs

Graphs are powerful mathematical tools for representing data with complex structures and detailed attributes, and they have been utilised for nearly fifty years in automatic object classification within pattern recognition and for deducing global properties through regression methods [1,3].

The well-known K-Nearest Neighbours algorithm can be employed for graph classification or regression, requiring a distance metric between graphs. The most common used distance is the Graph Edit Distance [11,13]. However, calculating graph distances is computationally intensive and increases the processing time as the number of data elements grows [8–10].

To circumvent the need for calculating graph distances, it is common to convert graphs into a Euclidean space using graph embedding techniques. Earlier techniques for graph embedding did not utilise trainable parameters [4]. Recently, Graph Convolutional Networks (GCNs) [5,17] and Graph Autoencoders [2,7,16] have been developed as tools to embed graphs, distinguishing themselves from older methods by their ability to select intrinsic graph features through learned parameters.

3 NanoFingerprints: A Graph Embedding for Almost Constant Structures

NanoFingerprint is a graph embedding for attributed graphs, which is represented as a vector of numbers, designed to encode attributed graphs with nearly constant small substructures. These graphs have nodes that are labelled into only two classes and they have unattributed edges. This type of graph appears in distinct scenarios, such as predicting the toxicity of chemical nanocompounds (where the graphs represent the three-dimensional structure of the compound) and social network analysis (for example, where nodes represent people labelled by their voting tendencies).

More formally. A graph $G = (V, E)$ is defined as a set of nodes V and a set of edges E. Where, G_i is the ith node in V and $G_{i,j}$ is the edge in E between the ith node and the jth node in V. Besides, the label or attribute of node G_i is $\gamma_i = \{N_k, N_p\}$. N_t is a cardinal attribute, being, $1 \leq t \leq T$. It is important to recall that a specific graph can only have two types of attributes, N_k and N_p and it is not possible to have more than two different types of attributes, such as, N_k, N_p and N_n, being $k \neq p \neq n$. For this reason, for the rest of the paper, we classify the nodes of a specific graph in two types: O and M, where class O are the nodes that have attribute N_k and class M are the nodes that have attribute N_p, being $k < p$.

In the following subsection, the input parameters of the algorithm that generates a NanoFingerprint are detailed. Then, the local substructures are also detailed, which are the basics of the NanoFingerprint. Finally, the definition of NanoFingerprint are specified.

3.1 Input Parameters of the Algorithm

The algorithm needs three basic parameters:

- Shell thickness: A positive real number representing the external radius of the compound, measured in angstroms (Å). Atoms (nodes in the graph) within this defined volume influence on the generation of the NanoFingerprint and are considered to be inside the shell. Only the most external nodes of the graph are used to generate the NanoFingerprints, as these atoms impact the toxicity level. If the entire graph needs to be converted into the NanoFingerprint, a larger value can be specified.
- Maximum number of bonds (MAX): A natural number specifying the maximum number of bonds per atom (edges per node) considered in generating the NanoFingerprint. This parameter ensures a fixed representation of the NanoFingerprint. Note that a larger number results in a larger NanoFingerprint and increases the likelihood of more values with zeros within it.
- 3D structure: The 3D structural information of the NanoFingerprint in the format of a file in XYZ format. An "emptied" 3D structure, an XYZ file containing only the atoms within the shell, can be used for a smaller file size and faster NanoFingerprint generation.

3.2 Local Substructures

A *local structure* refers to a usually small set of interconnected nodes within the graph. In this section, we introduce various *local structures*, which are systematically utilised to define the NanoFingerprint.

$O(x)$: It is a node (atom) type O that has x edges (bonds).

$M(x)$: It is any node (atom) of type M that have x edges (bonds).

$O(x,y)$: A small local structure composed of a central node (atom) of type O connected to (there is an edge) x nodes (atoms) of type O and y nodes (atoms) of type of M. Note these x O and y M could be connected (having other bonds) to other nodes (atoms).

$M(x,y)$: Similarly, a local structure composed of a central node (atom) of type M connected to x nodes (atoms) of type O and y nodes (atoms) of type M.

$O(x,y)-O(x',y')$: A larger structure that is composed of two of the previous ones. It is composed of an $O(x,y)$ and an $O(x',y')$ whose central O are connected by an edge (there is a bond).

$M(x,y)-M(x',y')$: In a similar way, this structure is composed of an $M(x,y)$ and an $M(x',y')$ whose central M are connected by an edge (there is a bond).

$O(x,y) - M(x',y')$: Finally, another similar structure but composed of an $O(x,y)$ and an $M(x',y')$ whose central nodes (atoms) are connected by an edge (bond).

3.3 Definition of the NanoFingerprint Embedding

A NanoFingerprint is composed of a vector of Natural numbers that tallies the occurrences of *local structures*. It is divided into four principal sections: 1) Global Information, 2) Node Information, 3) Edge Information, and 4) Structural Information. More information in [12,14].

- **Section 1: Global information**
  Six global properties are specified in this section.
- **Section 2: Node information**
  $2 * MAX$ values compose this section, where MAX is the maximum number of bonds per atom (edges per node). It includes all combinations of $O(i)$ and $M(j)$ being $0 \leq j \leq MAX$)
- **Section 3: Edge information**
  This section includes the counting of all combinations of the local structures $M(x,y)$ and $O(x,y)$. Its length is $2(MAX+1)^2$.
- **Section 4: Structural information**
  The information of the local structures $O(x,y)-O(x',y')$, $M(x,y)-M(x',y')$ and $O(x,y) - M(x',y')$ is included in this last section. It is composed of $3(MAX+1)^4$ values.

The length of the NanoFingerprint is $5 + 2MAX + 2(MAX+1)^2 + 3(MAX+1)^4$ and depends on the maximum accepted number of bonds MAX. Thus, the theoretical worst computational cost of generating NanoFingerprints is in fourth power of the maximum number of bonds MAX.

4 NanoFingerprint Computation

This section gives a general idea of our algorithm based on a regression method to generate an approximation of a NanoFingerprint. The exact computation is implemented in ATENA. Note the computational cost of the algorithm to compute the exact NanoFingerprint is to the power of four, with respect to the number of atoms, whereas our algorithm is linear with respect to the number of atoms.

The goal of the algorithm is to generate a NanoFingerprint with lower computational cost and without needing the XYZ file beforehand. This allows us to extrapolate a NanoFingerprint for a large nanocompound that has never been physically obtained. Consequently, properties of this nanocompound can be approximated, such as its toxicity level, without having to actually possess it. For instance, if we determine the toxicity of small chemical nanocompounds in the lab but cannot produce larger ones due to practical constraints, we can approximate their NanoFingerprints and predict their toxicity without generating or purchasing them.

This section is focused on the artificial generation of NanoFingerprints. As described in previous sections, it is a must to previously compute the three-dimensional structure of the NanoFingerprint (XYZ file) for the exact computation of the NanoFingerprint, which could be non realistic given some sizes of nanocompounds due to the computational cost. Therefore, this section proposes a machine learning model capable of artificially predicting new NanoFingerprints given only the original parameters of size, thickness and type of metal.

To select the appropriate machine learning model, the behaviour of various specific local structures across different sizes of $TiO2$ nanocompound has been analysed. As an example, Fig. 4 shows the number of the local structures $O(3)$ (Sect. 2), $M(10)$ (Sect. 2), $O(0,3)$ (Sect. 3) and $M(6,4)$ (Sect. 3) throughout sizes ranging from 20 Å to 200 Å and always considering the whole nanocompound (thickness infinitive). It appears to be a clear exponential tendency in the appearance of this structures in the NanoFingerprint. Therefore, the suggested model is an exponential regression.

Note that other more developed models could be explored, for instance artificial neural networks, but in the experimental section were discarded due to the small number of samples.

Moreover, it was noticed that many positions of the NanoFingerprints have null values. Thus, before proceeding to deduce the regression model, the positions that are null in all the samples are not considered and thus, assumed to always being null.

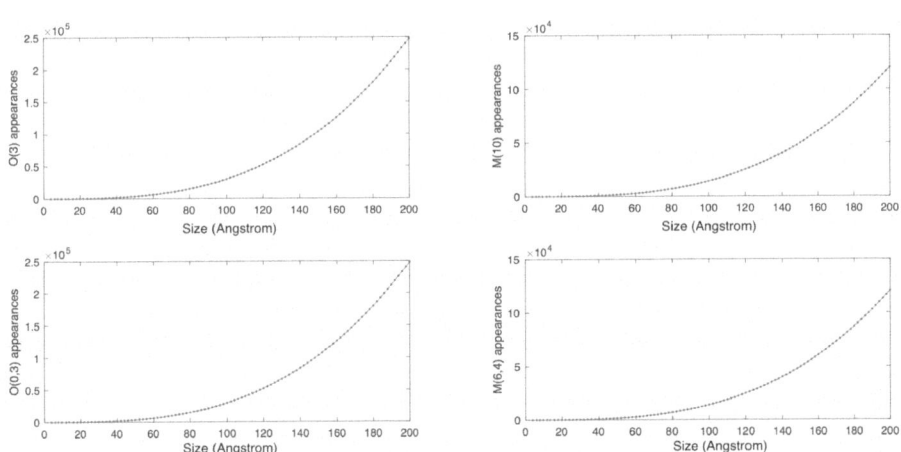

Fig. 4. Number of the Local Structures $O(3)$, $M(10)$, $O(0,3)$ and $M(6,4)$ over different sizes ranging from 0.4 nm to 20 nm for TiO_2.

4.1 Validating the NanoFingerprint Correctness

This section explains the process to validate the correctness of the NanoFingerprint generated by the approximate method. Note that, on the one hand, the original XYZ file is not accessible because the NanoFingerprint has been generated through the regression method and it is supposed the original structure is not available. And on the other hand, the fact of non detecting any error does not mean the NanoFingerprint being completely valid thus some non-correct combinations could be undetectable.

Validation of Section 1:

Any error that appears in Sect. 1 cannot be validated.

Validation of Section 2:

The sum of O(x) counts must be less or equal to the total number of oxygen atoms of the nanoparticle, present at the fifth position (Sect. 1) of the NanoFingerprint, as well as the sum of M(x) counts must be less or equal to the total number of metal atoms of the graph, present at the sixth position (Sect. 1) of the NanoFingerprint. The reason why it is less or equal instead of exactly equal is because the MAX parameter fixes the maximum number of edges that a node can have in order to be counted. If it is the case that the number of real edges is higher than MAX, there are nodes that do not figure in the count and thus, the sum of Sect. 2 nodes (oxygen and/or metal, and vice versa) would be lower than the total number of oxygen and/or metal atoms, and vice versa.

Validation of Section 3:

Three different validations are carried our for each Local Structure $O(x, y)$ or $M(x, y)$ whose count is different than zero:

• The number of edges to other nodes, this is, x+y, must be equal or lower than MAX, since the maximum number of edges is imposed as a parameter.

• The number of edges to other nodes, this is, x+y, must be different to 0 (it does not exist neither O(0,0) nor M(0,0), since it would mean having an isolate node).

• The sum of appearances of the Local Structures whose number of edges to other nodes must be equal to Sect. 2 Local Structure count whose central node and number of edges appear to the same. For example, the Local Structures $O(2,3)$ and $O(4,1)$, whose number of appearances are 3 and 4, respectively, would be compared to the Local Structure of Sect. 2 $O(5)$. Note that $O(2,3)$ and $O(4,1)$ have 5 edges to other nodes both, so their number of appearances would be summed, resulting in 7 (suppose there are no other $O(x, y)$ with 5 edges). Then, in case that $O(5)$ did not have 7 appearances, the NanoFingerprint would not be valid.

Validation of Section 4: For all appearances of structures $O(x,y) - O(x^{'}, y^{'})$, $M(x,y) - M(x^{'}, y^{'})$ and $O(x,y) - M(x^{'}, y^{'})$, it is validated that structures $O(x,y)$ and $O(x^{'}, y^{'})$, $M(x,y)$ and $M(x^{'}, y^{'})$ and finally $O(x,y)$ and $M(x^{'}, y^{'})$, respectively, appear in Sect. 3 with a positive value. Note it might happen to have more count appearances in the Sect. 4 local structures than the addition of the two forming ones in Sect. 3.

5 Practical Experiments

We conducted two classes of experiments. In the first ones (Sect. 5.1), we analysed the quality of the NanoFingerprints generated by the regression model. In the second ones (Sect. 5.2), we deduced the toxicity of nanocompounds based on NanoFingerprints generated by both algorithms: the exact and the regression one.

5.1 Quality of the NanoFingerprints Based on Regression Algorithm

The practical experiment was based on generating NanoFingerprints given both algorithms and then comparing the generated vectors. We computed the normalised mean square error, considering that the NanoFingerprint generated by the exact algorithm was the correct one. Moreover, we also computed the normalised number of detected errors using the validation method in Sect. 4.1. The normalisation process, in both cases, is based on dividing the value by the length of the NanoFingerprint.

To do so, we generated the crystallographic structure of 161 TiO_2 in XYZ format from sizes 40Å to 200Å through the crystallographic tool VI-SEEM[5]. From this initial pool, we randomly split 10 times the samples such that 40 samples were used to test and the other 121 ones were used to train the regression. The shown results are the mean of these 10 tests.

First of all, we wanted to visualise the predicted values of some elements in the NanoFingerprint. Considering Sect. 1 of the NanoFingerprint, Fig. 5 shows the real and predicted number of oxygen (fifth position) and metal (sixth position) over different sizes ranging from 40 Å to 200 Å for TiO_2. The continuous blue line represent the number of occurrences in the exact NanoFingerprint. The green and red dots are the predicted number of occurrences in the training and testing datasets, respectively. We realise the accuracy is specially good since the dots in both datasets accurately follow the blue line.

Considering Sect. 2 of the NanoFingerprint, we have selected the local structure that generated the lower mean square error and the one that generated the larger mean square error, $O(0, 3)$ and $M(3, 1)$, respectively. Similarly to Fig. 5, Fig. 6 shows the real and predicted values. We realise that in the second case, much more samples would be needed for properly modelling the NanoFingerprint.

Figure 7 shows the normalised validated errors. There is a clear tendency of exponentially increasing the number of errors and in the largest NanoFingerprint, 1/10 of the cells obtain non-correct structures.

---

[5] https://nanocrystal.vi-seem.eu.

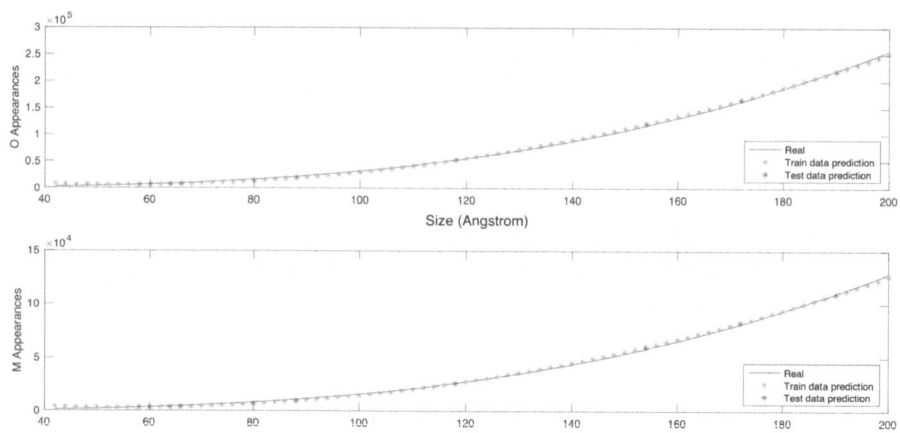

Fig. 5. Real (blue line) and predicted (red and green dots) number of oxygen and metal over different sizes ranging from 40 Å to 200 Å for TiO_2. (Color figure online)

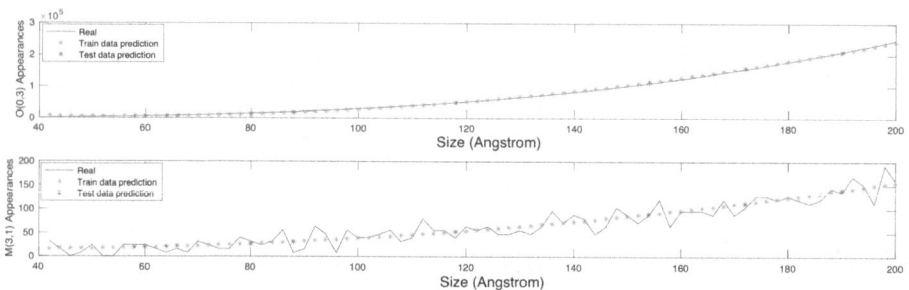

Fig. 6. Real and predicted number of $O(0,3)$ and $M(3,1)$ over different sizes ranging from 40 Å to 200 Å for TiO_2.

Finally, Fig. 8 shows the normalised mean square error. Again, there is a clear tendency of exponentially increasing the mean square error. Thus, it seems logical to think that when the size of the nanocompound increases, so it does the errors while generating the NanoFingerprint. Then, we could conclude there is a limitation on the generation of NanoFingerprints with respect to their desired size.

5.2 NanoFingerprint Regression for Toxicity Prediction

Table 1 shows the accuracy, precision and recall obtained in four different architectures. 1) Data and method in [15]; 2) Regression on the data in [15] concatenated to (represented by symbol '+') the embedding vector generated by [6]; 3) Regression on the NanoFingerprint generated by the 3D-structure of data in [15] and 5) Regression on the NanoFingerprint. In 3) and 5), the NanoFingerprint were generated by the method reported in [12]. Moreover, in bold appear the

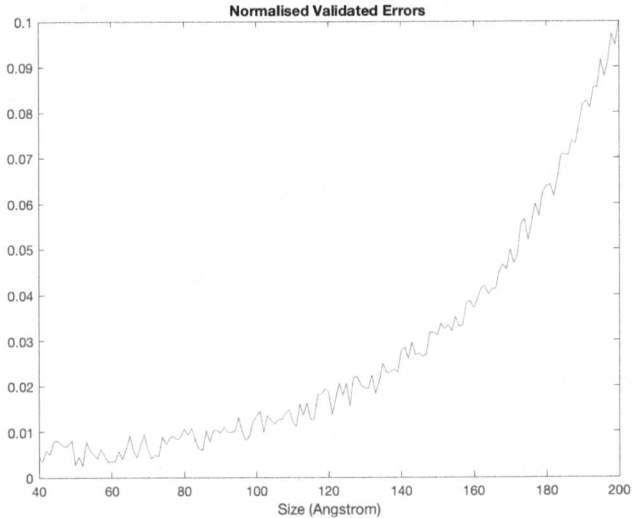

Fig. 7. Normalised validated errors over different sizes ranging from 40 Å to 200 Å for TiO_2.

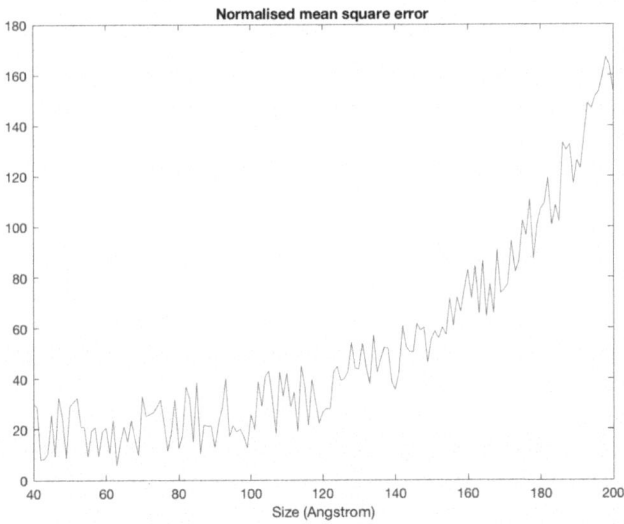

Fig. 8. Normalised mean square error over different sizes ranging from 40 Å to 200 Å for TiO_2.

classification metrics but the NanoFingerprints were replaced by the ones generated by our regression model. 4) The same architecture than 3) and 6) the same architecture than 5).

We realise adding the NanoFingerprint embedding reports higher quality in the classification process. But, this is not the aim of this paper, since NanoFingerprints were presented in [12]. What is in important is that using the NanoFingerprint generated by our new fast method makes a slight decrease of the accuracy, precision and recall, but keeps being the best method. Thus, we conclude have defined a method, which is fast to generate new samples and keeps a high quality. The original data was composed of 484 different nanocompounds with different sizes as reported in the supplementary material of [12].

Table 1. Precision, Recall and Accuracy obtained by: 1) Method in [15] (the same data has been used); 2) Regression on the data in [15] concatenated to the embedding vector defined in [6]; 3) Regression on the NanoFingerprint generated by the 3D-structure of data in [15]; 4) The same architecture than 3) but the NanoFingerprint generated by our new method; 5) Regression on the Fingerprint; 6) the same architecture than 5) but the NanoFingerprint generated by our new method. In all rows, symbol '+' represents concatenation.

	Precision	Recall	Accuracy
1) [15]	0.81	0.98	0.81
2) [15]+GCN	0.80	0.88	0.56
3) [15]+NanoFingerprint	0.92	0.98	0.93
4) [15]+**Generated NanoFingerprint**	**0.85**	**0.88**	**0.89**
5) Regression on NanoFingerprint	0.77	0.78	0.77
6) **Regression on Generated NanoFingerprint**	**0.76**	**0.71**	**0.75**

6 Conclusions

NanoFingerprints have been seen useful for toxicity prediction of metal-oxide nanocompounds and its computation can be done through the public website ATENA. Nevertheless, the computational cost of the algorithm is to the power of four with respect to the number of nodes. As an example, the NanoFingerprint computation of a nanocompound of size 200 Å might take 10 min in ATENA. Since in same cases this time might be excessive, we have presented an algorithm to approximate NanoFingerprint in linear time, which toke 10 s in ATENA. We have analysed the error and we concluded that the generated embedding vectors could be used for toxicity prediction. Thus, we have generated a tool that seems to be very practical for the toxicity prediction of large chemical compounds described by its 3D structure on a graphs. We are currently in the process of including this new algorithm in ATENA.

References

1. Conte, D., Foggia, P., Sansone, C., Vento, M.: Thirty years of graph matching in pattern recognition. Int. J. Pattern Recognit. Artif. Intell. **18**(3), 265–298 (2004). https://doi.org/10.1142/S0218001404003228
2. Fadlallah, S., Julià, C., Serratosa, F.: Graph regression based on graph autoencoders. In: Krzyzak, A., Suen, C.Y., Torsello, A., Nobile, N. (eds.) S+SSPR 2022. LNCS, vol. 13813, pp. 142–151. Springer, Cham (2022). https://doi.org/10.1007/978-3-031-23028-8_15
3. Foggia, P., Percannella, G., Vento, M.: Graph matching and learning in pattern recognition in the last 10 years. Int. J. Pattern Recognit. Artif. Intell. **28**(1) (2014). https://doi.org/10.1142/S0218001414500013
4. Gibert, J., Valveny, E., Bunke, H.: Graph embedding in vector spaces by node attribute statistics. Pattern Recogn. **45**(9), 3072–3083 (2012)
5. Kipf, T.N.: Deep Learning with Graph-Structured Representations. Ph.D. thesis, University of Amsterdam (2020)
6. Kipf, T.N., Welling, M.: Semi-supervised classification with graph convolutional networks. In: 5th International Conference on Learning Representations, ICLR 2017, Toulon, France, April 24-26, 2017, Conference Track Proceedings. OpenReview.net (2017). https://openreview.net/forum?id=SJU4ayYgl
7. Lin, M., Wen, K., Zhu, X., Zhao, H., Sun, X.: Graph autoencoder with preserving node attribute similarity. Entropy **25**(4) (2023). https://doi.org/10.3390/e25040567, https://www.mdpi.com/1099-4300/25/4/567
8. Serratosa, F.: Fast computation of bipartite graph matching. Pattern Recogn. Lett. **45**, 244–250 (2014)
9. Serratosa, F.: Speeding up fast bipartite graph matching through a new cost matrix. Int. J. Pattern Recogn. Artifi. Intell. **29**, 1550010 (2014). https://doi.org/10.1142/S021800141550010X
10. Serratosa, F.: A general model to define the substitution, insertion and deletion graph edit costs based on an embedded space. Pattern Recognit. Lett. **138**, 115–122 (2020). https://doi.org/10.1016/j.patrec.2020.07.010
11. Serratosa, F.: Redefining the graph edit distance. SN Comput. Sci. **2**(6), 1–7 (2021). https://doi.org/10.1007/s42979-021-00792-5
12. Serratosa, F.: Graph embedding of almost constant large graphs. CIARP2023, Iberoamerican Congress on Pattern Recognition (2023)
13. Serratosa, F., Cortés, X.: Graph edit distance: moving from global to local structure to solve the graph-matching problem. Pattern Recogn. Lett. **65**, 204–210 (2015)
14. Serratosa, F., Álvarez, S., Escorhiuela, L., Calatayud, M.: Subgraph nanofingerprint for modelling metal oxide nanoparticles based on connected atoms exploration. NanoWeek and NanoCommons Final Conference 2022 (2022)
15. Subramanian, N.A., Palaniappan, A.: Nanotox: development of a parsimonious in silico model for toxicity assessment of metal-oxide nanoparticles using physicochemical features. ACS Omega **6**(17), 11729–11739 (2021). https://doi.org/10.1021/acsomega.1c01076, pMID: 34056326
16. Wang, J., Liang, J., Yao, K., Liang, J., Wang, D.: Graph convolutional autoencoders with co-learning of graph structure and node attributes. Pattern Recogn. **121**, 108215 (2022). https://doi.org/10.1016/j.patcog.2021.108215, https://www.sciencedirect.com/science/article/pii/S0031320321003964
17. Wu, Z., Pan, S., Chen, F., Long, G., Zhang, C., Yu, P.S.: A comprehensive survey on graph neural networks. IEEE Trans. Neural Networks Learn. Syst. **32**(1), 4–24 (2021). https://doi.org/10.1109/TNNLS.2020.2978386

Impact of Agricultural Production on Climate Change in South America: Comparative Analysis Between 1990 and 2020

Carlos Miguel Aizaga and Rafael Melgarejo-Heredia(✉) [iD]

Pontificia Universidad Católica del Ecuador, 1076 Quito, Ecuador
rmelgarejo@puce.edu.ec

Abstract. This article comprehensively addresses the relationship between agricultural production and climate change in the South American region. Through a documentary study with quantitative approach and, historical data are analyzed to assess the impact of greenhouse gas emissions in agriculture and their consequent influence on climate. Correlation, principal component analysis and clustering techniques are used to identify significant patterns and trends. The results obtained provide valuable information on the dynamics of agricultural production and its relationship with climate change, highlighting the importance of adopting sustainable practices to mitigate adverse effects on the environment.

Keywords: agricultural production · GHG emissions · climate change · patterns · trends

1 Introduction

Climate change is one of the most critical challenges of the 21st century, with significant implications for global food security. As the world population is projected to increase by at least 20% by 2050, agricultural production will need to grow by at least 60% to meet rising food demands, driven by both population growth and changing dietary patterns (Smith & Gregory, 2012). This increased demand is occurring alongside the ongoing impact of climate change on food systems globally, including in South America—a region that, while not among the top contributors to greenhouse gas (GHG) emissions, is still significantly affected by these emissions.

Agriculture, although not the largest contributor to GHG emissions, is a key activity linked to climate change, primarily through the release of nitrous oxide (N2O), methane (CH4), and carbon dioxide (CO2) (Foodinsight, 2021). These gases are emitted through practices such as fertilizer use, irrigation, and the burning of crop residues. While agriculture represents a small proportion of the global economy, it is vital to the livelihoods of millions and plays a crucial role in food production, particularly in South America,

---

The original version of the chapter has been revised. A correction to this chapter can be found at https://doi.org/10.1007/978-3-031-76604-6_20

where major crops like rice, soybeans, and maize are threatened by climate change (Ferrero et al., 2017).

The impacts of climate change on food supply and security are significant, with potential decreases in agricultural productivity due to rising temperatures and increased pest and disease incidences (Rani and Reddy, 2023). The urgency to reduce emissions and achieve sustainable agriculture is clear, as climate change could have detrimental effects, particularly in developing regions (Smith and Gregory, 2012).

As farmers adapt to climate variations, they adjust their farming practices and technologies to manage the risks that accompany changes in crop yields. These adjustments can range from small modifications, such as adjusting crop insurance coverage, to significant investments, such as implementing irrigation systems. Their decisions are influenced by their risk preferences and their perception of how climate change will impact the risks they face (Loduca, 2024).

The impact of climate change on agriculture includes changes in crop yields, water availability, and the prevalence of pests and diseases. Farmers are facing challenges such as rising temperatures, altered precipitation patterns, and more frequent extreme weather events, which can lead to reduced productivity and increased production risks (Acharya & Saren, 2024).

This article examines the relationship between environmental and climatic variables and the production and yield of key crops in nine South American countries (Argentina, Bolivia, Brazil, Colombia, Chile, Ecuador, Peru, Paraguay, and Uruguay) over two periods (1990 and 2020). The focus is on identifying crops that contribute to GHG emissions, specifically N_2O and CH_4, to provide insights into their environmental impact through agricultural activities.

2 Methodology

Regarding the systematic design of the study, the study is framed in a documentary type, with a quantitative approach, using the deductive method and correlational scope. Upon The World Bank and FAO sites mainly, some statistical environmental variables query was made. In the study, continuous numerical environmental variables will be measured, such as the production and yield of agricultural crops, food and greenhouse gas emissions of each of them. The reasoning is deductive, since it makes use of general premises, from which it is intended to obtain conclusions, by methodologies related to data science to count, sort and categorize the data obtained from the observations, to build tables, graphs and/or models that allow simplifying the difficulty of the data involved. A correlational study is used, given that we want to see to what extent the production of the selected crops impacts on variables related to climate change, in terms of their corresponding greenhouse gas emissions.

Throughout this work we made use of the Python programming language using the IDE (Integrated Development Environment) Google Collaboratory, making use of the following packages: pandas, numpy, matplotlib, seaborn, sklearn.preprocessing (StandardScaler, MinMaxScaler), sklearn.decomposition (PCA), sklearn.model_selection (train_test_split), IPython, scipy.stats.

2.1 Data Processing Techniques

Pearson's correlation coefficient reveals how closely two variables are associated with each other, with values between -1 and 1. The closer it is to -1, the more it refers to a perfect negative correlation. Conversely, the closer it is to 1, the more it refers to a perfect positive correlation. On the other hand, values close to O (zero) imply that there is no association between the variables studied.

PCA (Principal component analysis) was applied to identify temperature change patterns across countries in the region. Subsequently, Agglomerative Clustering was used to group observations of selected countries and their crops based on three variables: production, yield, and gas emissions (CH4 and N2O). This clustering method was chosen for its ability to merge similar data points, starting with each point as an individual cluster.

In this case, the CRISP-DM (Cross-Industry Standard Process for Data Mining) methodology is applied, in a first stage, to understand the problem in terms of the impact generated by the variables related to climate change in the selected countries. Figure 1 shows the methodological process used in the analysis of data related to the environmental and economic impact of agriculture and emissions, upon CRISP-DM.

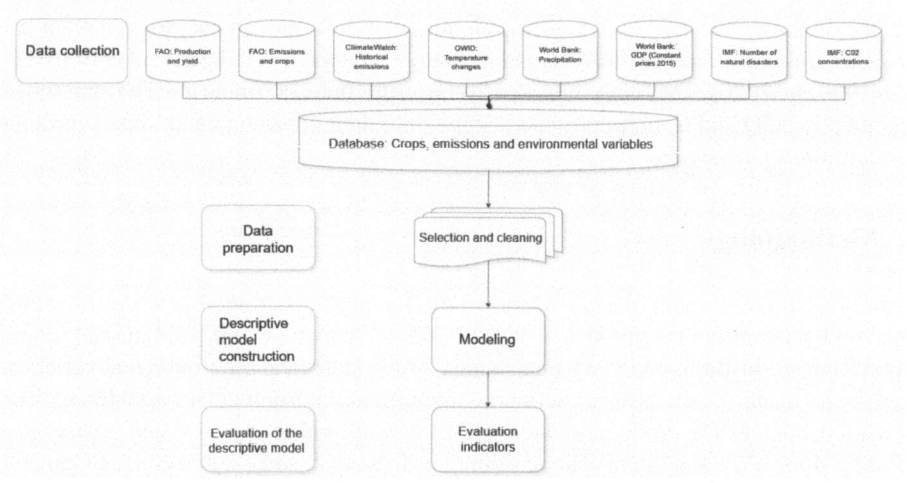

Fig. 1. Diagram for CRISP-DM methodology.

2.2 Data Sources

Using the Web as a Public Web (Melgarejo-Heredia et al., 2016) he first dataset used to obtain data for the clustering study was obtained from FAO (FAOSTAT, Emissions from Crops, n.d.b), with a dimension of 17289 records and 36 variables. The countries corresponding to the South American region were selected from this data, as well as their respective crops. For the section on yield and production of each of the chosen crops, the variables yield (Yield) measured in 100 g/h and production (Production) measured in tons were taken from the FAO website (FAOSTAT, Crops and livestock products, n.d.a), obtaining a total of 2458207 records and 6 variables.

By joining the datasets and after the necessary transformations to obtain the variables of interest related to each country, the type of crop, the type of unit, the type of gas emitted and the respective values for production and yield, we proceeded to filter the data to obtain the two years of interest for this study, obtaining a dataset with 84 records and 8 variables for each of the years. To compile the dataset used in the exploratory study through PCA, a first dataset was used, related to historical emissions by country, obtained from the ClimateWatch web page (ClimateWatch, n.d.) with an original total of 1881 records and 37 variables.

A dataset from the Our World in Data web page (Our World in Data, n.d.) was also used, which contains the annual changes in temperature resulting from the different greenhouse gases, obtaining a total of 50598 records and 79 variables. To further complement the environmental analysis, a dataset related to rainfall in each of the countries was also added, obtained from the World Bank web page (World Bank, n.d.), for a total of 16758 records and 6 variables. Similarly, the Gross Domestic Product variable (measured in constant 2015 dollars) was added in the same way from the World Bank website (World Bank, n.d.), with a total of 266 records and 67 variables.

A dataset related to the type and total number of natural disasters within countries was also added from the Climate Change Dashboard of the International Monetary Fund (IMF Climate Change Indicators Dashboard, n.d.), with a total of 970 records. Finally, also from the same website (IMF Climate Change Indicators Dashboard, n.d.), we obtained data related to CO2 concentrations in the atmosphere with a total of 279 records and 10 variables.

All these databases were previously filtered for the years of interest, 1990 and 2020, and for the selected countries, it was also necessary to homogenize the year and country columns, obtaining a unified data set with 17 variables and 9 observations for each year.

2.3 Variable Operationalization Matrix

The description of each of these variables can be seen in Table 1:

Table 1. Variable operationalization matrix

Dimension	Variable	Indicator	Item Source
Demographic	Population	Population size, number of inhabitants	Our World in Data
	All GHG	Million metric tons of carbon dioxide equivalent	Climatewatch ($mtCO_2e$)

(*continued*)

Table 1. (*continued*)

Dimension	Variable	Indicator	Item Source
	CH4	Million metric tons of carbon dioxide equivalent	Climatewatch (mtCO$_2$e)
Environmental	CO2	Million metric tons of carbon dioxide equivalent	Climatewatch (mtCO$_2$e)
	F-Gas	Million metric tons of carbon dioxide equivalent	Climatewatch (mtCO$_2$e)
	N2O	Million metric tons of carbon dioxide equivalent	Climatewatch (mtCO$_2$e)
	Temperature_change_from_ch4	Degrees Celsius, climate change, CH4 emissions	Our World in Data based on Jones et al. (2023)
	Temperature_change_from_co2	Degrees Celsius, climate change, C02 emissions	Our World in Data based on Jones et al. (2023)
	Temperature_change_from_ghg	Degrees Celsius, climate change, greenhouse gases	Our World in Data based on Jones et al. (2023)
	Temperature_change_from_n2o	Degrees Celsius, Climate Change, N2O Emissions	Our World in Data based on Jones et al. (2023)
	Average precipitation in depth (mm per year)	Millimeters of water	World Bank
	Number of Disasters: TOTAL	Number of natural disasters	International Monetary Fund
	Carbon Dioxide Concentrations (Annual mean)	Percentage	International Monetary Fund

(*continued*)

Table 1. (*continued*)

Dimension	Variable	Indicator	Item Source
Economic	GDP (constant 2015 US$)	Millions of dollars, Gross Domestic Product, economic performance	World Bank
	Production (tons)	Industrial production, measured in tons	Food and Agriculture Organization of the United Nations
	Yield (100 g/ha)	Industrial performance, measured in grams	Food and Agriculture Organization of the United Nations

3 Results

3.1 Correlation Analysis

The correlation matrix was constructed based on all the years of the dataset considered for the subsequent PCA, corresponding to the studied period from 1990 to 2020 (see Fig. 2).

One of the main findings of interest are the correlations of the variables corresponding to gas emissions (except for F-Gas) with the changes in temperatures resulting from these same gases, coefficients that in their totality are greater than 0.80 according to Pearson's correlation coefficient. This does not indicate anything other than the fact that, in the face of a greater emission of gases, the variations in temperature are greater because of these same gases; it is important to note that correlation does not indicate causation.

On the other hand, it may seem worrisome that the correlation between Gross Domestic Product and many of the environmental variables, related to both gas emissions and temperature changes caused by gases, are correlations that exceed 0.85 in all cases. Moreover, the interaction of the Gross Domestic Product with other economic variables, such as production or yield, where although the Gross Domestic Product has a direct and strong positive correlation with production of 0.96, the same does not happen with crop yield, where it is much weaker, with only 0.33. Additionally, it can also be seen that in the case of yield the correlations are much weaker with respect to the other variables (apart from production), showing that crop yield may depend on other variables beyond those considered in this study.

3.2 PCA Study Analysis

From the visualization in the Scree diagram, you may see a good part of the explained variance in the first main component, with 76.88% for 1990 and with 78.24% by 2020,

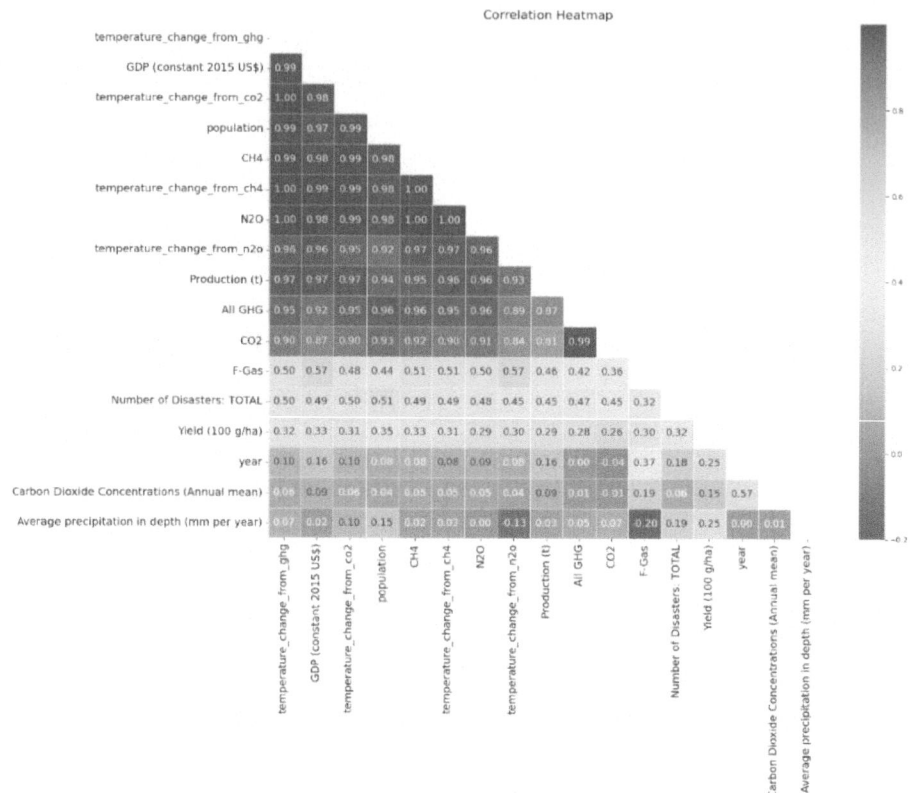

Fig. 2. Correlation matrix for the selected variables

while in the case of the second with only 13.86% for 1990 and 9.68% by 2020. A first analysis shows more than 85% of the variance of the data is in only two components for both, and little being what a third component can contribute to both periods.

Based on the above, and through the application of the PCA algorithm with 2 main components, a type of graph (MDS multidimensional scaling) is obtained that shows the position and distance each of the countries enters according to the newly created (see Figs. 3 and 4).

Through the application of this algorithm, a set of loadings were obtained, corresponding to each of the selected variables and their relationship with each of the main components.

For the PC1, those with positive loadings greater than 0.5 are considered. These include all variables related to greenhouse gases (GHG, CH4, CO2, F-GAS, N2O), as well as temperature changes associated with these gases, which suggests that PC1 is mainly associated with the variability in emissions of greenhouse gases and temperature change. In addition, the gross domestic product, production and population also have positive loadings, indicating a positive association with these emissions and temperature changes.

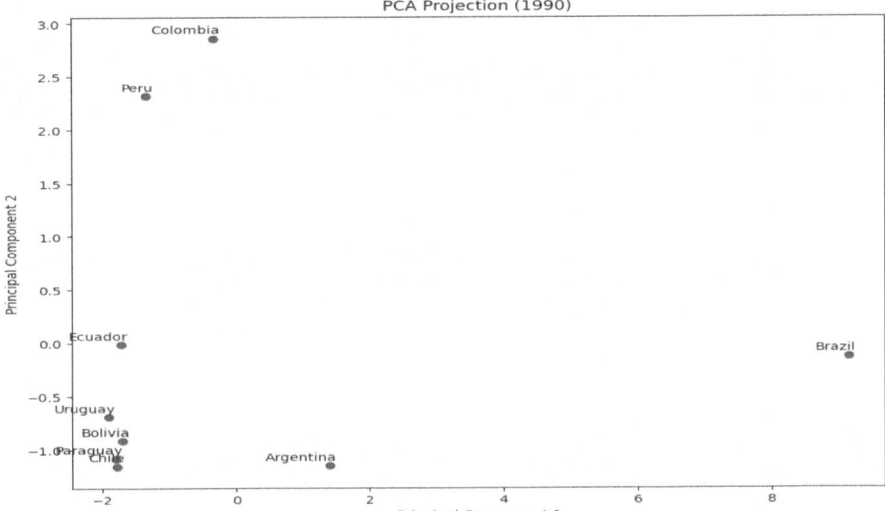

Fig. 3. PCA projection for 1990

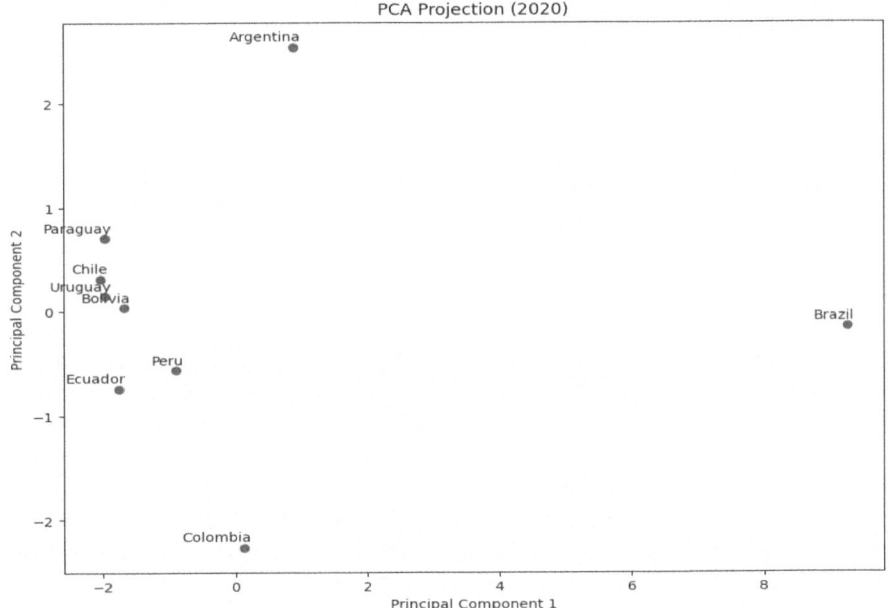

Fig. 4. PCA projection for 2020

For the PC2, the variables with positive loadings greater than 0.5 include the performance, average precipitation and the number of natural disasters, suggesting that PC2 is associated with these variables. Variables are also observed with negative loadings lower than -0.5, such as N2O emissions and temperature change product of greenhouse gases, indicating a negative association.

In summary, PC1 seems to be related to greenhouse gas emissions, economic growth and population, while PC2 is related to agricultural productivity, (Yield), precipitation and natural disasters (see Table 2).

Table 2. Loadings PCA 1990

All GHG	1,049	0,010
CH4	1,057	−0,027
CO2	1,038	0,025
F-Gas	1,038	−0,138
N2O	1,056	−0,056
GDP (constant 2015 US$)	1,058	−0,025
population	1,047	0,106
temperature_change_from_ch4	1,057	−0,074
temperature_change_from_co2	1,052	0,071
temperature_change_from_ghg	1,057	0,034
temperature_change_from_n2o	0,882	−0,255
Yield (100 g/ha)	0,178	0,830
Production (t)	1,048	0,000
Average precipitation in depth (mm per year)	0,033	0,830
Number of Disasters: TOTAL	0,137	0,921
Carbon Dioxide Concentrations (Annual mean)	0,000	0,000

For the PC1 all the variables with positive loadings are greater than 0.5. These include all the variables related to greenhouse gases (GHG, CH4, CO2, N2O) and the associated temperature changes, suggesting that PC1 is mainly associated with the variability in greenhouse gas emissions and temperature change. Also, GDP (in constant dollars of 2015) and the population have positive loadings, indicating a positive association with emissions and temperature change.

For the PC2, variables with positive loadings are observed greater than 0.5, such as the number of natural disasters, indicating that PC2 is associated with the variability in the disaster number. The average precipitation has a significant negative Loading on PC2, suggesting an inverse association with the disaster number.

In summary, PC1 seems to be related to greenhouse gas emissions, economic growth, and population, while PC2 with fishes" (Fluorinated gases "F-gases") such as sulfur hexafluoro (SF6), hydrofluorocarbons (HFC) or Perfluorocarbons (PFC that are not generated directly in agriculture, but mainly in other industries, such as electronics, cooling and the manufacture of chemicals (see Table 3).

Table 3. Loadings PCA 2020

Variables	PC1	PC2
All GHG	1,049	0,010
CH4	1,057	−0,027
CO2	1,038	0,025
F-Gas	1,038	−0,138
N2O	1,056	−0,056
GDP (constant 2015 US$)	1,058	-0,025
population	1,047	0,106
temperature_change_from_ch4	1,057	−0,074
temperature_change_from_co2	1,052	0,071
temperature_change_from_ghg	1,057	0,034
temperature_change_from_n2o	0,882	−0,255
Yield (100 g/ha)	0,178	0,830
Production (t)	1,048	0,000
Average precipitation in depth (mm per year)	0,033	0,830
Number of Disasters: TOTAL	0,137	0,921
Carbon Dioxide Concentrations (Annual mean)	0,000	0,000

Comparing the years 1990 and 2020, it is revealed that in both years the PC1 is related to greenhouse gas emissions and variables related to temperature change with slight variations. While in PC2, productivity, disaster number and precipitation are shown defining 1990 and while F: gas in 2020.

3.3 Cluster Analysis

To achieve an optimal number of K, you can use the silhouette metric, this metric indicates how well the clusters are defined according to data points within each conglomerate, providing information about how relatively similar they are similar between Yes at the same time in which they are different from the points of other conglomerates. The optimal number of K clusters obtained is 4.

N2O emissions for 1990: in Cluster 0, Brazil and Argentina stand out in N2O emissions, corn and soy being the main contributory crops. Brazil leads in corn emissions, while Argentina stands out in soybeans and wheat. The Cluster 1 shows lower emissions, with Argentina, Bolivia and Chile participating in crops such as barley, beans and corn. Cluster 2 highlights sugar emissions from Argentina, Bolivia and Paraguay, while in Cluster 3, Brazil is a leader in sugarcane emissions (see Fig. 5).

N2O emissions by 2020: In Cluster 2 of 2020, Brazil continues to lead corn and soy emissions, with significantly increased figures since 1990. In Cluster 0, Argentina stands out in the production of corn and wheat, while in the cluster 1, Argentina, Colombia and

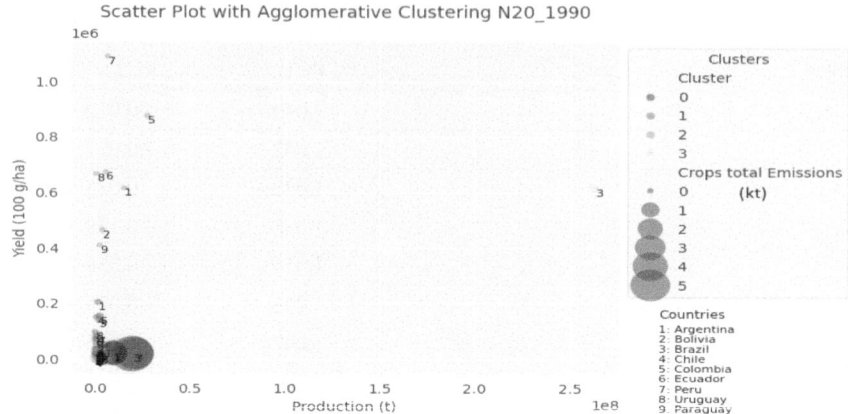

Fig. 5. Scatter plot with Agglomerative Clustering for N2O 1990

Brazil lead sugarcane emissions, Argentina being the main taxpayer. In Cluster 3, Brazil remains the leader in sugarcane emissions, with higher values than in 1990 (see Fig. 6).

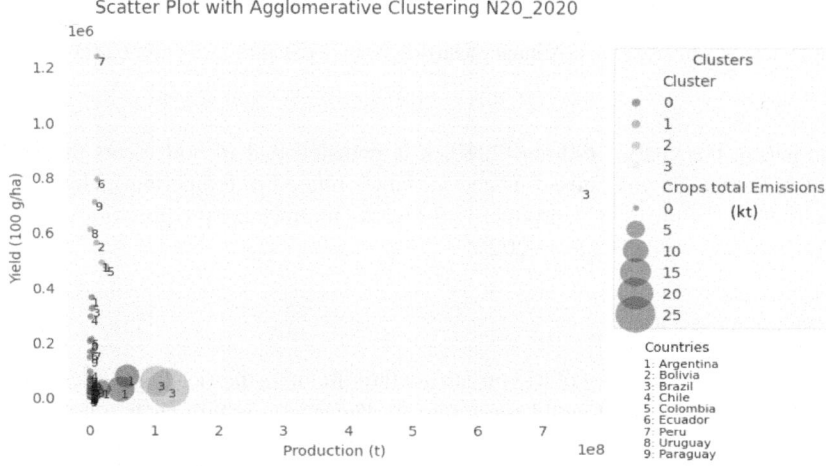

Fig. 6. Scatter plot with Agglomerative Clustering for N2O 2020

CH4 emissions for 1990: in Cluster 0 of 1990, it is observed that Argentina, Bolivia, Colombia, Ecuador, Paraguay and Uruguay are involved in the production of sugar cane, the main taxpayer being Colombia. Meanwhile, in Cluster 1, Brazil stands out significantly with high sugarcane emissions. In Cluster 2, Argentina leads corn and wheat emissions, followed by Chile, Colombia and Ecuador. Cluster 3 highlights corn and wheat emissions in Brazil, with notable figures (see Fig. 7).

CH4 emissions by 2020: In Cluster 0 of 2020, the trend of sugarcane emissions, with Argentina, Bolivia, Colombia, Ecuador, Paraguay and Uruguay, is maintained, Colombia

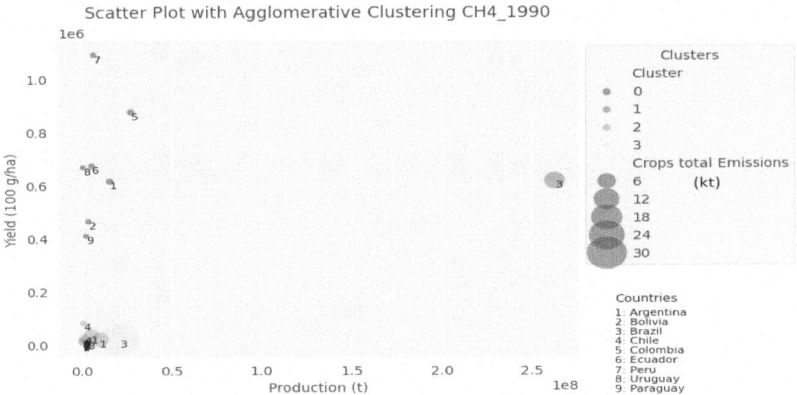

Fig. 7. Scatter plot with Agglomerative Clustering for CH4 1990

being the main issuer. The Cluster 1 shows a great contribution from Brazil in sugarcane emissions, with notoriously higher values than in 1990. In Cluster 2, Argentina leads corn and wheat emissions, followed by Colombia, Brazil and Chile. Finally, Cluster 3 stands out again the significant emissions of corn and wheat in Brazil (see Fig. 8).

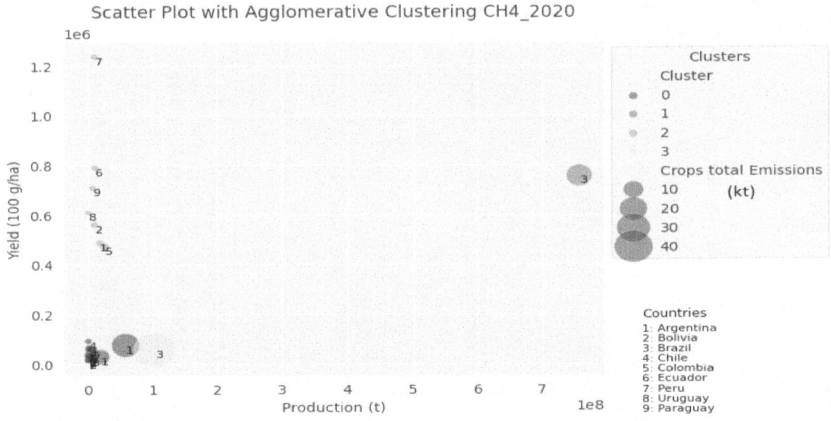

Fig. 8. Scatter plot with Agglomerative Clustering for CH4 2020

Through the results obtained, it can be observed that Brazil and Argentina are key actors in greenhouse gas emissions (N20 and CH4) related to agriculture, the production of sugar cane, corn and soybeans being the main sources of these emissions, this being also a product of its geographical extension. Another interest insight is to see the crop yield characteristic, where sugarcane stands out among the other crops in this regard.

Finally, there is a considerable increase in emission levels in 2020 compared to 1990, which highlights the need to consider these emissions to mitigate environmental impact, paying special attention to crops such as corn and soybeans.

Table 4. Summary table of clustering N20 1990 and 2020

Crops	N20 1990					N20 2020					Variations		
	Country	Crops total Emissions (KT)	Production (t)	Yield (100 g/ha)	Country	Crops total Emissions (Kt)	Production (t)	Yield (100 g/ha)			Emissions	Production	Yield
Sugar cane	Todos	6	8.989.135	683.714	Todos	9	11.976.000	695.198			65%	33%	2%
Barley, Beans, dry, Maize (corn), Oats, Potatoes, Rye, Sorghum, Soya beans, Wheat	Todos	105	512.445	28.661	Todos	502	2.511.778	49.831			377%	390%	74%
Sugar cane	Brasil	194	262.674.144	614.788	Brasil	456	757.000.000	756.043			134%	188%	23%
Soya beans, maize, wheat	Argentina Brasil	4.115	15.500.319	19.143	Brasil	23.249	113.000.000	44.854			465%	629%	134%

Table 5. Summary table of clustering N2O 1990 and 2020

Crops	CH4 1990				CH4 2020				Variations		
	Country	Crops total Emissions (Kt)	Production (t)	Yield (100 g/ha)	Country	Crops total Emisions (Kt)	Production (t)	Yield (100 g/ha)	Emissions	Production	Yield
Sugar cane	Todos	219	8.989.135	683.714	Todos	362	11.978.498	695.198	65%	33%	2%
Maize (corn), Wheat	Todos	1,276	1.576.268	20.453	Todos	2,357	5.948.653,4	38.897	85%	277%	90%
Sugar cane	Brasil	7,498	262.674.144	614.788	Brasil	17,575	757.116.900	756.043	134%	188%	23%
Maize (corn)	Brasil	30,765	21.347.770	18.735	Brasil	49,285	103.963.600	56.955	60%	387%	204%

When observing the clusters with the crops, the information can be summarized in the following tables:

Regarding N2O, there is a growth in the emissions of sugarcane crops in 65%, with an increase in the production of only 33%. While in the case of soy, corn and wheat grains, it happens in reverse, that is, an increase in emissions of 465%, with an increase in the production of 629% (see Table 4).

In relation to the CH4, something like the previous analysis happens, provoking growth in the emissions of the sugarcane crops of 65%, with an increase in the production of only 33%. However, corn crop emissions increased by 60% with a considerable increase in production of 387% (see Table 5).

4 Conclusions

In the first instance, the impact of each of the selected crops could be estimated in terms of their contribution measured in greenhouse gas emissions, observing how there are certain crops with greater contribution with respect to others, as is the case in the case of corn and soybeans in the case of N20, and corn and wheat in the case of CH4, maintaining these same trends in the face of their proportions in both periods studied, 1990 and 2020. In turn, the Performance of other crops, such as sugarcane, finding a lower level of emissions with respect to other crops and providing an outstanding hectare, when compared to other crops.

It is important to recognize the proportion of emissions that Brazil and Argentina have with respect to the other selected countries, being able to easily between the two to represent a proportion of more than 85% of the total emissions.

During the study, the intrinsic relationship between economic variables such as the gross domestic product, with environmental cutting variables, such as changes in temperature and levels of greenhouse gas emissions, has also been observed, as could be seen in the main component analysis. These elements must be considered to evaluate practices when producing, which in the case that is responsible for this research, is related to the search for less polluting agricultural practices.

Regarding the comparison of both periods, the fact of how they remain relevant and intensify the variables related to environmental change, as in the case of the number of natural disasters and changes in rainfall and rainfall are also remarkable, a situation that serves as a basis for generating "Awareness" on the gradual changes produced on the planet, the product of human action.

At present, the use of data science techniques applied to this type of problem represents a great advantage, mainly, due to the opening that allows working with opensource languages, such as Python, and availability of tools such as Google Collaboratory, being able to make deeper analysis raising awareness about what has been happening in our environmental context.

References

Acharya, S.K., Saren, P.: Climate Change, Uncertainty and Agriculture. PMW (2024)

ClimateWatch. (s.f.). Historical Emissions. https://www.climatewatchdata.org/dataexplorer/historical-emissions?historical-emissions-data-sources=climate-watch&historicalemissions-end_year=2020&historical-emissions-gases=All%20Selected&historicalemissions-regions=LAC&historical-emissions-sectors=

FAOSTAT. (s.f.). Crops and livestock products. https://www.fao.org/faostat/en/#data/QCL

FAOSTAT. (s.f.). Emissions from Crops. https://www.fao.org/faostat/en/#data/GCE

Ferrero, R., Lima, M., Gonzalez-Andujar, J.: Crop production structure and stability under climate change in South America. Academia (2017). https://www.academia.edu/45600016/Crop_production_structure_and_stability_under_climate_change_in_South_America%2010.1111/aab.12402

Foodinsight. La Agricultura y las Emisiones de Gases de Efecto Invernadero (2021). https://spanish.foodinsight.org/seguridad-alimentaria/la-agricultura-y-lasemisiones-de-gases-de-efecto-invernadero/

IMF Climate Change Indicators Dashboard. (s.f.). Atmospheric CO_2 Concentrations. https://climatedata.imf.org/datasets/9c3764c0efcc4c71934ab3988f219e0e/explore

IMF Climate Change Indicators Dashboard. (s.f.). Climate-related Disasters Frequency. https://climatedata.imf.org/datasets/b13b69ee0dde43a99c811f592af4e821_0/about

Loduca, N.R.: Understanding Farmer Decisions About Climate Change Adaptation and Conservation Decisions in Agriculture. ProQuest (Michigan State University ProQuest Dissertations & Theses), 2 (2024). https://www.proquest.com/openview/5ec940c9b04bde2fa02b3ad22c2bbd4f/1?pqorigsite=gscholar&cbl=18750&diss=y

Melgarejo-Heredia, R., Carr, L., Halford, S.: The public web and the public good. In: Proceedings of the 8th ACM Conference on Web Science (WebSci2016). Association for Computing Machinery, New York (2016). https://doi.org/10.1145/2908131.2908181

Our World in Data. (s.f.). https://ourworldindata.org/co2-and-greenhouse-gas-emissions. https://ourworldindata.org/co2-and-greenhouse-gas-emissions

Rani, P., Reddy, R.: Climate change and its impact on food security. Int. J. Environ. Climate Change **13**(3), 104–108 (2023). https://doi.org/10.9734/ijecc/2023/v13i31687

Smith, P., Gregory, P.J.: Climate Change and Sustainable Food Production. Cambridge University Press (2012). https://www.cambridge.org/core/journals/proceedings-of-the-nutritionsociety/article/climate-change-and-sustainable-foodproduction/DE02043AE462DF7F91D88FD4349D38E7

World Bank. (s.f.). Average precipitation in depth (mm per year). https://data.worldbank.org/indicator/AG.LND.PRCP.MM

World Bank. (s.f.). GDP-Gross domestic product - 2015 US$-Constant. https://data.worldbank.org/indicator/NY.GDP.MKTP.KD

VAVnets: Retinal Vasculature Segmentation in Few-Shot Scenarios

Idris Dulau[1,2(✉)], Benoit Recur[1,3], Catherine Helmer[1,4], Cecile Delcourt[1,4], and Marie Beurton-Aimar[1,2]

[1] Bordeaux University, Talence, France
[2] LaBRI UMR 5800, Talence, France
idris.dulau@labri.fr
[3] INSERM LAMC U1029, Pessac, France
[4] INSERM U1219, Bordeaux, France

Abstract. The structure of the retinal vasculature can indicate various health issues. Quantitatively measuring changes in retinal arteries and veins offers significant potential for disease prevention and management. We propose VAVnets: three variants of a deep-learning network to generate Vessels, Arteries and Veins binary segmentations. Training is conducted in a few-shot and cross-dataset manner using images from four different fundus datasets, collectively comprising 137 images: DRIVE, DUMO, HRF, and LESAV. Training occurs with 3-fold cross-validation. Each fold employs a cross-dataset-build of 40 images (10 from each dataset), with testing on the remaining 97 images. In this article, we discuss our experiments involving architectural choices, transfer-learning, and data augmentation. We assess performances using the dice score as we aim to achieve the best possible pixel-wise segmentations. Our dice scores for each dataset are, for vessels: 0.81, 0.83, 0.81, 0.86; for veins: 0.78, 0.81, 0.78, 0.79; and for arteries: 0.73, 0.78, 0.74, 0.77. To the best of our knowledge, VAVnets demonstrate superior performances compared to existing few-shot methods across these datasets.

Keywords: Deep-learning · Segmentation · Retinal vasculature

1 Introduction

Changes in retinal vascular morphology are associated with various ocular and systemic diseases. Quantitative measurements of retinal arteries and veins could thus serve as biomarkers for managing such diseases. Performing measurements requires a segmentation step to differentiate the retinal vasculature from other fundus structures. However, the scarcity of available data for arteries and veins makes it challenging to train accurate segmentation networks. We propose VAVnets: three variants of a deep-learning network to generate vessels, arteries and veins binary segmentations. To train and test the networks, we use images from four different fundus datasets that are available with the retinal arteries and veins segmentations: DRIVE [1], DUMO [2], HRF [3] and LESAV [4]. Having access to a total of solely 137 images, we perform the training using a cross-dataset-build of 40 images, 10 from each dataset (which is a few-shot scenario), and the testing on the remaining 97 images. Training occurs with 3-fold cross-validation to compensate for this small amount of data. In Sect. 2, we present the datasets and summarize their characteristics. In Sect. 3, we survey the state-of-the-art few-shot retinal vasculature segmentation of vessels, veins, and arteries. In Sect. 4, we discuss our experiments involving architectural choices, transfer-learning, and data augmentation. In Sect. 5, we analyze our obtained results and then compare them to the state of the art in Sect. 6. Finally, in Sect. 7, we outline future applications of the methods in the field.

2 Datasets

Datasets of fundus images with manual segmentation of veins and arteries are scarce due to the labor-intensive and costly nature of the annotation process, which requires specialized expertise. Ethical and privacy concerns limit data sharing, while variability in image quality and lack of standardized protocols further complicate the creation of comprehensive datasets. Additionally, the technical challenges in accurately distinguishing between veins and arteries contribute to the scarcity of such annotated datasets. Still, some datasets including veins and arteries manual segmentation are publicly available: *DRIVE* [1], *DUMO* [2], *HRF* [3] and *LESAV* [4]. Figure 1 shows a sample of each dataset.

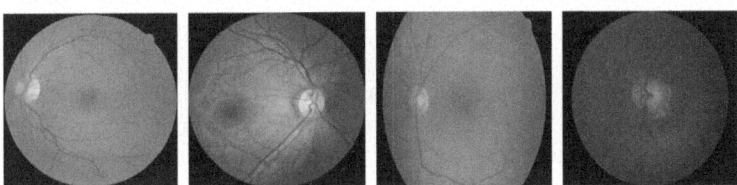

Fig. 1. Samples of the four datasets used in this study. From left to right: DRIVE [1], DUMO [2], HRF [3], LESAV [4].

The datasets characteristics are sum-up in Table 1. Collectively, the four datasets comprise 137 fundus images. These fundus images have been captured using different materials (row. 1), with different resolutions (row. 2), and comprise various pathological signs such as: hemorrhages, glaucoma and exudates.

Table 1. Datasets characteristics

	DRIVE	DUMO	HRF	LESAV
Material	Canon CR5	OT-110M	Canon CR-1	NA
Resolution	565*584	1024*1024	3504*2336	1620*1444
Images	40	30	45	22
Pathologies	7	NA	30	11
Focus	macula	macula	macula	optic disc

3 Related Works

Numerous methods have been developed for retinal vasculature segmentation in fundus images. For this task, deep-learning-based methods produce the best results [5], and the most effective methods are built by modifying the U-Net [6] architecture However, few have addressed the challenge of retinal vasculature segmentation in few-shot scenarios. This gap has recently been explored in seven articles, five for vessels segmentation, and two for veins and arteries segmentation. All these methods, published between 2022 and 2023, underscore the growing interest in tackling real-world application problems within this field.

For vessels segmentation, Xu Jianguo et al. in [7] propose architectural modifications to U-Net, focusing primarily on enhancing the skip-connections. Similarly, Hao-Chiang Shao et al. in [8] also make architectural changes but concentrate mainly on the loss function. Junyan Lyu et al. in [9] introduce an automatic data augmentation method. Hu Dewei et al. in two works [10,11] employ prior knowledge of the vasculature to modify U-Net, emphasizing the "tubular" shapes of vessels.

In veins and arteries segmentation, Shi Danli et al. in [12] leverage 220 fundus images from an in-house dataset to perform cross-training with public datasets. Notably, in contrast to most related work (not limited to few-shot), they generate veins and arteries as two separate binary outputs. A year later, in [13], they adopted a different approach which implements a multi-class output strategy, relying heavily on transfer-learning from fundus fluorescein angiography images.

4 Methods

The primary objective of the proposed VAVnets is to perform the best pixel-wise segmentations of vessels, veins and arteries in few-shot scenarios. To do so, we investigate various aspects of the deep-learning training processes, including architecture, transfer-learning, and data augmentation.

4.1 Architecture

As mentioned in Sect. 3, the most effective retinal vasculature segmentation methods are built by modifying the U-Net [6] architecture. Thus, the proposed

VAVNets perform in a similar fashion, as illustrated in Fig. 2. In addition to the original design, we first include new blocks into the network such as: (i) attention, (ii) normalization, and (iii) dropout; we adjusted: (iv) the feature map sizes and depths of the networks; and we replaced U-Net building blocks by: (v) others blocks that perform similar functions but in different ways. Attention [14] (i) enable more focus on relevant parts of the input data. Normalization [15] (ii) ensure consistent feature distribution across different inputs. Dropout [16] (iii) reduce reliance on specific features. Adjustment of feature maps size and depth [17] (iv) enable the networks to capture more intricate features and patterns in the down-sampled input. Other blocks (v) and related experiments are discussed in Sect. 5.

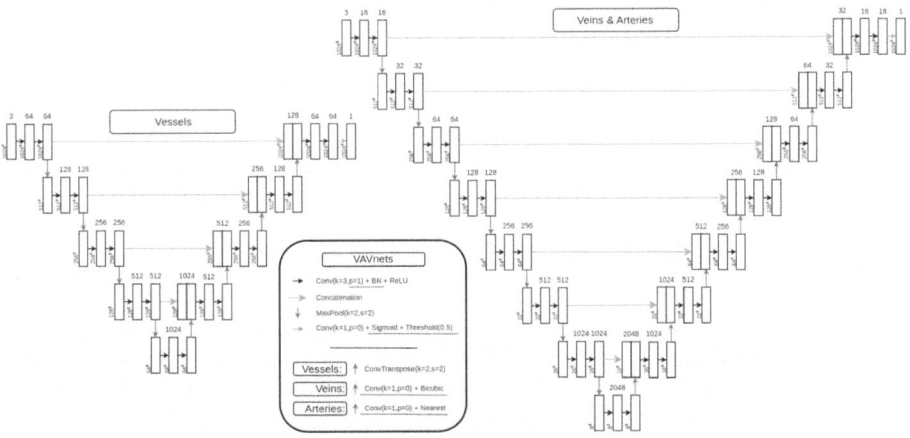

Fig. 2. VAVnets configurations: vessels (left), veins and arteries (right). (Color figure online)

The proposed **VAVnets**, share common characteristics between themselves as illustrated in Fig. 2. Batch-normalization layers are added after each convolution and before each ReLU. Feature maps are concatenated between up and down paths Maximum-pooling layers are used for down-sampling. Convolution followed by a sigmoid and a threshold is used to generate a single class output. Vessels, arteries and veins networks have also distinct characteristics between themselves in term of depth, feature map size and up-sampling. The vessels segmentation network is built with transpose-convolution for up-sampling, and a depth of five with feature maps size ranging from 64 to 1024 as in the U-Net. The veins network performs the up-sampling with a convolution followed by a bicubic interpolation, while the arteries network up-sample with a convolution followed by a nearest neighbors interpolation. Both share a depth of eight and a feature map size adjusted from 16 to 2048. Some modifications of the U-Net [6] leading to VAVnets are underlined in red in Fig. 2. First, padding is applied to the convolutions to maintain the size of the convolved input across consecutive layers, eliminating the need for cropping before concatenating skip connections. Also, padding in par with changes in the input size now being 1024*1024 enable to adjust the feature maps size and to go deeper in the architecture. Batch-normalization added after each convolution lead to interesting findings for few-shot scenarios (see in Sect. 5). The output is a single channel processed through a sigmoid function and then thresholded to produce the binary segmentations. Finally, the up-sampling operators are modified for arteries and veins networks.

4.2 Transfer-Learning

To cope with a lack of data, we experiment transfer-learning with additional fine-tuning using several publicly available datasets as well as pre-trained networks. We define various configurations utilizing: common objects, curvilinear structures, and retinal vessels from other datasets. Figure 3 shows samples of the data involved in the transfer-learning experiments.

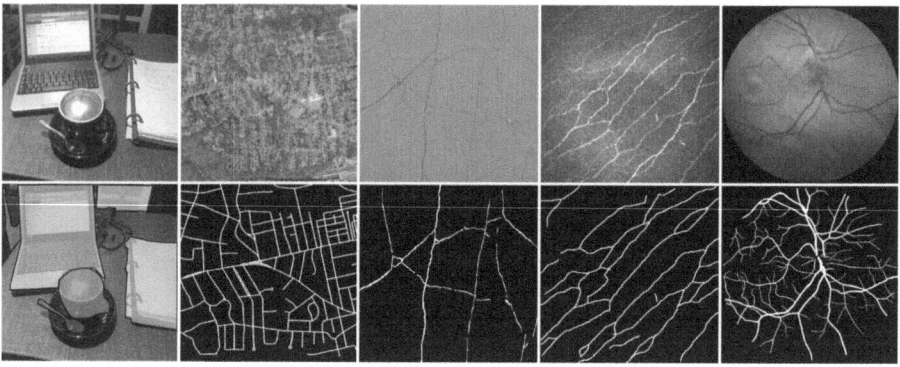

Fig. 3. Samples of the data involved in the transfer-learning experiments. From left to right: common objects [18], roads [19], cracks [20], corneal nerves [21], and retinal vessels [22–24].

First, fine-tuning a network pre-trained on a huge dataset of common objects enable to extract consequent amount of features. By leveraging ResNet-50 [25] abilities to extract rich features from COCO [18] images, we benefit from fundamental low-level features like edges, textures, and colors. Then, curvilinear structures such as corneal nerves [21], wall cracks [20] and aerial roads [19] are good candidates to benefit from more targeted features like tortuosity and connectivity. Up to 10k images can be collected, making these smaller yet sizable datasets but some more targeted choices of external knowledge. Finally, retinal vessels from other datasets can serve to fine-tune our networks. The 1.5K images collected from public datasets [22–24] makes it the less sizable but more targeted option.

4.3 Data Augmentation

Lastly, we experiment many data augmentation techniques to compensate for the data scarcity of few-shot configurations. Figure 4 shows samples of the data augmentation techniques we experiment during the development process of the VAVnets.

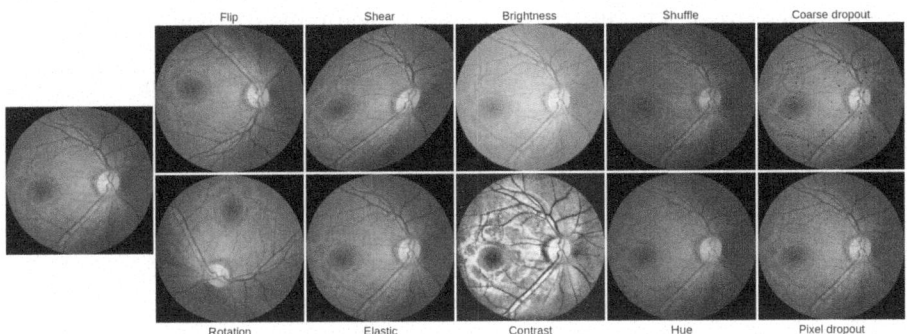

Fig. 4. Samples of the data augmentation techniques

We selected augmentation techniques to mimic the theoretical new unseen data. The different perspectives and orientations that a camera could have taken to obtain the fundus image is reproduced by flips and rotations. The variety of potential retinal vascular shapes that could be present in other fundus is approximated by shearing and elastic transforms. Lighting conditions during the image acquisition are mimicked by non-linear transformations of pixel intensities leading to changes in brightness and contrast. The presence of pathologies or masking luminous artifacts caused during acquisition is represented by masking out small to very small regions of the image, referred to as coarse and pixel dropout.

Also, we observed many manual segmentations performed on identical images by different annotators. Some annotators tend to rely heavily on colors to differentiate arteries and veins. This leads to huge classification differences between them on smaller vasculature, which thus tend to be way less differentiable by solely their color aspect. To steer the deep-learning process away from focusing too much on color, we shuffle the color channels and modify the hue of the image.

5 VAVnets Experiments

For the experiments on the **architecture**, we found that the use of normalization techniques proves to be most beneficial for our specific tasks. In particular, batch normalization [15] when added after each convolution layer. However, we utilize it in a non-standard manner during inference. After training on input training data, we can perform inference on test data using either (Learn) or (Pred). When using (Learn), the batch normalization layers' expectation and variance are computed over the training data. Conversely, when using (Pred), the batch normalization layers' expectation and variance are computed over the testing data. Applying (Pred) on the generated weights and biases during inference, as mentioned in [15], allows us to recover the original activations by setting $\gamma^{(k)} = \sqrt{\text{Var}[x^{(k)}]}$ and $\beta^{(k)} = \text{E}[x^{(k)}]$. It involves no cost, neither in time nor in training resources. This utilization of normalization is particularly important in domain shift scenarios, considering the performance gap observed between (Pred) and (Learn), as shown in Table 2, where (pred) outperform (Learn) at all tasks. The performance gap is observed at the very first stage of the network development (First results) as well as in the very last stage (Last results). The first modification to U-Net is the addition of batch-normalization layers after each convolution. The last development step is the use of data augmentation describe afterwards. Results are reported as the average of each segmentation dice score over all datasets.

Table 2. Inference using the learned data statistics (Learn) compared to inference using the testing data statistics (Pred). **Bold green** is the best dice score.

	Vessels	Veins	Arteries
First results			
Pred	0.813 ± 0.001	0.716 ± 0.002	0.689 ± 0.006
Learn	0.754 ± 0.005	0.591 ± 0.017	0.512 ± 0.027
Last results			
Pred	0.821 ± 0.001	0.788 ± 0.003	0.746 ± 0.005
Learn	0.765 ± 0.003	0.677 ± 0.012	0.626 ± 0.017

Less impactful but still valuable insights are utilized to construct the current configurations. For the vessels segmentation task, U-Net depth proves to be the optimal choice. Deeper or shallower layers compared to U-Net result in a performance decrease. Such cases may introduce either too little or unnecessary complexity for vessels segmentation. In contrast, for veins and arteries, deeper layers yield significantly better results. This suggests that deeper layers are advantageous for capturing the intricate patterns that help differentiate between veins and arteries in the vascular tree. Experiments involving downsampling layers and skip connections do not lead to any improvement compared to U-Net. Regarding up-sampling layers, while variations in performance are observed among the interpolation methods tested, the reasons behind these differences are not entirely clear. The distinctions are subtle, particularly for veins, where the superiority of one method over another is weakly discerned. However, the chosen changes still result in improved results.

For the experiments on **transfer-learning**, performances gradually increase with the proximity of the used data to the vascular structure: common objects, curvilinear structures, and retinal vessels. Initially, the use of COCO-trained weights probably under-performed because vessels, veins, and arteries are too thin and precise structures to benefit from common objects' gathered information. Subsequently, for the use of curvilinear structures, the results were higher than with common objects, but still insufficient. Finally, the experiments exploring the utilization of new vessels datasets to enhance performances lead to the most interesting findings. We conducted three approaches: transfer-learning with no fine-tuning (referred to as: TL), transfer-learning with complete fine-tuning (referred to as: FT), and single-pass cross-dataset training (referred to as: CD) with 1493 images from new vessels datasets combined with the 40 images in the cross-dataset-build training subset. TL results in performance reduction. Examination of the predicted segmentations reveals that without fine-tuning, the network only generates larger structures. While this approach creates very few false positives, it also misses the details of the retinal structures, explaining the diminished performance. In contrast, both FT and CD demonstrate comparable results with the 40 images cross-dataset-build training from scratch. This indicates that using the pre-trained network with complete fine-tuning or incorporating cross-dataset training put the networks performances to a level on par with the network trained from scratch. Despite the multiplication of training data by 37, it does not yield additional performance improvements.

In contrast to the use of external datasets, **data augmentation** significantly improves performance, particularly for veins and arteries segmentation. Initially, the increase in data quantity through data replication led to a slight decrease in performance, indicating slight over-fitting to the training set. The increase in performances that we observed afterwards is thus solely related to the data augmentation techniques employed. For all three tasks, as shown in Table 3 the most significant performance enhancements are consistently achieved through the flips & rotations, showing growth that correlates with the data replication scaling. Shearing & elastic transforms augmentations exhibit a more ambiguous impact on performance, with slight enhancements for veins and arteries while showing a performance decrease for vessels. In opposition, brightness & contrast, coarse dropout & pixel dropout, and channel shuffle & hue modifications augmentations consistently result in performances inferior to those without any augmentation.

Finally, as shown in Table 4 we scaled further the data replication factor (r) during the experiments. For vessels, the replication was stopped at $r = 6$ due to consecutive performance decreases from $r = 3$. For arteries and veins, replication was extended to $r = 15$ to ensure a thorough understanding of the behavior, as performance continued to improve with increasing replication factors. The vessels segmentation performances are already high without any augmentation which limits the impact of scaling to provide additional enhancements. The arteries and veins, as for them, benefits a lot more than the vessels. Results are reported as the average of each segmentation dice score over all datasets.

In conclusion, retinal vessels in fundus images are unique structures that vary significantly between datasets, influenced by acquisition material, illumination

Table 3. Data augmentation (DA) scaling through replication factor (r). **Bold green** is the best dice score. **Bold blue** signals that the dice score increased compared to (No DA). **Bold red** signals that the dice score decreased compared to (No DA). **Bold black** signals equivalent performances to (No DA).

r	Vessels	Veins	Arteries
	No DA		
1	**0.813 ± 0.001**	**0.732 ± 0.002**	**0.696 ± 0.007**
	Flip & Rotation		
1	**0.813 ± 0.004**	**0.736 ± 0.001**	**0.697 ± 0.014**
2	**0.819 ± 0.003**	**0.758 ± 0.008**	**0.720 ± 0.008**
3	0.821 ± 0.001	0.776 ± 0.003	0.737 ± 0.005
	Shear & Elastic		
1	0.810 ± 0.003	0.719 ± 0.012	**0.696 ± 0.007**
2	0.808 ± 0.004	**0.734 ± 0.006**	**0.705 ± 0.005**
3	0.805 ± 0.004	**0.742 ± 0.002**	**0.696 ± 0.002**
	Brightness & Contrast		
1	0.809 ± 0.001	0.721 ± 0.002	0.680 ± 0.002
2	0.799 ± 0.002	0.721 ± 0.004	0.669 ± 0.001
3	0.799 ± 0.002	0.710 ± 0.003	0.649 ± 0.004
	Channel Shuffle & Hue		
1	0.804 ± 0.004	0.719 ± 0.001	0.661 ± 0.007
2	0.800 ± 0.001	0.719 ± 0.007	0.659 ± 0.012
3	0.793 ± 0.002	0.713 ± 0.001	0.652 ± 0.008
	Coarse dropout & Pixel dropout		
1	0.811 ± 0.002	0.730 ± 0.001	0.686 ± 0.009
2	0.802 ± 0.001	**0.732 ± 0.002**	0.688 ± 0.004
3	0.801 ± 0.002	0.724 ± 0.002	0.684 ± 0.004

conditions, and the presence of diseases. A well-suited architecture is essential to maximize performances in each sub-task of vasculature segmentation: vessels, veins, and arteries. Counterintuitively, external data do not enhance performance in this cross-dataset and few-shot scenario. This is also true for most data augmentation techniques that attempt to replicate the appearance of new unseen data. However, some data augmentation techniques, as flips and rotations, work efficiently and demonstrate a notable ability to scale with the quantity of data. Figure 5 shows the results of VAVnets segmentations alongside corresponding groundtruth.

Table 4. Flips and rotations augmentations response to scaling. Bold green is the best dice score. **Bold blue** is the second best.

r	Vessels	Veins	Arteries
1	0.813 ± 0.004	0.736 ± 0.001	0.697 ± 0.014
2	**0.819 ± 0.003**	0.758 ± 0.008	0.720 ± 0.008
3	0.821 ± 0.001	0.776 ± 0.003	0.737 ± 0.005
4	**0.819 ± 0.002**	0.781 ± 0.002	0.733 ± 0.008
5	0.818 ± 0.001	0.778 ± 0.007	0.743 ± 0.001
6	0.815 ± 0.002	0.782 ± 0.001	**0.745 ± 0.002**
7	-	**0.783 ± 0.006**	0.743 ± 0.006
8	-	0.782 ± 0.004	0.746 ± 0.005
9	-	0.784 ± 0.007	0.737 ± 0.001
10	-	0.780 ± 0.007	0.742 ± 0.003
12	-	0.780 ± 0.005	0.742 ± 0.006
15	-	0.781 ± 0.006	0.730 ± 0.009

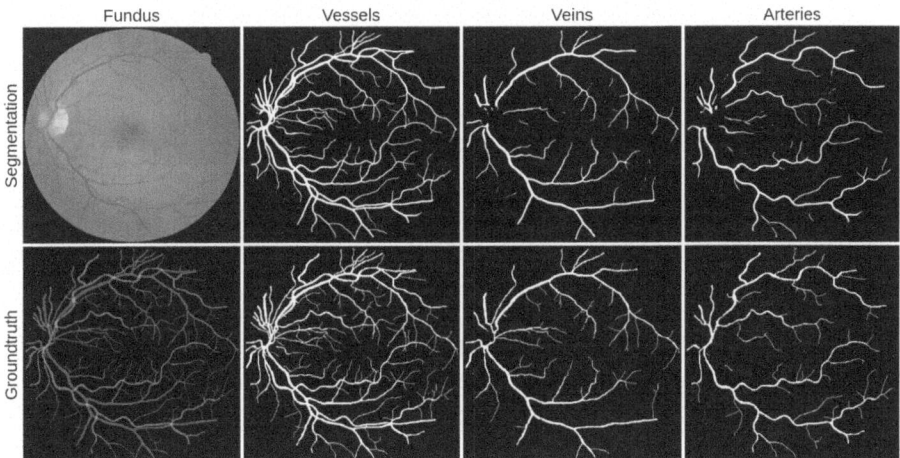

Fig. 5. Segmentation and corresponding groundtruth for vessels, veins and arteries.

6 Comparison to Related Works

The performances of VAVnets are compared to the state-of-the-art methods for vasculature segmentation in few-shot scenarios. Table 5 shows vessels segmentation and Table 6 shows veins and arteries segmentation. The values are reported from the referenced articles and rounded to two decimal places when more were given. If the third decimal is five or higher, the value is rounded up.

Vessels Segmentation
We have shown in Sect. 3 that [7,8] propose architectural modifications, [10,11]

employ prior knowledge, and [9] leverage on data augmentation. For the development of VAVnets, we leverage on architecture, transfer-learning, and data augmentation all together. Table 5 shows the comparisons of VAVnets performances with the state-of-the-art on vessels segmentation. On the DRIVE dataset composed of only seven pathological images out of 40 and a small resolution of 565*584, the performances of VAVnets are superior, but not by a large margin. For HRF, the performances of VAVnets are consequently superior. HRF is composed of 30 pathological images out of 45, and an high resolution of 3504*2336 thus each image present more details due to the visibility of more depth in the vessels trees. We attribute the performances gap to the leverage on architecture, transfer-learning, and data augmentation all together. The performances of VAVnets on DUMO and LESAV are referenced for comparison with other methods that may be developed in the future.

Table 5. Comparison between vessels segmentation methods in few-shot scenarios. Bold green is the best dice score. **Bold blue** is the second best.

date	method	DRIVE	DUMO	HRF	LESAV
2022	[11]	–	–	0.64	–
2022	[7]	< **0.80**	–	–	–
2022	[9]	0.79	–	**0.74**	–
2023	[10]	–	–	0.68	–
2023	[8]	0.77	–	0.67	–
2024	**VAVnets**	0.81	**0.83**	0.81	**0.86**

In [7], the F1 scores are not explicitly provided in the text and can only be roughly estimated from a chart. From the chart, it appears that the average F1 score is slightly below 0.8. For the purposes of reported tables, we will assume an average F1 score of approximately 0.8. We want to mention that dice and $F_\mathcal{B}$ scores are numerically equal in binary cases.

Veins and Arteries Segmentation. The most difficult but also most interesting tasks are the arteries and the veins segmentations. Table 6 shows the results of VAVnets compared to related works for the segmentation of veins and arteries in few-shot scenarios.

As mentioned in Sect. 3, Shi Danli *et al.* propose a binary-output approach in [12] and a multi-class output strategy in [13]. The performance of the binary-output approach in [12] is consistently lower than that of the multi-class approach in [13]. However, VAVnets, despite using a binary-output strategy, demonstrate superior performances. The only exception is for artery segmentation on the DRIVE dataset, where VAVnets perform almost equivalently.

First, comparing VAVnets to [12] (binary-output approach), we attribute the performance gap to our use of batch-normalization layers during inference,

Table 6. Comparison between veins and arteries segmentation methods in few-shot scenarios. Bold green is the best dice score. **Bold blue** is the second best.

date	method	Veins				Arteries			
		DRIVE	DUMO	HRF	LESAV	DRIVE	DUMO	HRF	LESAV
2022	[12]	0.63	–	0.50	0.61	0.57	–	0.46	0.58
2023	[13]	**0.71**	–	**0.70**	**0.67**	**0.72**	–	**0.69**	**0.72**
2024	**VAVnets**	0.78	**0.81**	0.78	0.79	0.73	**0.78**	0.74	0.77

our data augmentation choices, and proper scaling application. Table 2 presents the results of VAVnets at a developmental stage without data augmentation. The (Pred) describe our use of the normalization where (Learn) describe the absence of it. The veins and arteries segmentations performances reported in this table are on par with those of [12]. Remaining performances differences may be attributed to architectural differences.

Second, comparing VAVnets to [13], we mainly attribute the performance gap to the use the binary-output strategy instead of the multi-class strategy. The binary-output strategy is better than the multi-class strategy because of the complexity of the segmentation task. The structures are uncommon and consist of complex patterns. The task involves segmenting two tortuous trees of various sizes, depths, and branching patterns. Moreover, the trees overlap in a 2D view and can be obscured by pathologies, making some parts invisible. Nonetheless, overlapping areas are crucial, and should not be classified as only veins or arteries. Multi-class segmentation in the literature typically uses only three classes: arteries, veins, and background. It is the case here in [13]. Methods that attempt to detect overlaps as a separate class often fail because overlaps represent a small and disparate percentage of the data. However, combining binary segmentations of arteries and veins leads to more accurate overlap segmentations.

Another issue with the multi-class approach is that it introduces discontinuities in the structures. We referred to them as intra-vascular and inter-vascular miss-classification. Inter-vascular miss-classification is the close alternation between arteries and veins along a single vascular path. Intra-vascular miss-classification is the presence of veins pixels within a portion that is dominantly classified as arteries, and vice versa. However, both inter-vascular and intra-vascular miss-classifications are less common when performing binary segmentations.

Additionally, the veins are typically better predicted than the arteries in most state-of-the-art methods, as they are more easily differentiable by their appearance. In contrast, in [13], the opposite is true. This may seem unusual, but the explanation lies in both: i) the anatomical specificities of the eye, and ii) the multi-class output strategy without considering the overlap class.

i) In fundus images, at overlapping positions, arteries tend to be over veins, meaning that veins are not visible in the 2D view. In [13], as mentioned in

Sect. 3, the method rely heavily on transfer-learning from fundus fluorescein angiography images (referred to as *FFA*). The *FFA* groundtruth are acquired by fluorescein dye. Such *FFA* do not have overlap labels, one pixel correspond to only one class. Thus mostly arteries are classified at overlap points. Prioritizing arteries classification at overlap positions makes the network learn this behavior.

ii) The choice of a multi-class output strategy without considering the overlap class leads to the detection of only arteries at overlap points. On the other hand, DRIVE, HRF, and LESAV datasets utilized for performances assessment have overlap labels. Therefore, veins are never detected at overlaps points, but are compared to a groundtruth that has veins labels at overlap points, resulting in decreased performances. Conversely, overlaps are easily detected at a convolutional level, thus all the crossings are well predicted as arteries, enhancing the performances. This could justify the better performances for arteries than veins.

Finally, across all datasets, VAVnets demonstrate significantly superior performances for both veins and arteries, with one exception: arteries segmentation on the DRIVE dataset, where results are on par. This can be attributed to two main factors. The aforementioned impact of the multi-class strategy and the utilization of *FFA* groundtruth. The nature of the DRIVE dataset, which consists of only seven pathological images out of 40 and has a relatively small resolution of 565*584, making the segmentation process easier. Comparatively, HRF comprises 30 pathological images out of 45, with an higher-than-DRIVE resolution of 3504*2336, providing more detailed information. Similarly, LESAV includes 11 pathological images out of 22 and has an higher resolution of 1620*1444 compared to DRIVE, also providing more details.

7 Conclusion

In this article, we propose VAVnets: three variants of a deep-learning network designed to generate binary segmentations of vessels, veins, and arteries. We focus our experiments on few-shot scenarios and cross-dataset training. Through architectural changes, transfer-learning, and data augmentation experiments, we optimize VAVnets. Our results demonstrate that VAVnets outperform state-of-the-art methods for vessels, veins, and arteries segmentation across multiple datasets: DRIVE, DUMO, HRF, and LESAV. In conclusion, VAVnets significantly enhance the reliability of retinal vasculature segmentation in few-shot scenarios.

Segmentation alone cannot indicate health issues, so quantitatively measuring changes in retinal veins and arteries is still necessary. Such measurements need to be performed on coherent structures. In the future, we plan to further enhance the segmentation of veins and arteries at both pixel and structural levels. Additionally, we aim to ensure a fully connected structure through reconnection, ultimately enabling quantitative measurements of the retinal vasculature.

References

1. Staal, J., Abràmoff, M.D., Niemeijer, M., Viergever, M.A., van Ginneken, B.: Ridge-based vessel segmentation in color images of the retina. IEEE Trans. Med. Imaging **23**, 501–509 (2004)
2. Zhang, S., et al.: Simultaneous arteriole and venule segmentation of dual-modal fundus images using a multi-task cascade network. IEEE Access **7**, 57561–57573 (2019)
3. Odstrcilík, J., et al.: Retinal vessel segmentation by improved matched filtering: evaluation on a new high-resolution fundus image database. IET Image Process. **7**, 373–383 (2013)
4. Orlando, J.I., Breda, J.B., van Keer, K., Blaschko, M.B., Blanco, P.J., Bulant, C.A.: Towards a glaucoma risk index based on simulated hemodynamics from fundus images. In: International Conference on Medical Image Computing and Computer-Assisted Intervention (2018)
5. Khandouzi, A., Ariafar, A., Mashayekhpour, Z., Pazira, M., Baleghi, Y.: Retinal vessel segmentation, a review of classic and deep methods. Ann. Biomed. Eng. **50**(10), 1292–1314 (2022)
6. Ronneberger, O., Fischer, P., Brox, T.: U-Net: convolutional networks for biomedical image segmentation. In: Navab, N., Hornegger, J., Wells, W.M., Frangi, A.F. (eds.) MICCAI 2015. LNCS, vol. 9351, pp. 234–241. Springer, Cham (2015). https://doi.org/10.1007/978-3-319-24574-4_28
7. Xu, J., Shen, J., Wan, C., Jiang, Q., Yan, Z., Yang, W.: A few-shot learning-based retinal vessel segmentation method for assisting in the central serous chorioretinopathy laser surgery. Front. Med. **9**, 821565 (2022)
8. Shao, H.-C., Chen, C.-Y., Chang, M.-H., Yu, C.-H., Lin, C.-W., Yang, J.-W.: Retina-transnet: a gradient-guided few-shot retinal vessel segmentation net. IEEE J. Biomed. Health Inform. **27**, 4902–4913 (2023)
9. Lyu, J., Zhang, Y., Huang, Y., Lin, L., Cheng, P., Tang, X.: AADG: automatic augmentation for domain generalization on retinal image segmentation. IEEE Trans. Med. Imaging **41**, 3699–3711 (2022)
10. Hu, D., Li, H., Liu, H., Yao, X., Wang, J., Oguz, I.: Vesselmorph: domain-generalized retinal vessel segmentation via shape-aware representation. arXiv preprint arXiv:2307.00240 (2023)
11. Hu, D., Li, H., Liu, H., Oguz, I.: Domain generalization for retinal vessel segmentation with vector field transformer. In: Proceedings of The 5th International Conference on Medical Imaging with Deep Learning (2022)
12. Shi, D., et al.: A deep learning system for fully automated retinal vessel measurement in high throughput image analysis. Front. Cardiovasc. Med. **9**, 823436 (2022)
13. Shi, D., He, S., Yang, J., Zheng, Y., He, M.: One-shot retinal artery and vein segmentation via cross-modality pretraining. Ophthalmol. Sci. 100363 (2023)
14. Ruan, D., Wang, D., Zheng, Y., Zheng, N., Zheng, M.: Gaussian context transformer. In: Proceedings of the IEEE/CVF Conference on Computer Vision and Pattern Recognition, pp. 15129–15138 (2021)
15. Ioffe, S., Szegedy, C.: Batch normalization: accelerating deep network training by reducing internal covariate shift. In: International Conference on Machine Learning, pp. 448–456. PMLR (2015)
16. Srivastava, N., Hinton, G., Krizhevsky, A., Sutskever, I., Salakhutdinov, R.: Dropout: a simple way to prevent neural networks from overfitting. J. Mach. Learn. Res. **15**(1), 1929–1958 (2014)

17. Eldan, R., Shamir, O.: The power of depth for feedforward neural networks. In: Conference on Learning Theory, pp. 907–940. PMLR (2016)
18. Lin, T.-Y., et al.: Microsoft COCO: common objects in context. In: Fleet, D., Pajdla, T., Schiele, B., Tuytelaars, T. (eds.) ECCV 2014. LNCS, vol. 8693, pp. 740–755. Springer, Cham (2014). https://doi.org/10.1007/978-3-319-10602-1_48
19. Mnih, V.: Machine learning for aerial image labeling. Ph.D. thesis (2013)
20. König, J., Jenkins, M., Mannion, M., Barrie, P., Morison, G.: What's cracking? a review and analysis of deep learning methods for structural crack segmentation, detection and quantification. arXiv preprint arXiv:2202.03714 (2022)
21. Mou, L., et al.: Cs2-Net: deep learning segmentation of curvilinear structures in medical imaging. Med. Image Anal. **67**, 101874 (2021)
22. Chalakkal, R.J., Abdulla, W.H., Sinumol, S.: Comparative analysis of university of Auckland diabetic retinopathy database. In: International Conference on Signal Processing Systems (2017)
23. Popovic, N., Vujosevic, S., Radunović, M., Radunović, M., Popović, T.: Trend database: Retinal images of healthy young subjects visualized by a portable digital non-mydriatic fundus camera. PLoS ONE **16** (2021)
24. Jin, K., et al.: Fives: a fundus image dataset for artificial intelligence based vessel segmentation. Sci. Data **9** (2022)
25. He, K., Zhang, X., Ren, S., Sun, J.: Deep residual learning for image recognition. In: Proceedings of the IEEE Conference on Computer Vision and Pattern Recognition, pp. 770–778 (2016)

Remote-Sensing Based Precipitation Detection Using Conditional GAN and Recurrent Neural Networks

Pablo Negri[1,2]( ), Alejo Silvarrey[3], Sergio Gonzalez[4,5,7], Juan Ruiz[4,5,6], and Luciano Vidal[7]

[1] Instituto de Investigación en Ciencias de la Computación (ICC) UBA-CONICET, Buenos Aires, Argentine
pnegri@dc.uba.ar
[2] Departamento de Computación, FCEyN, UBA, Buenos Aires, Argentine
[3] Universidad Católica del Uruguay, Punta del Este, Uruguay
[4] Departamento de Ciencias de la Atmósfera y los Océanos, FCEyN, UBA, Buenos Aires, Argentine
[5] Centro de Investigaciones del Mar y la Atmósfera (CIMA) UBA-CONICET, Buenos Aires, Argentine
[6] Institut Franco-Argentin d'Études sur le Climat et ses Impacts (IRL IFAECI / CNRS-IRD-CONICET-UBA), Buenos Aires, Argentine
[7] Servicio Meteorológico Nacional (SMN), Buenos Aires, Argentine

Abstract. Precipitation detection using infrared (IR) brightness temperature (BT) temporal flux data is a challenging problem. Other sensors, such as microwave (MW), have reliable and more robust predictive performance, but lack land coverage and temporal availability. IR-BT provides high-frequency data (from half an hour to 10 min) at very low resolution (4 km). However, automatic precipitation detection frameworks should face the simple nature of this variable on the one hand, and the very low number of rain events occurring in nature on the other hand. This paper addresses this challenge by proposing a conditional GAN framework using recurrent neural networks, which transforms the unbalanced problem into a small (short) pattern detection algorithm. Several tests allow the identification of robust architectures and useful loss functions that enable promising results, minimize false alarms, and improve the overlap of positive events.

Keywords: Remote-Sensing · Precipitation Detection · cGAN · Recurrent Neural Networks · Invert Dice Loss

1 Introduction

This study proposes the use of machine learning models to detect precipitation over land areas that are not covered by rain gauges. It aims to map a spatio-temporal distribution of precipitation near water resources to analyze the local climatic impact in the development of cyanobacterial harmful algal blooms (CyanoHABs). The vulnerability of water resources to the effects of global warming is influenced by several factors, and some of them may produce CyanoHABs as increases in temperature, and changes in wind and rainfall patterns [19]. Both aspects increase the frequency of blooms and can lead to CyanoHABs due to enhanced stratification conditions and increased nutrient loading to water sources [3,4]. Figure 1 is an example of precipitation detection around Laguna Del Sauce, in the state of Marldonado, Uruguay, where this type of bloom is habitual in the months of December and January. As the image shows, the rain was detected in some places around the lake, but not in the whole region. This local information at precise times could help to understand some variables that can trigger the bloom.

Fig. 1. Precipitation detection around Laguna del Sauce, in Uruguay. Red boxes represent local places where the model detects precipitation using remote-sensing signals based on IR and BT. (Color figure online)

1.1 Detecting Precipitation Using Remote Sensors

In many regions of the world, precipitation variability across spatial and temporal scales cannot be fully resolved by conventional rain gauges. Geostationary satellites (GEO) have large spatial coverage and real-time availability with high spatio-temporal resolution in the order of 2–4 km and 10–30 min, respectively. These characteristics are highly relevant for detecting and monitoring the life cycle of precipitating clouds. For example, many previous studies have demonstrated the relationship between infrared (IR) brightness temperature (BT) and precipitation by providing information on cloud top properties [2,9,10,17]. Other sensors, such as microwave (MW) data, which capture the internal structure of clouds, provide a better quantitative estimate of precipitation, but have very poor coverage and observation frequency.

Satellite quantitative precipitation estimate (SQPE) methodologies can be classified based on how they use precipitation dynamics information. Those using instantaneous information (or snapshots) will be referred to as static methods. Others, namely dynamic methods, integrate available information on a time window and explicitly consider the evolution of the precipitation system.

Static SQPE methods are characterized by a single or a set of satellite data corresponding to the same time, in which the objective is to integrate such information to obtain precipitation estimation. Among them, there are most of the products of the Precipitation Estimation from Remotely Sensed Information Using Artificial Neural Networks (PERSIANN) family [10].

More current versions, such as PERSIANN-CNN [22], implements an Deep Convolutional Neural Network (CNN), and use IR-BT and water vapor (WP) channels. The use of CNNs improves the extraction of 2d features such as shape, texture, and extent of precipitating clouds, aspects that ANNs cannot capture directly.

Dynamic methods use a sequence of satellite data with temporal order, which can be composed of one or more input variables. Among these methods, we can name those based on optical flow, such as the Climate Prediction Center morphing method (CMORPH) [13], Global Satellite Mapping of Precipitation (GSMaP) [14], and IMERGE [11], which are characterized by propagating the precipitation field estimated with MW sensors through the IR data. [25] uses conditional GANs to integrate estimations based on IR and MW over a time window. Recurrent neural networks (RNNs) are particularly suited for extracting information and predicting patterns in sequential data such as satellite observations. In [1] they propose a PERSIAN architecture and Long Short Time Memory (LSTM), a widely employed RNN, for estimating accumulated precipitation from a sequence of images over a time window.

1.2 Adversarial Learning on Sequential Time-Series

This paper combines an adversarial learning framework and recurrent network architectures inspired by the work of Rezaei et al. [20,21]. Their approach, called RNN-GAN, addresses the semantic segmentation of medical images that face a highly imbalanced data problem. The tumor segmentation system applies a Conditional Generative Adversarial Network (cGAN) [15] architecture to two cardiac magnetic resonance (MR) images and one abdominal computed tomography (CT) dataset. The input stream consists of a 2D sequence of images and slices where the tumor and lesions are manually segmented at the pixel level. In such cases, the number of pixels representing a tumor is significantly lower than the number of pixels representing the background. The effect of unbalanced pixel distribution is mitigated by using complementary mask classes, some of which may overlap.

In this work, the precipitation detection system evaluates a remote sensing based IR-BT input signal of length L to diagnose a rainfall event. This signal is sampled at a frequency of 30 min, and the detection output corresponds to a binary stream of length L, where 1 is the value if it is raining and 0 otherwise. Since precipitation events do not usually last more than a few hours or less, the binary target is mostly populated with zeros even for a rainy day (see Fig. 1).

A traditional prediction model that forecasts rainfall at time $t+1$ from a signal going from 0 to t have the problem of positive collection of samples, and the dataset can finally be highly unbalanced. On the other hand, facing the problem in a similar way as Rezaei et al., the proposed system trains a conditional Generator following a sequence-2-sequence approach that estimates the most probably rainfall output from aht IR-BT input signal. Working with cGAN on target segments with at least one precipitation event within the stream, changes the learning approach, correctly detecting rainfall and minimizing false alarms.

The objective of this paper involves precipitation detection using satellite IR-BT information streaming data using a cGAN framework and recurrent neural networks. It proposes:

- A study of Generator and Discriminator architectures to tackle this particular problem,
- Evaluate different losses to improve regularization and convergence,
- Analyze the neighbor influence to detect precipitation.

Next section introduces materials and methods, Sect. 3 present different experiments and results. Conclusions and perspective are proposed at Sect. 4.

2 Materials and Methods

The Global full-resolution infrared dataset is provided by the Climate Prediction Center (CPC)[1] It merges IR-BT data between 10μ to 11μ from all available GEO satellites (GOES-8/9/10/11/12/13/14/15/16, METEOSAT-5/7/8/9/10, and GMS-5/MTSat-1R/2/Himawari-8) every 30 min with a spatial resolution of 4 km [12]. Spatial coverage is between 60S-60N and 180W-180E.

Brazil's INMET[2] provides the precipitation event targets from more than 350 automated rain gauges deployed throughout their country. The dataset consists of hourly automated rain gauge measurements of accumulated precipitation. The location (latitude and longitude) of each rain gauge station is used to retrieve the corresponding IR-BT data from GEO satellites. Precipitation values are interpolated to half-hourly measures to obtain the same length of IR-BT values.

The dataset comprises measurements collected during the years 2010 and 2011. The data set has been divided into four semesters, with the initial semester of 2010 and the subsequent two semesters of 2011 designated for training and validation purposes, while the second semester of 2010 has been reserved for testing. The IR-BT and observed rainfall sequence for the years 2010 and 2011 are 8,832 and 17,520, respectively. The number of valid rain gauges for each year is 376 for 2010 and 368 for 2011.

Figure 2 shows an example of the rain gauge accumulated precipitation in mm/h as the blue line and the IR-BT data in Kelvins degrees (K) at the rain gauge site as the red line. The figure below shows a sequence binary target consisting of the instantaneous points where the precipitation was greater than 5 mm/h. It can be seen in the Fig. 2 that there is a correlation between precipitation and low temperature values, but this is not a rule that can be solved with

---
[1] https://search.earthdata.nasa.gov/
[2] https://portal.inmet.gov.br/

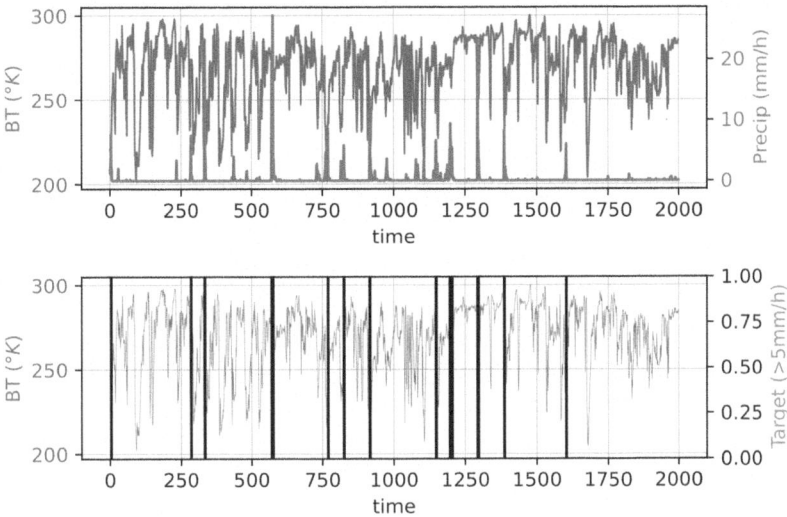

Fig. 2. Examples of IR-BT at the location of the rain gauges and the accumulated precipitation. The red line shows the IR-BT sensor data, and the blue line shows the accumulated rainfall. In the bottom figure, we produce a binary target sequence of those instants where precipitation was greater than 5 mm/h. (Color figure online)

a simple threshold. At the same time, it is clear that we face a problem where precipitation events are rare in the input sequence. In fact, on average, 7.6% of the points in the 2011 dataset have precipitation greater than 0 mm/h, and only 0.6% of the points have precipitation greater than 5 mm/h. In the following, we fix $T_{pr} = 5mm/h$ as the threshold for precipitation events, because this amount of precipitation can be considered significant for the generation of CyanoHABs.

2.1 Recurrent Neural Networks and Conditional GAN Framework

Recurrent Neural Networks (RNN) are a potentially accurate prediction model for precipitation detection. RNNs are neural networks that compute current variables based on their prior states, giving them a "dynamic memory" [6]. This is extremely useful for prediction within a time series, where each element fed into the model is related to the previous and next values.

Temporal series denoted as $(\mathbf{x}^{(1)}, \mathbf{x}^{(2)}, ..., \mathbf{x}^{(T)})$ are usually the inputs of RNN models. Similarly, the target sequences corresponding to precipitation detection are given as a binary sequence $(\mathbf{y}^{(1)}, \mathbf{y}^{(2)}, ..., \mathbf{y}^{(T)})$, where $\mathbf{y}^{(t)} = 1$ if the accumulated precipitation is greater than T_{pr}, and $\mathbf{y}^{(t)} = 0$ otherwise (see Fig. 2). The predictions produced by the recurrent model are denoted as $\hat{\mathbf{y}}^{(t)}$.

We express the estimation of target **y** at time t as a dependent function R with internal parameters θ_{RNN}:

$$\hat{\mathbf{y}}^{(t)}, \theta_{RNN}^{(t)} = R(\mathbf{x}^{(t)} | \theta_{RNN} = \theta_{RNN}^{(t-1)}) \tag{1}$$

Modern RNN architectures introduce several improvements to overcome traditional training problems. The Long-Short Term Memory model [8] (LSTM) is one of the most successful networks, widely used in several applications such as natural language processing. We will use LSTM in the architecture of our precipitation diagnosis framework.

Generative adversarial networks consist of a generator G and a discriminator D model, which are both trained simultaneously according to the two-player min-max game with value function $V(G, D)$:

$$\min_G \max_D V(G, D) = \mathbf{E}_{\mathbf{y} \sim p_{data}(\mathbf{y})}[log(D(\mathbf{y}))] + \mathbf{E}_{\mathbf{z} \sim p_z(\mathbf{z})}[log(1 - D(G(\mathbf{z})))] \tag{2}$$

D estimates whether the data is real or generated by G (fake). On the other hand, G learns how to fool D improving the quality of the output. Formally, $D(\mathbf{y}; \theta_d)$ outputs the probability that **y** comes from the training data instead of p_y, a generator distribution over the data **y** that G maps from a prior noise distribution $p_y(\mathbf{z})$ to the data space as $G(\mathbf{y}, \theta_g)$. The framework adjusts the parameters of D to minimize $log(D(\mathbf{y}))$, the first part of Eq. 2, and simultaneously adjusts the parameters of G to minimize $log(1 - D(G(\mathbf{z})))$.

GANs are extended to a conditional model when G and D are conditioned with additional data **x**. For G, the input noise $p_y(\mathbf{z})$ and **x** are combined in a joint hidden representation, and for D, **y** and **x** are presented as inputs to the discriminator. The objective functions $\mathcal{L}_{adv}$ would be [15]:

$$\min_G \max_D V(G, D) = \mathbf{E}_{\mathbf{y} \sim p_{data}(\mathbf{y})}[log(D(\mathbf{y}|\mathbf{x}))] + \mathbf{E}_{\mathbf{z} \sim p_z(\mathbf{z})}[log(1 - D(G(\mathbf{z}|\mathbf{x})))] \tag{3}$$

In our framework, the **y** distribution is the binary stream of the precipitation diagnosis, and the **x** data is the normalized IR BT in Kelvin degrees. To normalize **x**, we subtract the overall mean and divide it by the standard deviation of the IR BT values. The framework is trained to approximate $G(\mathbf{z}|\mathbf{x}) = \hat{\mathbf{y}}$ to **y**.

Figure 3 shows the cGAN framework. The generator G follows a sequence-2-sequence framework, arranged in a deep network composed of LSTM cells, followed by fully connected layers with a single sigmoid output. The discriminator D uses a one-dimensional convolutional neural network architecture with a single sigmoid output.

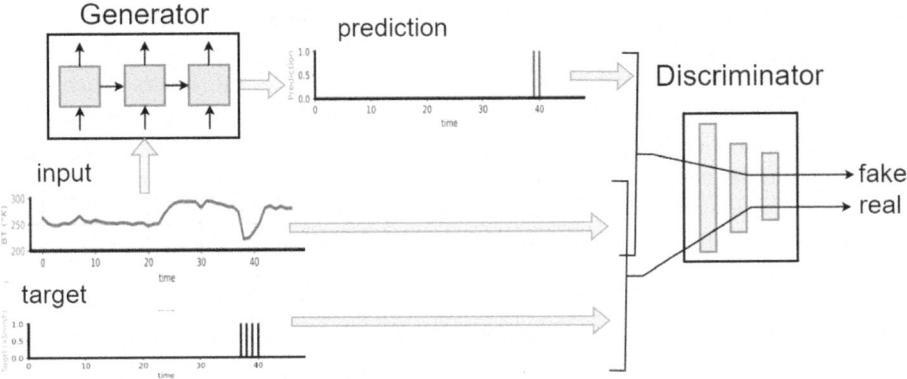

Fig. 3. Proposed framework for precipitation detection cGAN.

2.2 Loss Functions

In this work, we evaluate the following loss functions as regularizers to train G:

$$\mathcal{L}_{L1} = ||\mathbf{y}_{target} - \hat{\mathbf{y}}|| \tag{4}$$

$$\mathcal{L}_{Ldice} = 1 - \frac{2 \cdot \sum_{t=1}^{L} \left(\mathbf{y}_{target}^{(t)} \cdot \hat{\mathbf{y}}^{(t)} \right) + \epsilon}{\sum_{t=1}^{L} \left(\mathbf{y}_{target}^{(t)} + \hat{\mathbf{y}}^{(t)} \right) + \epsilon} \tag{5}$$

$$\mathcal{L}_{Linvdice} = \frac{\sum_{t=1}^{L} \left(\mathbf{y}_{target}^{(t)} + \hat{\mathbf{y}}^{(t)} \right) + \epsilon}{2 \cdot \sum_{t=1}^{L} \left(\mathbf{y}_{target}^{(t)} \cdot \hat{\mathbf{y}}^{(t)} \right) + \epsilon} - 1. \tag{6}$$

were ϵ is a low value number to avoid division by zero.

Table 1. Table evaluating Dice loss functions for a target sample and $\epsilon = 0.1$.

	target (0,0,1,1,0)	description	L1 loss	Dice Loss	Inverted Dice Loss
predictions	(0,0,0,0,0)	2 fn	2	0.95	20.0
	(0,0,0,1,0)	1 fn	1	0.32	0.47
	(1,0,1,1,0)	1 fp	1	0.19	0.24
	(1,1,1,1,0)	2 fp	2	0.32	0.48
	(1,1,0,0,1)	all wrong	5	0.98	50.0

$\mathcal{L}_{L1}$ refers to the ℓ_1 distance and measures the deviation of $G(z|y) = \hat{\mathbf{y}}$ from a binary output. This loss not only highlights prediction errors, but also forces G to produce binary values close to 0 and 1, which improves the quality of the prediction. This step is very important because this output is later evaluated by the discriminator D.

The Eqs. 5 and 6 describe the dice function loss and the inverse dice loss, respectively. A dice score is usually used to measure the overlap between distributions. Table 1 compares the results of the three loss functions on a target sample similar to the goal and predictions of our framework. As can be seen, Inverse Dice (iDice) generally produces larger loss values if the prediction $\hat{y}$ does not trigger a rainfall event, which is considered a false negative (fn). It is expected that the generator G learns to triggers a rainfall event at a very precise position to avoid this negative reward. Empirically results showed that only using $\mathcal{L}_{L1}$ cGAN train a generator G that does not trigger rainfall events. This is because the loss of false negatives for this score is not important. Dice and inverse Dice losses, on the other hand, forces the generator to produce a distribution similar to the binary target.

The final objective function $\mathcal{L}_{pd-gan}$ of the precipitation detection generator is: $\mathcal{L}_{pd-gan} = \mathcal{L}_{adv} + \mathcal{L}_{L1} + \mathcal{L}_{xdice}$.

3 Experiments and Results

3.1 Evaluation Scores

The evaluation of the precipitation detection task is similar to the tumor segmentation in [20]. The idea is to obtain a measure of the overlap between the ground truth and the model prediction distributions. We then define: tp, true positives, which is the number of points correctly classified as precipitating; fn, false negatives, the number of points incorrectly classified as non-precipitating; and fp, false positives, those points classified as precipitating but it is not.

We propose the following scores:

$$F1 = \frac{2 \cdot precision \cdot recall}{precision + recall} \quad (7)$$

$$IoU = \frac{intersection}{union} \quad (8)$$

where $precision = \frac{tp}{(tp+fp)}$, $recall = \frac{tp}{(tp+fn)}$, $intersection = tp$, and $union = tp + fn + fp$. The **F1** score (Eq. 7) measures the exact overlap between the target distribution and the prediction distribution. The **IoU** (Intersection over Union) score (Eq. 8) calculates the ratio of true rainfall events detected to the total number of events triggered and the times the system should detect rainfall. **F1** and **IoU** range from 0 to 1, with 1 being optimal. Since precipitation events are usually very short, the system sometimes predicts precipitation at time $t \pm 1$ instead of time t. For these cases, we also define a $F1 - ext$ score, which is calculated as the correct detection of predictions at times $t \pm 1$ of the correct positive prediction.

3.2 BT-PR-Dataset

To train and validate the PD-GAN framework, we set tuples $(bt_k, pr_k)_{k=1,...,N}$. Each $bt_k^{(t)}$ is a sequence of IR BT, and $pr_k^{(t)}$ is the corresponding sequence of precipitation values at the same location, both of length L: $t = 1,...,L$. We fix the length of the sequences to $L = 48$.

This set is populated by **positive** and **negative** tuples. The **positive** ones have at least one position $pr_k^{(t)}$ equal to 1. This means that the precipitation accumulated by the rain gauge was greater than $T_{pr} = 5$ mm/h. **negative** tuples have all values of $pr_k^{(t)}$ equal to 0, which means that the accumulated precipitation was less than T_{pr} or did not rain at all. We will refer to this dataset as **BT-PR-dataset** and will use it later to train the PD-GAN framework. For $T_{pr} = 5$ mm/h, **BT-PR-dataset** collects 64,755 tuples, with 27,969 **positive** tuples and 36,786 **negative** tuples.

3.3 Naive Detection

Naive precipitation detection classifies pixels as raining or non-raining by applying a fixed threshold to the IR BT. From **BT-PR-dataset** we obtain the distribution of those IR BT values $bt_k^{(t)}$ that correspond to effective precipitation $pr_k^{(t)} = 1$. We also compute the distribution of IR BT values without precipitation. Both distributions are then plotted on cumulative histograms, as shown in Fig. 4, to obtain two IR BT thresholds. The first threshold $T_{naive-cross} = 256K$ indicates the point where the two curves coincide. The other threshold $T_{naive-otsu} = 262K$ follows the idea of the Otsu methodology [18].

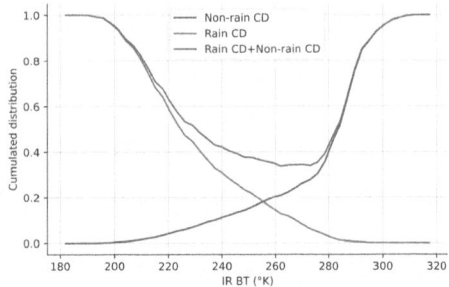

Fig. 4. Cumulative Distribution (CD) of BT for raining and non-raining events. The green curve defines two thresholds: $T_{naive-cross} = 256K$ and $T_{naive-otsu} = 262K$. (Color figure online)

Table 2 presents the result of applying this threshold to the whole 2010 dataset.

3.4 Precipitation Detection GAN

This section develops an ablation study on the different hyperparameters of the system and different architectures. We use a 5-fold cross-validation evaluation. The reported results are computed as the average within the folds of the performance scores. The models were trained for up to 50 epochs with a batch size of 64. We choose the Adam optimizer with a learning rate of $2e-4$. The length L of the sequence of training tuples in **BT-PR-training** is set to $L = 48$, which means a 1-day history.

As explained in Sect. 2.1, the architecture of the discriminator D is fixed. It consists of two branches of 1d CNN networks:

- br1(bt_k) = [1dConv(L, 32, 5), MaxPool(9), 1dConv(32, 64, 5], BN(), MaxPool(1), BN(), FC(64, 32)]
- br2(pr_k) = [1dConv(L, 32, 3), MaxPool(9), 1dConv(32, 32, 3], BN(), MaxPool(1), BN(), FC(32, 32)]

where 1dConv(A, B, K) is a 1d convolutional layer with input A, output B, and 1d kernel length K, MaxPool(C) is a max-pooling computed to length C, BN() is a batch normalization layer, and FC(D, E) is a fully connected layer with input size D and output size E. The input of br1 is the BT sequence, and the input of br2 is the precipitation binary target. Then the outputs of both branches br1 and br2 are concatenated and feed the output branch: ob = [FC1(64, 32), FC(32, 1)]. ob has a binary output that indicates whether the input is real or fake.

Table 2 shows the results using **BT-PR-dataset**, where each bt_k is the stream of IR BT at the location of the rain gauge that gives the target pr_k. In this test, bt_k is a one-dimensional variable of normalized IR BT values that feeds the training and validation framework.

Table 2. Results of naive thresholding and different PD-GAN architectures. Best results are highlited in bold.

Naive threshold		F1	IoU	precision	recall	F1-ext	FAs
$T_{naive-cross}$= 256 K		0.0750	0.0394	0.0399	0.7551		704.05
$T_{naive-otsu}$= 262 K		0.0631	0.0328	0.0331	**0.8006**		899.4
Hidden Units	dLoss	F1	IoU	precision	recall	F1-ext	FAs
8	d	0.0088	0.0044	0.0044	0.7716	0.0188	5184.6
16		0.0362	0.0308	0.0409	0.0709	0.0667	126.6
32		0.0074	0.0037	0.0037	0.7739	0.0158	6405.9
64		0.0341	0.0238	0.0224	0.0899	0.0690	118.2
8	id	0.1299	0.0731	0.0959	0.2642	0.2213	**79.1**
16		0.1344	0.0755	0.0901	0.3434	0.2323	112.3
32		0.1334	0.0749	0.0902	0.3292	0.2300	105.5
64		**0.1423**	**0.0806**	**0.0976**	0.3405	**0.2455**	103.4

Table 2 compares different numbers of hidden units in the LSTM cells and the use of Dice Loss and Inverted Dice Loss. As can be seen, the use of Dice Loss does not provide the framework for correct learning of precipitation detection. High values of FAs and low values of F1 show that some of the K-fold models fall into non-optimal working states. In the former case, the generator predicts ones most of the time, or in the latter case, the generator misses precipitation events and often predicts zeros.

The inverted dice loss, on the other hand, shows robust learning for the precipitation detection framework. The best results are obtained with 64 hidden units in the LSTM cell. However, using 8 hidden units produces a minimum number of FAs.

F1-ext is another interesting score that shows a large increment compared to F1. In general, using the extended target, the number of TPs doubles, while the number of FPs decreases in the same proportion. This clearly shows that the detection framework is sensitive to the temporal evolution of the IR BT associated with the precipitation, but triggers the signal once after or before the actual precipitation time.

Table 3. Results of stacked LSTM architectures using inverse dice loss.

Nr. Layers	Hidden Units	F1	IoU	precision	recall	F1-ext	FAs
2Lay	8	**0.1395**	**0.0787**	0.0924	0.3646	**0.2427**	116.4
	16	0.1320	0.0740	0.0847	**0.3789**	0.2309	131.8
	32	0.1189	0.0658	0.0748	0.3695	0.2087	144.0
	64	0.1151	0.0635	0.0716	0.3705	0.2025	150.8
3Lay	8	0.1388	0.0782	**0.0927**	0.3577	0.2407	114.1
	16	0.1373	0.0773	0.0773	0.3749	0.2387	123.4
	32	0.1309	0.0737	0.0922	0.3252	0.2268	**113.5**
	64	0.1193	0.0662	0.0769	0.3550	0.2093	138.2

We also use a stacked LSTM network [26] for the generator. In practice, an easy way to increase the depth of the recurrent network is to stack the cells into D layers, trying to better capture the dynamics of the time-dependent signal. This architecture has been shown to improve efficiency and performance in problems such as vehicle-to-vehicle communication [5], regional commodity price estimation [16], and French-English translation [23]. However, the results in Table 3 do not show a significant improvement in precipitation detection.

Next experiments increase the context of the information around the position of interest. Instead of only using the IR-BT value at the rain gauge location, we extract a grid from the EarthData around that point, thus the samples $bt_k^{(t)}$ became an $R \times R$ grid centered on that rain gauge location. The horizontal resolution of these grids is 4 km, which means that a 3×3 grid covers a field of 12 km^2. Thus, the framework now has the neighborhood information that can be used to better estimate a precipitation event, but, at the same time, we have increased the number of parameters in the network.

Table 4 shows the results using a 1-layer architecture and two grid sizes: 3×3 and 5×5. As expected, additional information about IR-BT at neighboring locations allows us to improve the results in both F1 and FAs for 8 hidden units architecture.

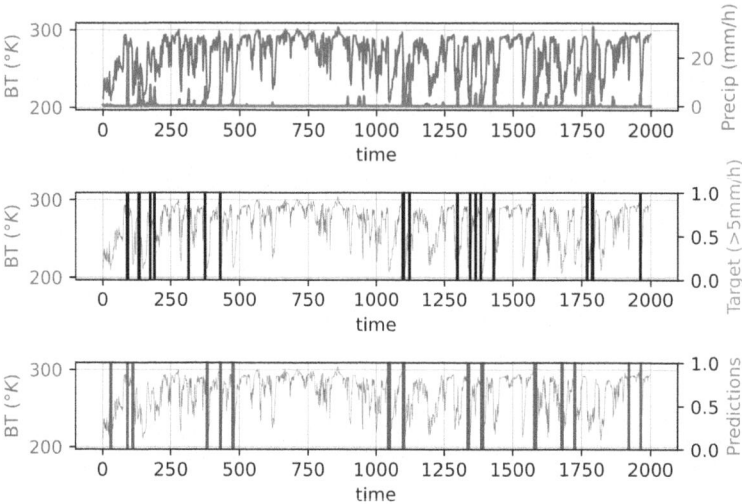

Fig. 5. Precipitation detection on the testing dataset 2010, 8 hidden units architecture.

Table 4 shows the results of the framework using the bidirectional LSTM [7] in the generator, which is also used in [20]. However, the performance of this modified LSTM cell drops in F1 and FAs scores.

The final test involves a different choice for the precipitation threshold. Using a lower value of 1 mm/h, the new binary target represents a higher number of precipitation events. Table 5 shows the results of the cGAN framework using a naive BT threshold. The performance shows a better F1 score compared to a threshold of 5 mm/h, but more than twice as many FAs. This is the expected behavior. In the training loop, the generator learns to trigger many more precipitation events, even if the BT corresponds to warm values. In this way, the framework loses generalization power.

The Figs. 5 and 6 show the results of the predictions for the two best architectures with $T_{pr} = 5$ and $T_{pr} = 1$ respectively.

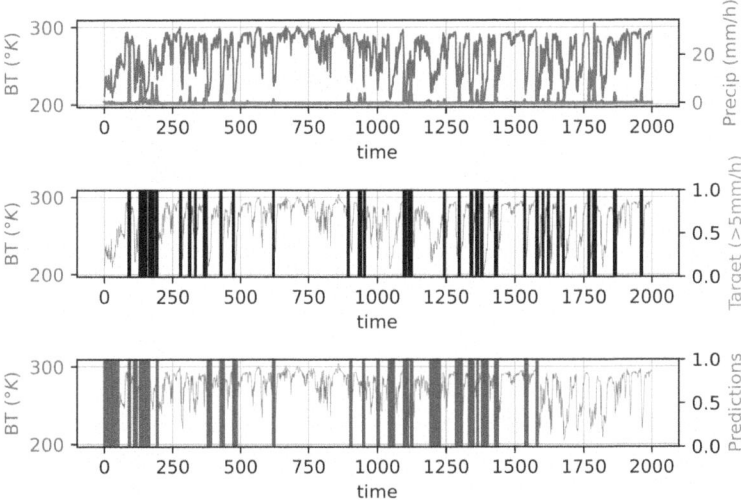

Fig. 6. Precipitation detection 2010 dataset using a $T_{pr}=1\,\text{mm/h}$, 16 hidden units architecture.

Table 4. Results using a grid region around pluviometer position.

Hidden Units	R	F1	IoU	precision	recall	F1-ext	FAs
8	3	0.1443	0.0828	0.1338	0.2549	0.2433	63.1
16		0.1415	0.0800	0.0988	0.3360	0.2429	97.3
32		0.1211	0.0672	0.0775	0.3531	0.2113	125.1
64		0.1214	0.0673	0.0783	0.3474	0.2123	123.5
8	5	0.1258	0.0702	0.0842	0.3410	0.2180	118.2
16		0.1141	0.0630	0.0723	0.3441	0.2010	136.5
32		0.1123	0.0618	0.0709	0.3384	0.1976	133.1
64		0.1123	0.0618	0.0697	0.3647	0.1976	143.9
Bidirectional LSTM							
8	3	0.1141	0.0632	0.0786	0.2690	0.1964	91.5
16		0.1250	0.0695	0.0796	0.3770	0.2184	136.2
32		0.1166	0.0644	0.0751	0.3346	0.2023	123.2
64		0.1174	0.0650	0.0743	0.3791	0.2057	154.6
8	5	0.1202	0.0666	0.0809	0.2964	0.2088	99.1
16		0.1244	0.0695	0.0845	0.3263	0.2154	114.6
32		0.1203	0.0666	0.0752	0.3771	0.2106	139.2
64		0.1200	0.0666	0.0787	0.3221	0.2091	115.7

Table 5. Results using a precipitation threshold of 1 mm/h.

Naive threshold	F1	IoU	precision	recall	F1-ext	FAs
257 K	0.2080	0.1193	0.1278	0.6687		736.5
Hidden Units	F1	IoU	precision	recall	F1-ext	FAs
8	0.2036	0.1208	0.1509	0.3785	0.2862	255.4
16	0.2120	0.1267	0.1606	0.3821	0.2964	239.5
32	0.2102	0.1255	0.1548	0.3989	0.2954	264.1
64	0.2009	0.1188	0.1431	0.4200	0.2840	310.6

4 Conclusions

This paper develops a framework to tackle local detection of precipitation using IR-BT temporal sequence signal, a cGAN framework and recurrent neural networks to produce the output sequence. The generator G of the cGAN is particularly suitable for producing binary temporal sequence outputs where the target has small (short) positive events, while it is mostly populated by zeros, using the IR-BT as input signal.

An inverse Dice function allows correct learning in the adversarial game, improving the performance of the traditional $L1$ norm applied to binary outputs. Considering the simple nature of the IR-BT data signal, the results are promising and the system in its current form is ready to be incorporated into the framework of CyanoHABs prediction and to help set social policies for water resource management.

Further research could be directed at optimizing the forecast diagnosis by regionalizing the rain gauges, using categorical precipitation with multiple thresholds, or other architectures for the generator and discriminator, such as transformers [24].

Acknowledgements. This research was supported by The National Research and Innovation Agency of Uruguay (ANII), International Development Research Centre of Canada (IDRC) and National Scientific and Technical Research Council of Argentina (CONICET). Grant Number: ANII-IA_2021_1_1010782. Sergio Gonzalez's scholarship is supported by CONICET and the National Meteorological Service of Argentina.

References

1. Akbari Asanjan, A., et al.: Short-term precipitation forecast based on the Persiann system and LSTM recurrent neural networks. J. Geophys. Res. Atmospheres **123**(22), 12543–12563 (2018)
2. Arkin, P.A., Meisner, B.N.: The relationship between large-scale convective rainfall and cold cloud over the western hemisphere during 1982–84. Mon. Weather Rev. **115**(1), 51–74 (1987)

3. Barruffa, A.S., Sposito, V., Faggian, R.: Climate change and cyanobacteria harmful algae blooms: adaptation practices for developing countries. Mar. Freshw. Res. **72**(12), 1722–1734 (2021)
4. Burford, M., et al.: Perspective: Advancing the research agenda for improving understanding of cyanobacteria in a future of global change. Harmful Algae **91**, 101601 (2020)
5. Du, X., Zhang, H., Van Nguyen, H., Han, Z.: Stacked LSTM deep learning model for traffic prediction in vehicle-to-vehicle communication. In: 2017 IEEE 86th Vehicular Technology Conference (VTC-Fall). pp. 1–5. IEEE (2017)
6. Elman, J.L.: Finding structure in time. Cogn. Sci. **14**(2), 179–211 (1990)
7. Graves, A., Schmidhuber, J.: Framewise phoneme classification with bidirectional LSTM and other neural network architectures. Neural Netw. **18**(5–6), 602–610 (2005)
8. Hochreiter, S., Schmidhuber, J.: Long short-term memory. Neural Comput. **9**(8), 1735–1780 (1997)
9. Hong, Y., Hsu, K.L., Sorooshian, S., Gao, X.: Precipitation estimation from remotely sensed imagery using an artificial neural network cloud classification system. J. Appl. Meteorol. **43**(12), 1834–1853 (2004)
10. lin Hsu, K., Gao, X., Sorooshian, S., Gupta, H.V.: Precipitation estimation from remotely sensed information using artificial neural networks. J. Appl. Meteorol. **36**(9), 1176–1190 (1997)
11. Huffman, G.J., et al.: Integrated multi-satellite retrievals for the global precipitation measurement (GPM) mission (imerg). Satellite Precipit. Measurement **1**, 343–353 (2020)
12. Janowiak, J.E., Joyce, R.J., Yarosh, Y.: A real-time global half-hourly pixel-resolution infrared dataset and its applications. Bull. Am. Meteor. Soc. **82**(2), 205–218 (2001)
13. Joyce, R.J., et al.: Cmorph: a method that produces global precipitation estimates from passive microwave and infrared data at high spatial and temporal resolution. J. Hydrometeorol. **5**(3), 487–503 (2004)
14. Kubota, T., et al.: Global satellite mapping of precipitation (gsmap) products in the gpm era. Satellite Precipitation Measurement **1**, 355–373 (2020)
15. Mirza, M., Osindero, S.: Conditional generative adversarial nets. arXiv preprint arXiv:1411.1784 (2014)
16. Negri, P., Ramos, P., Breitkopf, M.: Regional commodities price volatility assessment using self-driven recurrent networks. In: Tavares, J.M.R.S., Papa, J.P., González Hidalgo, M. (eds.) CIARP 2021. LNCS, vol. 12702, pp. 361–370. Springer, Cham (2021). https://doi.org/10.1007/978-3-030-93420-0_34
17. Nguyen, P., et al.: Persiann dynamic infrared-rain rate model (PDIR) for high-resolution, real-time satellite precipitation estimation. Bull. Am. Meteor. Soc. **101**(3), E286–E302 (2020)
18. Otsu, N.: A threshold selection method from gray-level histograms. IEEE Trans. Syst. Man Cybern. **9**(1), 62–66 (1979)
19. Paerl, H.W., Hall, N.S., Calandrino, E.S.: Controlling harmful cyanobacterial blooms in a world experiencing anthropogenic and climatic-induced change. Sci. Total Environ. **409**(10), 1739–1745 (2011)
20. Rezaei, M., Yang, H., Meinel, C.: Recurrent generative adversarial network for learning imbalanced medical image semantic segmentation. Multimedia Tools Appl. **79**(21–22), 15329–15348 (2020)
21. Rezaei, M., et al.: Conditional generative refinement adversarial networks for unbalanced medical image semantic segmentation. arXiv:1810.03871 (2018)

22. Sadeghi, M., et al.: Persiann-CNN: precipitation estimation from remotely sensed information using artificial neural networks-convolutional neural networks. J. Hydrometeorol. **20**(12), 2273–2289 (2019)
23. Sutskever, I., Vinyals, O., Le, Q.V.: Sequence to sequence learning with neural networks. arXiv preprint arXiv:1409.3215 (2014)
24. Vaswani, A., et al.: Attention is all you need. In: Advances in Neural Information Processing Systems, vol. 30 (2017)
25. Wang, C., et al.: Precipgan: Merging microwave and infrared data for satellite precipitation estimation using generative adversarial network. Geophys. Res. Lett. **48**(5), e2020GL092032 (2021)
26. Yu, Y., et al.: A review of recurrent neural networks: LSTM cells and network architectures. Neural Comput. **31**(7), 1235–1270 (2019)

Data-Driven Genetic Algorithm for the Optimization of Water Distribution Networks: A New Surrogate Model for Estimating Investment and Operational Costs in Pumping Stations

Nicolás Gajardo-Sepúlveda[1](✉), Thalía Faúndez-Lizama[1], Jimmy H. Gutiérrez-Bahamondes[2], Daniel Mora-Melia[3], and César A. Astudillo[2]

[1] Master's in Operations Management, Faculty of Engineering, Universidad de Talca, 3340000 Curicó, Chile
nigajardo18@alumnos.utalca.cl

[2] Department of Computer Science, Faculty of Engineering, Universidad de Talca, 3340000 Curicó, Chile
jgutierrezb@utalca.cl

[3] Department of Engineering and Construction Management, Faculty of Engineering, Universidad de Talca, 3340000 Curicó, Chile

Abstract. The increase in water consumption and demand has highlighted the need to improve the efficiency of water distribution networks (WDNs). Pumping stations (PS) represent a significant challenge due to their high energy consumption and associated operational costs. This study presents a data-driven evolutionary optimization methodology focused on predicting cost and penalties in the design and operation of PS. The methodology integrates the functioning of genetic algorithms with machine learning techniques, developing a surrogate model capable of predicting associated costs and increasing the computational efficiency of optimization models. A case study is presented to validate the methodology, showing a significant reduction of 31.8% in evaluation times and a decrease in optimization time compared to traditional models. The results indicated that despite the reduction in computational effort, the increase in the final cost of the network was minimal, with only a 2.42% increase compared to the baseline model. These findings underscore the effectiveness of combining machine learning with genetic algorithms for the optimization of PS in WDNs, improving computational efficiency while maintaining high standards in solution quality.

Keywords: Water Distribution Networks · Data-driven evolutionary optimization · Offline optimization · Pump Stations design · Cost estimation

1 Introduction

The increasing demand for water and the significant impact of climate change on water resources underscore the crucial importance of water distribution networks (WDN) to ensure the supply of water to the population. These networks, made up of thousands of elements, present great complexity in their design and operation, requiring efficient and safe solutions (Sarbu, 2021). The design of WDN involves solving many nonlinear equations, and their optimization becomes more complicated with water pumping, which is the largest energy consumer in distribution systems, representing a significant portion of network operating costs and municipal budgets (Dadar et al., 2021; Pulido-Calvo & Gutiérrez-Estrada, 2011). Therefore, it is important to focus on pumping stations (PS) to reduce costs and save energy through two key areas: design and operations (Mala-Jetmarova et al., 2018).

The design of PS includes the location of the stations in the network, establishing the number of necessary stations, selecting the appropriate pump type and defining the required accessories such as valves and pipes (Gil et al., 2018). Studies have focused on these aspects in pump selection, station design and operational validation (Müller et al., 2021; Pulido-Calvo & Gutiérrez-Estrada, 2011).

On the other hand, the operation of PS centers on pump scheduling, defining the activation periods of the pumps and the flows supplied in each period (Mala-Jetmarova et al., 2018). This scheduling must be highly efficient to meet demand at different times of the day and achieve significant energy savings in operation. Relevant literature addresses station scheduling and the improvement of operational efficiency (Dadar et al., 2021; Dong & Yang, 2020).

Although planning and designing PS are critical aspects of network optimization, addressing them simultaneously presents a significant challenge due to their complexity. Recent advances have implemented operational validation systems in PS design (Müller et al., 2021) and optimization models that minimize capital and operating costs (Gutiérrez-Bahamondes et al., 2021). Some studies have incorporated controllers for pump capacity and continuous flow monitoring to detect faults and adjust PS designs (Wang et al., 2020).

Continuous hydraulic simulations are essential for verifying the feasibility of pump design, scheduling and evaluating energy costs, but they are complex and time-consuming (Maier et al., 2014; Sarbu, 2021). Metaheuristics, such as evolutionary and PSO algorithms, have proven to be efficient tools in addressing PS design and operation problems, supporting the hydraulic design of WDN to meet pressure and demand requirements, and improving operational efficiency (Garzón et al., 2022; Sangroula et al., 2022).

Genetic algorithms (GA) are notably useful for pump scheduling in networks (Dadar et al., 2021), providing near-optimal solutions and evaluating both operational and design costs (Maier et al., 2014). However, these algorithms face challenges with high computational costs during exhaustive searches, even though they reduce the search space. These design studies use simplified networks very often to facilitate the analysis, but real networks are much more complex due to their numerous components and requirements (Sarbu, 2021).

The main identified challenges include increasing the computational efficiency of optimization models, seeking evaluation methods that replace hydraulic simulations and designing hybrid models that combine different optimization techniques (da Costa Oliveira et al., 2023; Jin et al., 2019). In this context, data-driven evolutionary algorithms present a promising solution by combining evolutionary algorithms with machine learning models, increasing computational efficiency, and partially replacing hydraulic simulations in network problems (Garzón et al., 2022).

This study aims to explore and develop novel methods in data-driven evolutionary optimization, particularly focusing on optimizing pumping stations (PS). The main proposed contributions are as follows:

1. Development of a Data-Driven Evolutionary Optimization Methodology: We propose a new methodology that combines evolutionary algorithms with machine learning models to create a surrogate model capable of predicting costs from a historical data set.
2. Increasing Computational Efficiency: We will use data-driven algorithms to significantly reduce solution evaluation times in PS design and operation. Based on the problem of solution evaluation for PS design and operation described by Gutiérrez-Bahamondes et al. (2021), our methodology will seek to optimize these processes more effectively.
3. Implementation and Validation in a Case Study: We will present a case study applying the proposed methodology, demonstrating its effectiveness in accelerating optimization processes in WDN problems and validating the methodology to enable its application to larger-scale networks (da Costa Oliveira et al., 2023; Jin et al., 2019).
4. Evaluation and Comparison with Conventional Methods: We will conduct a thorough evaluation of the proposed methodology compared to conventional optimization methods, such as traditional hydraulic simulations. This analysis will identify the advantages and limitations of our methodology, as well as its potential to replace or complement existing methods.

2 Materials and Methods

This research builds on the work of Gutiérrez-Bahamondes et al. (2021), which proposes a mathematical optimization model for the design and operation of pumping stations (PS) in water distribution networks (WDN). The model calculates the optimal flow distribution for each PS should supply to meet the network's pressure and flow demands, minimizes investment costs (CAPEX) and operational expenses (OPEX) using an evolutionary algorithm.

In the proposed problem, as the network size increases, the hydraulic simulations required to evaluate each solution become increasingly complex, significantly increasing analysis time. This can make the massive execution of evaluations unfeasible, and therefore, the application of the method is limited by the network size. To overcome this limitation, this work proposes the use of data-driven evolutionary algorithms, and the stages involved in the design and implementation of the algorithm are described below.

2.1 Data-Driven Evolutionary Optimization

As described in the Introduction section, it is possible to use ML tools to enhance the performance of evolutionary algorithms during the optimization process. This work proposes a genetic algorithm that uses a pre-trained ML model to replace the evaluation process of the objective function that requires hydraulic simulations. The surrogate model, based on historical data, learns and evaluates the quality of the solutions found in the evolutionary algorithm, guiding the selection of solutions in each population of the algorithm and thereby replacing the original, expensive evaluation processes.

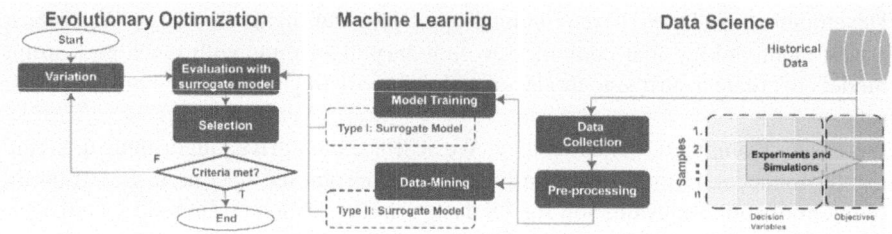

Fig. 1. Framework for data-driven offline evolutionary optimization (Jin et al., 2019).

In (Jin et al., 2019), the fundamental methodology of data-driven evolutionary optimization, along with the necessary tools to address each optimization problem are detailed. Figure 1 illustrates the framework and steps of the proposed evolutionary optimization, corresponding to an offline data-driven evolutionary optimization methodology. Offline models use a pre-existing and static dataset to guide the optimization process, without the need to acquire or process new information during the optimization. This approach aims to enhance the computational efficiency of traditional models, enabling the optimization of large-scale distribution networks.

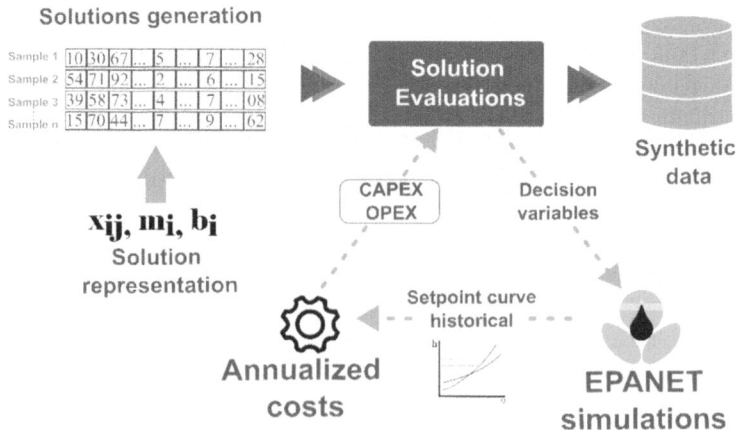

Fig. 2. Framework for Synthetic Data Generation.

The problem presents high dimensionality, and historical records are inadequate for forming an adequate training database. Therefore, the initial stage of the proposal involves generating a synthetic database, evaluating solutions generated through simulations, and assigning objective values to these solutions according to the methods described in (Jin et al., 2019). In this work, a continuous representation for the decision variables is used, considering the assignment of continuous values to the decision variables. The solutions are generated in such a way that the database contains evaluations of feasible and infeasible designs, ensuring the model's learning variety. Following this, the generated database is preprocessed with a principal component analysis (PCA) (Kelleher et al., 2015). Figure 2 represents the process flow for generating the synthetic database.

After generating the synthetic database, a surrogate model is trained to estimate and approximate the costs for the design and operation of PS. At the same time, it must detect penalty costs for not complying with network requirements. The model is a neural network for regression (multi-layer perceptron), for which its hyperparameters are optimized through halving grid search cross-validation (Kelleher et al., 2015).

The next stage is the data-driven genetic algorithm (DDGA). The crossover and mutation operators are adapted to the new representation, and the evaluation process implements the trained surrogate model as the unique evaluator of the algorithm. Parameters such as mutation probability, crossover probability, and population size, are modified for efficient optimization performance. The proposed algorithm is coded in Python, using metaheuristic tool libraries (JmetalPy) and machine learning methods from ScikitLearn and River. This process concludes when no significant improvement is detected in population quality, providing the best design and operation solution that the model can estimate. Figure 3 presents the operational steps of the algorithm.

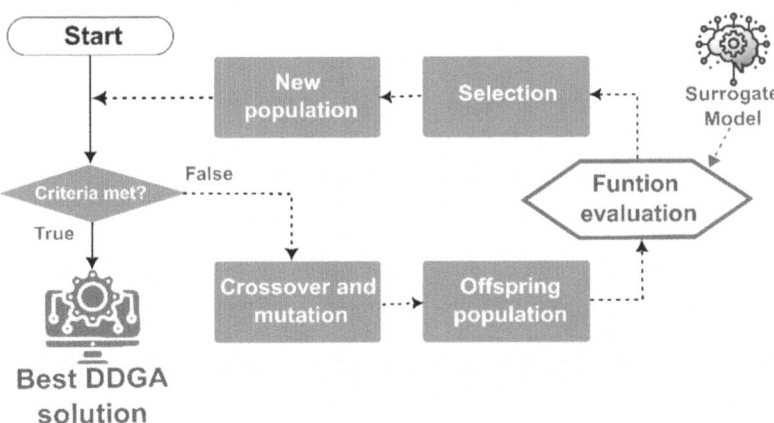

Fig. 3. Framework for the Data-Driven Evolutionary Optimization.

Finally, a local search is implemented to explore the neighboring solutions of the best solution obtained by the DDGA. This search explores and exploits regions close to the provided solution to find a local optimum with reduced costs. Implementing a refinement method improves the convergence and quality of the final solutions of the proposed algorithm. To ensure the consistency of the algorithm, 450 iterations have been performed and analyzed.

2.2 Case of Study

To verify the performance of the proposed methodology, a case study was developed. The case corresponds to the WDN called MTF. This network was presented in the article by Gutiérrez-Bahamondes et al. (2021), consisting of 3 PSs labeled as PS1, PS2, and PS3, 15 demand nodes, and 22 pipeline lines connecting the nodes and stations. Figure 4 illustrates this network.

For the case study analysis, the scope of the study includes the operation of the network for 24 h, considering one period per hour. Each node in the network must be supplied with a pressure of 20 m, while the average network demand is 100 L/s. All information related to the characteristics of the nodes, pipes, demand pattern, and electrical costs are specified in (Gutiérrez-Bahamondes et al., 2021).

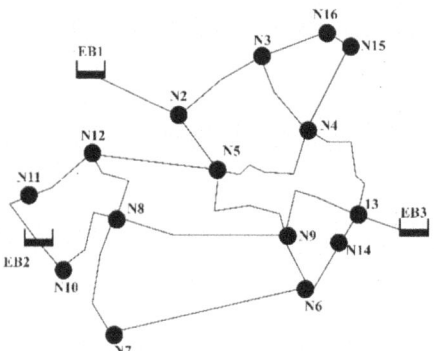

Fig. 4. MTF network.

3 Results

The surrogate model is trained with the synthetic dataset and then implemented in the 450 described evolutionary experiments, following the offline methodology. The genetic algorithm was executed until a population with a better-quality index could not be found. At this point, the best solution obtained was subjected to additional refinement through the local search algorithm, which included 10,000 additional evaluations. This section presents the results of the best solution found during the experimentation.

The distribution of flow supplied by each pumping station (PS) varies according to the network demand pattern in each period. In this context, PS1 must be activated in each operating period. On the other hand, PS2 is turned off only from period 2 to period 7, while PS3 is activated in periods 8, 9, 10, 11, 12, 15, and 22. Regarding the design of

Table 1. Pump Models for MTF Network Stations.

PS	Model	H0	A	Q_{max}
PS1	GNI 50-16/10	38,1	−0,0209	42,63
PS2	GNI 50-16/10	38,1	−0,0209	42,63
PS3	GNI 50-26/30	81,4	−0,0394	45,41

the stations, Table 1 presents the pump models to be installed in the stations and their operating characteristics.

Table 2 presents the performance indicators of the best result achieved by the genetic algorithm and the local search.

Table 2. DDGA indicators for best solution.

Surrogate Model	DDGA			Local Search	
	Populations	Estimated cost	Real cost	Evaluations	Final cost
MLP ANN	65	€40.044	€ 95.682,05	10.000	€41.232,7

Figure 5 compares the performance of the offline methodology's best solution with the baseline optimization model in terms of the number of evaluations with SH executed. The trained model is used for each experiment and is not re-trained either during the process or when conducting a new experiment. Therefore, the simulations of the synthetic data generation do not affect the solution search or the convergence of the genetic algorithm. Simultaneously, the DDGA does not include simulations in the evaluation; all computational effort is replaced by the surrogate model. Considering this, Fig. 5

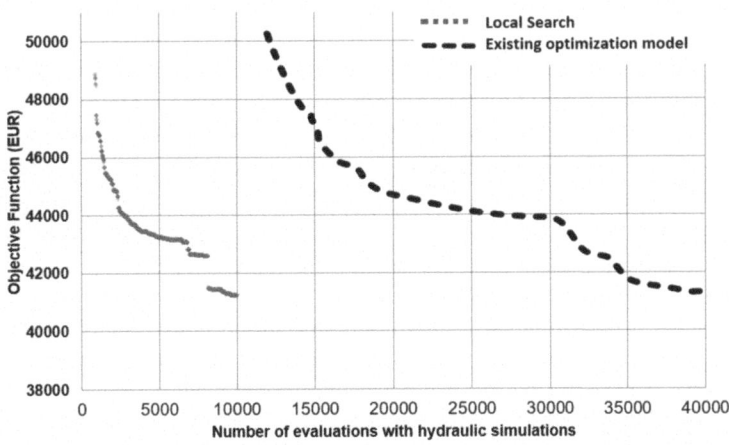

Fig. 5. Contrast between the Convergence of the Proposed Methodology and the Baseline Model.

contrasts the performance and convergence of the offline methodology and the original model in terms of evaluations involving hydraulic simulations.

The only evaluations to be compared are those included in the local search, which require simulations to evaluate the solutions. The local search functions as a refinement of the solution found in the evolutionary stage, showing convergence in the optimization, reaching up to 10,000 evaluations to find the best solution identified by the local search.

4 Discussion

Implementing the data-driven evolutionary methodology aims to reduce computational processes that prolong the optimization of PS design and operation. By reducing simulation costs and processes, this methodology seeks to achieve results close to the optimum found by the baseline model. The surrogate model is crucial in guiding the search for solutions, just as the synthetic data generation needs to be precise and reliable. The generated data must provide a set of information representative of feasible solutions and penalized designs. In this scenario, the simulated instances include infeasible solutions with corresponding penalties, high-cost feasible solutions, and viable solutions with reduced costs. This approach strengthens the model's ability to identify degrees of penalizations and potential feasible solutions, ensuring variability in training.

The traditional methodology effectively optimizes networks with a limited number of nodes; however, their efficiency decreases with more complex networks. Evaluating solutions in simplified networks involves high computational costs, and seeking an optimal solution in larger distribution systems leads to a significant increase in execution time and computational resources. The proposed methodology offers an alternative to achieve efficient solutions and reduce both execution resources and optimization times.

For each experiment, the DDGA leads to a low-cost solution detected by the surrogate. However, only 23 experiment solutions were penalty-free, the rest of the solutions present penalties for non-compliance with network demands. For this reason, the local search complements and refines the best solution that the model can detect. The final solution obtained through the local search is feasible and low-cost for most of the experiments conducted, although in 47 experiments, the local search was unable to find a penalty-free solution for the network.

The best solution presented in the previous section increases costs by 2,42% compared to the solution obtained in the baseline model (€40.259). However, this model required about 44.000 evaluations involving simulations processes. In contrast, the proposed methodology reduces simulations to 10.000, representing a 77,3% reduction in computational processing. According to (Mora-Melia, 2012), a good solution to the design and operation problems in WDNs does not exceed the baseline optimal solution by more than 3%. In this context, the obtained result meets the criteria for a good solution to the PS design and operation problem.

This proposal demonstrates a positive outcome in terms of resource reduction for optimization. However, the implemented tools present certain limitations related to data availability. The performance of the learning model heavily depends on the quality and variety of the synthetic data generated. Additionally, the effectiveness of the genetic algorithm relies on the surrogate model's ability to identify costly and beneficial solutions.

5 Conclusions

The increase in water demand, climate change concerns, and water scarcity challenges have driven the search for new strategies and methods to design efficient WDNs. Optimizing the design and operation of PS within these networks is complex and costly, especially when considering traditional methods and evaluation processes.

The literature presents few traditional models that focus simultaneously on the design and planning of PS. Such studies usually require a large number of evaluations and SH processing, which is inefficient for larger networks. This work proposes a methodology that combines evolutionary optimization tools with machine learning models to replace costly processes in optimization algorithms and enhance computational efficiency.

The implementation of the methodology in the case study shows a significant reduction in the number of SH evaluations compared to the baseline optimization model. Although the total costs slightly increased, the solution remains efficient for the design and operation of the network.

The proposal presents certain limitations when implementing a machine learning model. These are observed in the model's sensitivity to the synthetic data and in the preprocessing of this database. Although the dataset already incorporates some variety, implementing data augmentation techniques could further expand and diversify the data, enhancing the model's accuracy. Additionally, the representation of solutions plays a crucial role in the model's learning process and the convergence of the evolutionary algorithm. Although the experiments are conducted on a simplified network, the methodology aims to extend its application to more complex networks with a greater number of nodes.

References

da Costa Oliveira, A.L., Britto, A., Gusmão, R.: Machine learning enhancing metaheuristics: a systematic review. Soft. Comput. **27**(21), 15971–15998 (2023). https://doi.org/10.1007/s00500-023-08886-3

Dadar, S., Đurin, B., Alamatian, E., Plantak, L.: Impact of the pumping regime on electricity cost savings in urban water supply system. Water (Switzerland) **13**(9), 1141 (2021). https://doi.org/10.3390/w13091141

Dong, W., Yang, Q.: Data-driven solution for optimal pumping units scheduling of smart water conservancy. IEEE Internet Things J. **7**(3), 1919–1926 (2020). https://doi.org/10.1109/JIOT.2019.2963250

Garzón, A., Kapelan, Z., Langeveld, J., Taormina, R.: Machine learning-based surrogate modeling for urban water networks: review and future research directions. In: Water Resources Research, vol. 58, no. 5. Wiley (2022). https://doi.org/10.1029/2021WR031808

Gil, A., Antonio, F., Javier, F., Cortes, L., Vanessa, J.: Methodology for Projects of Pumping Stations Directly Connected to the Network Considering the Operation Strategy (2018)

Gutiérrez-Bahamondes, J.H., Mora-Meliá, D., Iglesias-Rey, P.L., Martínez-Solano, F.J., Salgueiro, Y.: Pumping station design in water distribution networks considering the optimal flow distribution between sources and capital and operating costs. Water (Switzerland) **13**(21), 3098 (2021). https://doi.org/10.3390/w13213098

Jin, Y., Wang, H., Chugh, T., Guo, D., Miettinen, K.: Data-driven evolutionary optimization: an overview and case studies. IEEE Trans. Evol. Comput. **23**(3), 442–458 (2019). https://doi.org/10.1109/TEVC.2018.2869001

Kelleher, J.D., Mac Namee, B., D'Arcy, A.: Fundamentals of Machine Learning for Predictive Data Analytics Algorithms, Worked Examples, and Case Studies. MIT Press (2015)

Maier, H.R., et al.: Evolutionary algorithms and other metaheuristics in water resources: current status, research challenges and future directions. Environ. Model. Softw. **62**, 271–299 (2014). https://doi.org/10.1016/j.envsoft.2014.09.013

Mala-Jetmarova, H., Sultanova, N., Savic, D.: Lost in optimisation of water distribution systems? a literature review of system design. Water **10**(3), 307 (2018). https://doi.org/10.3390/w10030307

Mora-Melia, D.: Diseño de redes de distribución de agua mediante algoritmos evolutivos. Universitat Politécnica de Valéncia, Análisis de eficiencia (2012)

Müller, T.M., Leise, P., Lorenz, I.S., Altherr, L.C., Pelz, P.F.: Optimization and validation of pumping system design and operation for water supply in high-rise buildings. Optim. Eng. **22**(2), 643–686 (2021). https://doi.org/10.1007/s11081-020-09553-4

Pulido-Calvo, I., Gutiérrez-Estrada, J.C.: Selection and operation of pumping stations of water distribution systems. **5**, 1–20 (2011). https://www.researchgate.net/publication/228451972

Sangroula, U., Han, K.H., Koo, K.M., Gnawali, K., Yum, K.T.: Optimization of water distribution networks using genetic algorithm based SOP–WDN program. Water (Switzerland) **14**(6), 851 (2022). https://doi.org/10.3390/w14060851

Sarbu, I.: Optimization of urban water distribution networks using deterministic and heuristic techniques: comprehensive review. J. Pipeline Syst. Eng. Pract. **12**(4) (2021). https://doi.org/10.1061/(asce)ps.1949-1204.0000575

Wang, G., Wang, H., Kang, Z., Feng, G.: Data-driven optimization for capacity control of multiple ground source heat pump system in heating mode. Energies **13**(14), 3595 (2020). https://doi.org/10.3390/en13143595

Gene Regulatory Network for the Tryptophanase Operon Under the Threshold Boolean Network Model

Felipe Encina-Chacana[1(✉)] and Gonzalo A. Ruz[1,2,3]

[1] Faculty of Engineering and Sciences, Universidad Adolfo Ibáñez, Santiago, Chile
fencinachacana@gmail.com
[2] Center of Applied Ecology and Sustainability (CAPES), Santiago, Chile
[3] Data Observatory Foundation, Santiago, Chile

Abstract. This paper presents an evolutionary computation framework that uses genetic algorithm and particle swarm optimization to infer a threshold Boolean network of the Tryptophanase operon. A unique feature of this network is that it exhibits bistability, converging to two fixed points under certain conditions. We proposed a fitness function to achieve this, ensuring the network showed the desired dynamics, particularly the bistability property. Additionally, we explored and analyzed the results obtained from 500 simulations conducted by each algorithm. The genetic algorithm could infer 23 different networks with perfect scores, but particle swarm optimization could not infer any. The results showed that, in general, the genetic algorithm could explore the search space more effectively, obtaining networks with more edges than particle swarm optimization, thus allowing it to find networks satisfying the biological restriction of the model inferred.

Keywords: Boolean Networks · Genetic Algorithm · Particle Swarm Optimization · Bistability

1 Introduction

Boolean networks (BNs) are widely used to model various phenomena. This work will focus mainly on modeling gene regulation processes, with extensive applications in medicine and neurology. BNs, as their name suggests, work with binary data in their nodes (1 and 0), which can be interpreted as a gene being "on" or "off." The edges represent interactions between nodes. While BNs may not precisely capture the intricate internal processes that genes perform, they are still effective at a larger scale and thus widely used. Threshold BNs are a particular case of BNs; the main difference is that, instead of using logical operators, we use weights on the edges and a threshold for each node.

To infer threshold BNs, one approach is to use evolutionary computing algorithms such as particle swarm optimization (PSO) or genetic algorithm (GA), which are widely used in the study of this subject as in [28,29]. So there is an opportunity to optimize the inference process through certain restrictions on the parameters to be estimated.

In this work, the weight matrix was limited to the set $[-1, 0, 1]$, following the same rules as [26]. We used two evolutionary computing algorithms such as (GA) and (PSO) to obtain the weight matrix representing the edge connections and the thresholds. This approach reduces the computational space and optimizes resources, following the methodology presented in [18].

The rest of the paper is organized as follows: in Sect. 2 we will provide a brief explanation of threshold BNs and their practical applications. Subsequently, we will explain the key concepts to understand the Tryptophanase operon from [9], in Sect. 3 we present the methods used to infer the network, and finally, the results are shown in Sect. 4, and the conclusions of this work in Sect. 5.

2 Background

To understand how threshold BNs work, we must start from the basics. This type of model seeks to represent how various nodes interact with each other. Each node has only two possible states (1 and 0), indicating whether the node is activated or not. Consider a set of n nodes, $\mathbf{x} = \{x_1, \ldots, x_n\}$, where $x_i \in \{0, 1\}$. This results in a set of 2^n initial configurations which we also refer to as input matrix. This approach is adapted from [27].

Threshold BNs models seek to observe how the input matrix changes over time. Normally, this matrix represents the initial state of a cell, which transforms with each iteration. To understand the transition that the matrix undergoes, the following concepts are needed:

1. Weight Matrix ($w_{i,j}$): represents the interactions between the nodes.
2. Activation threshold θ_i: in a biological sense, it represents that a certain potential is needed for the activation or deactivation of a node.
3. Activation function $H(x)$: mathematically relates the input matrix, weight matrix, and threshold to model the evolution of the network. It is important to highlight that there are various models of this function and they usually lead to different results. An example of this can be found in [12].
4. Fixed Points: Fixed points are essential in BNs, as they typically represent cell specialization or the expression of certain characteristics. As their name suggests, they appear when, over several iterations, the network configuration does not change [10].
5. Limit Cycle: Limit cycles, in contrast to fixed points, consist of points that repeatedly evolve cyclically, which prevents them from stabilizing in a single configuration, instead forming a series of configurations that repeat infinitely.
6. Basins of Attraction: consist of the configurations that lead to a specific fixed point or limit cycle.

The update function presented in [23], as shown in (1), utilizes the activation function presented in (2). It is important to highlight that this is relevant for understanding the behavior of these type of models.

$$x_i(t+1) = H(\sum_{j=1}^{N} w_{i,j} x_j - \theta_i) \quad (1)$$

$$H(x) = \begin{cases} 1 \ x > 0 \\ 0 \ x \leq 0 \end{cases} \quad (2)$$

In our case we are interested to infer a network that represents the behavior of the Tryptophanase operon, for that reason we used GA and PSO to get the parameters $w_{i,j}$ and threshold θ_i.

GA is based in the Darwinian principles [11] and is a sub-area from evolutionary computing. More details about their functionality can be found in [19]. Additionally, [2] details certain fundamental concepts about GA, demonstrating that, depending on the problem, the type of crossover can be modified to improve performance like in [24].

PSO is based on the social behavior of animals like certain species of birds and fishes, like GA, it is used to solve optimization problems [20]. Different variants of PSO are also studied in combination with other algorithms to evaluate performance differences. In [1], it is emphasized that the main distinction between PSO and fuzzy PSO lies in the fact that, in the latter, multiple neighbors influence the search for the optimum, leading to superior results compared to the original model. Finally, in the experimental section, fuzzy PSO was employed to conduct the experiments.

Several works have used only binary weights to optimize the space, for this they only use -1 and 1 to infer networks, as seen in [14,16]. In this work we adopted the same approach, i.e., we only used binary parameters, which is useful to understand dynamical systems. Some studies talk about this topic [3,4,7,15].

Nowadays, BNs are widely used in medicine. An example of this is the CABEAN software, which makes it possible to identify targets for intervention to modify cellular programming. In this way, specialization can be manipulated. This could lead to the generation of new treatments to cure various diseases and boost regenerative medicine [8]. Another application in the area of health consists of the study of basins of attraction, when certain gene configurations lack sufficient robustness, this results in the appearance of cancer [17]. Also, BNs are used in the study of the effects of certain drugs on the genetic regulation that subsequently leads to the development of cancer [13]. We can see applications of BNs in decease's like sepsis [21] or Parkinson [6], and these models have been used in the development of pharmacology [22] and in neuroscience to study the brain dynamics as discussed in [5].

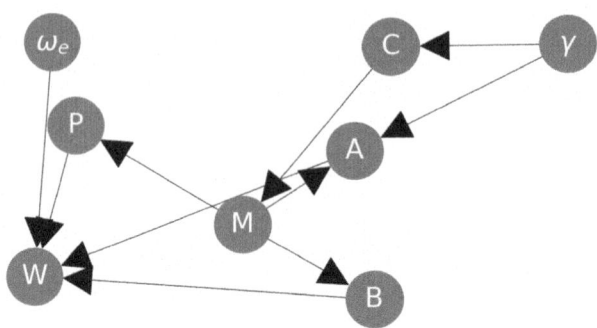

Fig. 1. Tryptophanase operon network [9]

Table 1. Tryptophanase operon data.

parameter vector			fixed points (Tryptophanase operon genes)							operon
γ	ω_e	ω_{em}	A	B	C	M	P	W	W_m	state
1	1	1	0	0	0	0	0	0	1	OFF
1	0	0	0	0	0	0	1	0	0	OFF
1	0	1	0	0	0	0	1	0	0	OFF
0	1	1	1	1	1	1	0	1	1	ON
0	0	0	0	0	1	0	1	0	0	OFF
0	0	1	0	0	1	0	1	0	0	OFF
			1	1	1	1	0	0	1	ON

3 Methods

3.1 Tryptophanase Operon

Now we give a brief explanation of the original Tryptophanase operon network from [9]. In this paper the authors propose the network in Fig. 1, but at the time of using the updating function (1), limit cycles appeared that did not correspond. The current challenge is to correctly infer the network.

The network consists of 3 special nodes known as parameter vector γ, ω_e, ω_{em}, which are artificial, i.e., they influence the other nodes but the rest of the nodes in the network cannot influence them, thus, they do not change between iterations. It is important to note that in Fig. 1 the parameter ω_{em} and W_m are omitted in the figure in order to simplify the illustration, which was obtained from [9]. The nodes in Fig. 1 represents,

- A = TryptophanaseA above metabolic tryptophan
- B = TryptophanaseB permease, responsible for transforming tryptophan
- C = cAMP- CAP protein complex that initiates transcription

- M = Tryptophanase mRNA
- P = TryptophanaseC sequence is bound to Rho protein
- W = high intracellular tryptophan levels
- W_m = intracellular tryptophan

3.2 Boolean Networks Inference

As shown in Table 1, there are special nodes called parameter vector (γ, ω_e, ω_{em}). These nodes remain invariant over time and initiate the gene regulation process. For example, consider the case where the parameter vector is γ, ω_e, $\omega_{em}(1,1,1)$. In this case, the 2^7 possible combinations of the nodes (A, B, C, M, P, W, W_m) will reach the fixed point $(0,0,0,0,0,0,1)$, which biologically indicates that the operon is off.

In Table 1, each parameter vector reaches a certain fixed point, except for the parameter vector γ, ω_e, $\omega_{em} = (0,0,1)$. In this case, it reaches the values of $(1,1,1,1,0,0,1)$ and $(0,0,1,0,1,0,0)$, which means that for that parameter vector, the network has the option of reaching two fixed points which indicates bistability.

3.3 Fitness Function

Now we will explain the creation of the input matrix X. There are 10 nodes in total: γ, ω_e, ω_{em}, A, B, C, M, P, W, and W_m. Therefore, there are $2^{10} = 1024$ possible configurations. However, we are only interested in combinations with biological relevance, meaning that the first 3 nodes, which refer to the parameter vector (γ, ω_e, ω_{em}), have a configuration that belongs to Table 1. Therefore, we only consider a total of 768 possible combinations.

We already have the initial matrix X, but we need to arrange it in a way that is convenient to work with. In this case, we group the first 128 combinations that correspond to the first parameter vector and follow the same pattern with the rest. The most important thing is to respect this order.

Prediction consists of taking X and the parameters from GA and PSO. The algorithms provide a set of parameters that are subsequently ordered to obtain the weight matrix w with dimensions 10×10 and the threshold θ of length 10. Since the nodes in the parameter vector are not updated, their weights in the weight matrix and their thresholds are set to 0. This means that in all inferred networks, the first 3 rows of the weight matrix and the first 3 values of the threshold vector are 0. Therefore, each algorithm needs to provide a total of $7 \times 10 + 7 = 77$ parameters to infer the network.

With this data, we use the equations (2) and (1) to iterate the network for a time t equal to 30. It is important to mention that in each iteration, the nodes corresponding to the parameter vector are ignored and remain constant over time, regardless of the update function. Finally, we obtain the resulting prediction matrix.

The resulting matrix, "Result", ensures that for a given parameter vector, the network reaches its corresponding fixed point. This matrix does not

Table 2. Experimental Metrics

Metrics	GA	fuzzy PSO
Total Execution Time [Hours]	154	87
Average Execution Time [Minutes]	18	10
Minimum Execution Time [Minutes]	5	8
Maximum Execution Time [Minutes]	49	38
Networks $E_{\text{Total}} = 0$	23	0
Average E_{Total}	217	1511
Minimum E_{Total}	0	1388
Maximum E_{Total}	240	2156

include the fixed points of the bistable parameter, which will be handled separately and referred to as $Fixed_1 = (0,0,1,1,1,1,1,0,0,1)$ and $Fixed_2 = (0,0,1,0,0,1,0,1,0,0)$. In (3), the parameter $sec = (2^7) * 5 = 640$ is used as a reference such that : sec includes the first 640 values corresponding to non-bistable cases, while sec : represents the 128 possible bistable cases.

$$E_1 = |\text{Prediction}[: \text{sec}] - \text{Result}| \qquad (3)$$

$$E_2 = \begin{cases} |\text{Prediction}[\text{sec} :] - \text{Fixed}_1| = e_1 \\ |\text{Prediction}[\text{sec} :] - \text{Fixed}_2| = e_2 \end{cases} \qquad (4)$$

$$E_3 = E_1 + \sum_{i=1}^{128} \min(e_{1i}, e_{2i}) \qquad (5)$$

$$E_{\text{Total}} = E_3 + \begin{cases} 1000 & \text{if Fixed}_1 \notin \text{Prediction} \\ 1000 & \text{if Fixed}_2 \notin \text{Prediction} \end{cases} \qquad (6)$$

It is important to note that in (3), only the absolute error of the non-bistable cases is calculated. In (4), two vectors, e_1 and e_2, are created to quantify how close the solution is to each of the possible bistable fixed points. As seen in (5), the error E_1 is added to the minimum error per record between e_1 and e_2, thus, giving the algorithm the flexibility to choose the proportions of the fixed points. Finally, (6) creates the necessary conditions to prevent the algorithm from converging to a unique solution, ensuring that it does not result in only fixed point 1 or only fixed point 2.

In summary, we can construct our fitness or score function using (3), (4), (5), and (6). This mathematical reasoning allowed us to model the fitness function correctly and perform the inference accurately. We used the evolutionary computing libraries of both GA and fuzzy PSO [25,30], the inference can be performed using the present methodology.

In this case, by conducting 500 simulations for each algorithm, we obtained the weight matrix $w_{i,j}$ and the threshold θ_i for each simulation. We also recorded

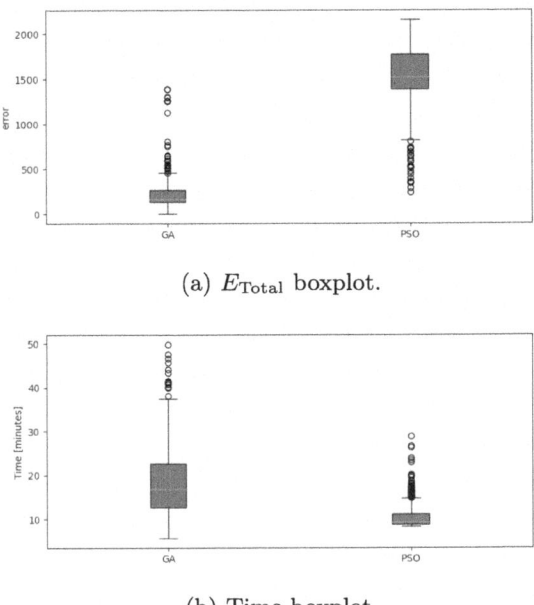

(a) E_{Total} boxplot.

(b) Time boxplot.

Fig. 2. Error and Time boxplots.

the computation times and the score associated with the inferred network. We used Python 3.11.4 on a laptop with an AMD Ryzen 9 5900HX processor with Radeon Graphics (3.30 GHz) and 16 GB of installed RAM (15.4 GB usable).

4 Results

In Table 2 we can observe the performance of each algorithm, focusing in the execution times and the score assigned by the fitness function to the network, which is reflected as error in the table. It is important to note that we have boxplots in Fig. 2 for a better visualization.

In summary, based on Table 2, we can conclude that the algorithm with the best performance is GA, as it successfully inferred 23 networks with an error of 0. This means that for these networks, all possible combinations for each parameter vector reach their respective fixed point, and the combinations belonging to the bistable parameter vector reach both possible fixed points.

In Fig. 2, we can clearly see that GA has an excellent performance in reducing the error, in general, the average error of GA is approximately 7 times smaller than the average error of fuzzy PSO. When evaluating the execution times of both algorithms, GA took longer to infer the network compared to fuzzy PSO.

To explain the results in greater detail, one of the 23 resulting networks is taken as an example, which can be visualized in Fig. 3a. It is important to note that all connections are $[-1, 1]$, making these networks binary. In Fig. 3b, we can

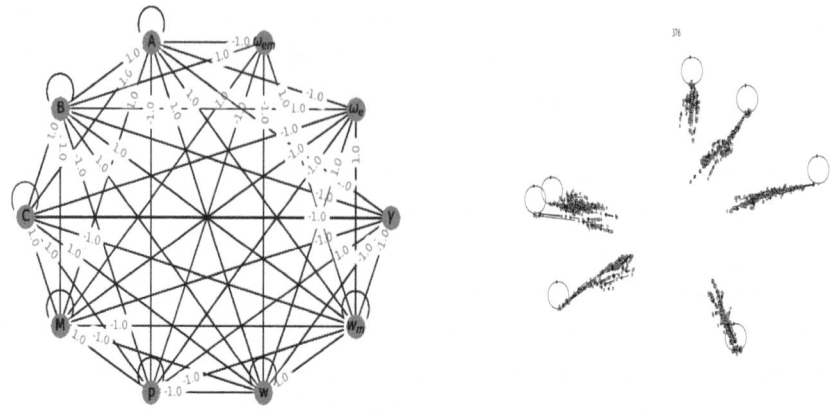

(a) Example of an inferred network. (b) State transition graph of the inferred network.

Fig. 3. Visualization of an inferred network of the Tryptophanase operon.

Table 3. Weight Matrix and Threshold Vector

Weight Matrix										Threshold Vector
0	0	0	0	0	0	0	0	0	0	0
0	0	0	0	0	0	0	0	0	0	0
0	0	0	0	0	0	0	0	0	0	0
-1	-1	0	0	1	0	0	-1	1	1	0
0	-1	1	1	0	1	0	-1	1	0	1
-1	-1	0	0	0	0	0	0	1	1	-1
0	-1	0	0	1	1	0	-1	0	0	0
0	-1	0	-1	0	0	1	0	0	-1	-1
-1	1	-1	0	1	1	-1	-1	0	0	0
-1	1	1	0	1	-1	-1	0	0	1	0

visualize its state transition diagram. This diagram shows the convergence of the network's points, and as we can see, there are 6 points where all the network's points converge. This means that the 6 possible fixed points are represented, and more details can be visualized in Fig. 4.

We also have the same network represented in Fig. 3a, where we illustrate the weight matrix and the obtained threshold vector. These parameters can be seen in Table 3. The first three rows of the weight matrix and the first three parameters of the threshold vector are null, as they represent the nodes of the parameter vector. This means they do not update and are predetermined to the value 0. The remaining parameters in this particular case come from the GA. It is important to note that the possible values are only $(-1, 0, 1)$, demonstrating that

Gene Regulatory Network for the Tryptophanase Operon   169

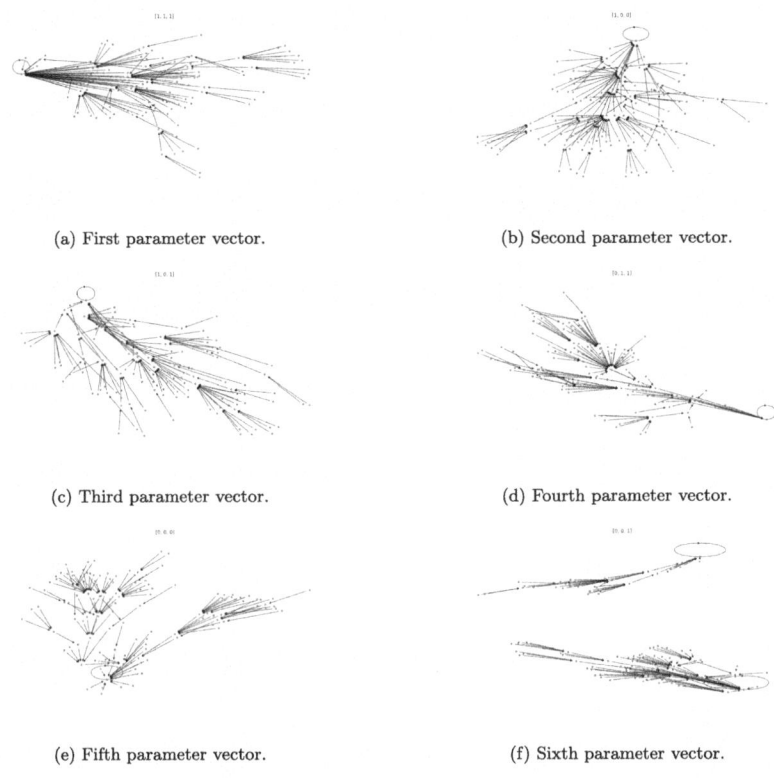

(a) First parameter vector.

(b) Second parameter vector.

(c) Third parameter vector.

(d) Fourth parameter vector.

(e) Fifth parameter vector.

(f) Sixth parameter vector.

Fig. 4. Detailed State transition graph.

we are using binary values $(-1, 1)$, but allowing the value 0 to avoid unnecessary connections.

In Table 4, each fixed point vector exhibits 128 possible combinations. Among the bistable configurations, their combined frequencies sum up to 128, indicating the absence of outliers and confirming the correct inference of the network. We also evaluated whether any of the inferred networks were repeated and whether all the networks were unique.

In Fig. 4, we can see in greater detail how the networks converge. Figures 4a, 4b, 4c, 4d, and 4e correspond to the parameter vectors $(1, 1, 1)$, $(1, 0, 0)$, $(1, 0, 1)$, $(0, 1, 1)$, and $(0, 0, 0)$, respectively. These five follow a similar pattern; we can observe a trend in the figure towards a central point, representing that each parameter vector effectively reaches its previously stipulated fixed point. In the last case, shown in Fig. 4f, the parameter vector is $(0, 0, 1)$. Here, we can visually confirm that there are two trends instead of just one. This corresponds to the fact that, for this particular parameter vector, there are two fixed points it can reach. For this reason, the same trend as in the previous figures does not exist.

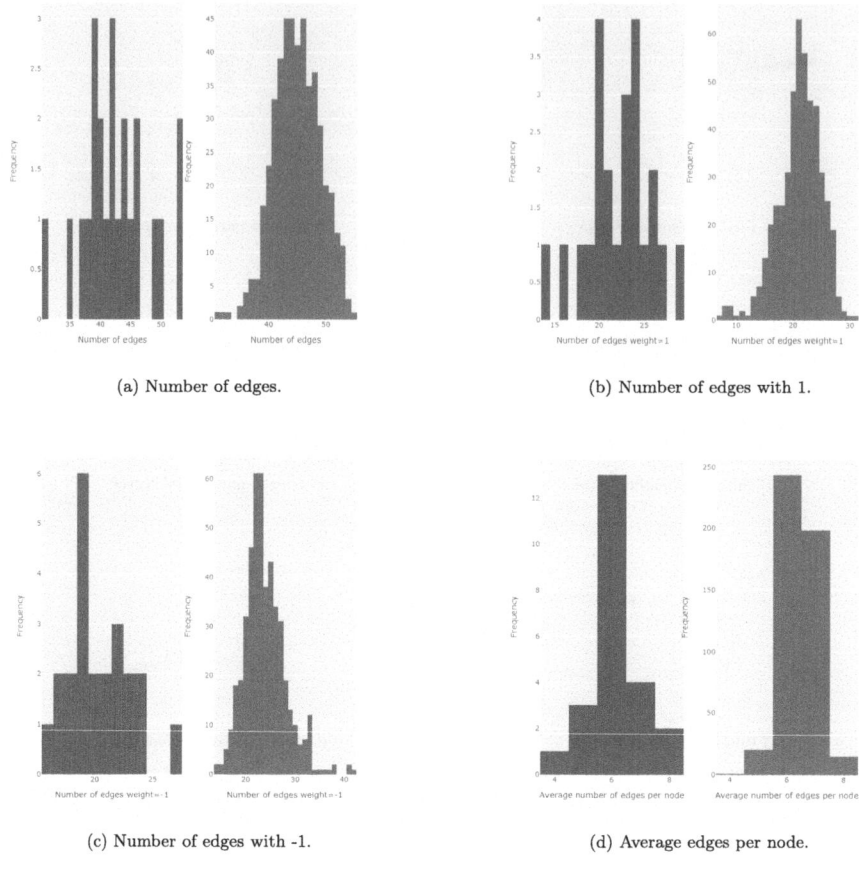

Fig. 5. GA metrics. (Color figure online)

In Fig. 5, we present certain metrics of the networks inferred by each algorithm. Each figure contains two histograms: green represents networks with zero error, and red represents the remaining networks. In this case, we are focusing on the networks from the GA. For the performance evaluation of the networks inferred by fuzzy PSO, refer to Fig. 6, where only red is used because this algorithm did not reach any network with a score of zero.

Figure 5 and Fig. 6 allows us is to study the behavior of the nodes by examining the number of positive and negative connections, as well as the average number of incoming and outgoing edges per node. This analysis provides a better understanding of the network's structure. In Fig. 5a and Fig. 6a, we observe the total number of edges per network, without considering whether they are positive or negative. Finally, we calculate the average number of connections per node, excluding the nodes of the parameter vector, as shown in Figs. 5d and 6d. One of the main differences that we can point out between working with GA and

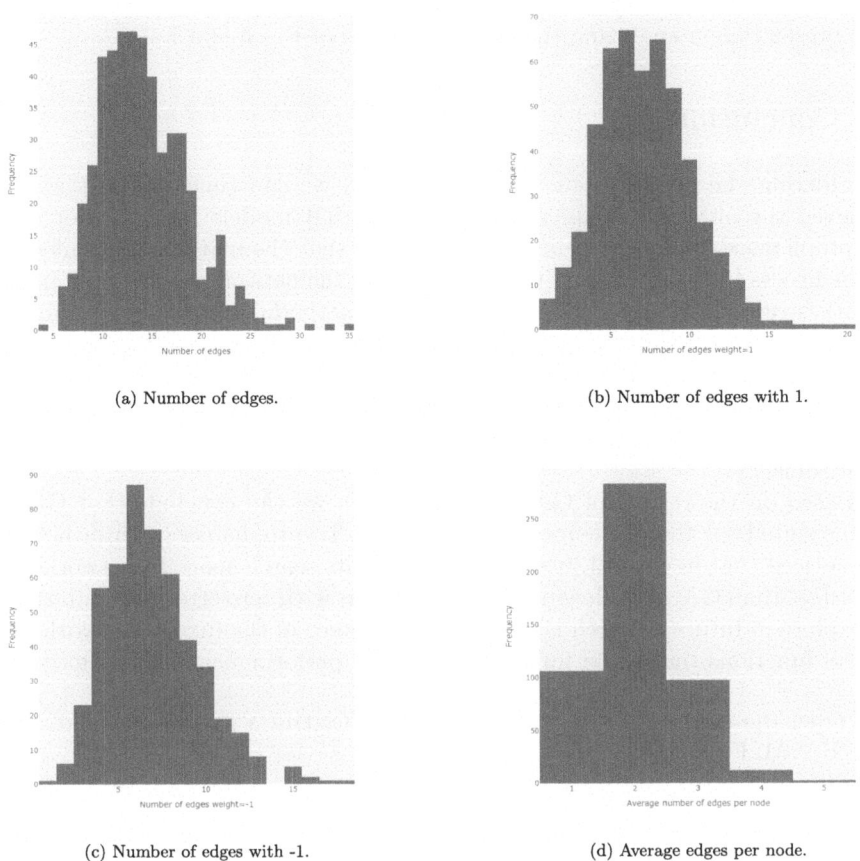

(a) Number of edges. (b) Number of edges with 1.

(c) Number of edges with -1. (d) Average edges per node.

Fig. 6. Fuzzy PSO metrics. (Color figure online)

Table 4. Results of the simulation

parameter vector	Fixed points	Quantity
[1, 1, 1]	[1, 1, 1, 0, 0, 0, 0, 0, 0, 1]	128
[1, 0, 0]	[1, 0, 0, 0, 0, 0, 0, 1, 0, 0]	128
[1, 0, 1]	[1, 0, 1, 0, 0, 0, 0, 1, 0, 0]	128
[0, 1, 1]	[0, 1, 1, 1, 1, 1, 1, 0, 1, 1]	128
[0, 0, 0]	[0, 0, 0, 0, 0, 1, 0, 1, 0, 0]	128
[0, 0, 1]	[0, 0, 1, 0, 0, 1, 0, 1, 0, 0]	36
	[0, 0, 1, 1, 1, 1, 1, 0, 0, 1]	92

fuzzy PSO, is that GA obtained solutions with more edges than fuzzy PSO. We can see in Fig. 5a that the mode for the networks with zero error (green color) is about 40 and the mode for the networks with non-zero solutions (red color) is about 45. Whereas, in Fig. 6a, the mode is about 12. This type of behavior is also present when analyzing the other bar plots of Fig. 5 and Fig. 6.

5 Conclusions

Considering the points discussed in the paper, we can conclude that we have achieved the objective of inferring a network that models the behavior of the Tryptophanase operon. It is important to note that the mathematical modeling of the fitness function was key to correctly infer the network since it contains the main restrictions of the inference process. The methodology used in this function can be extrapolated to infer other gene regulatory networks that exhibit the bistable property in their parameter vector. In summary, we were able to infer the network correctly and design a methodology that can solve the problem of bistability in genetic regulatory networks, and study the performance of both algorithms.

Based on the results of GA and fuzzy PSO, we can conclude that GA was better suited for the challenge of inferring the Tryptophanase operon network. We noticed that in general, fuzzy PSO limited its search space to networks with less edges than GA, not allowing to find solutions with zero error. One alternative to explore in future research is try the performance of the fuzzy PSO with other fitness functions that could improve the search performance of the algorithm.

Acknowledgements. G.A.R. thanks ANID FONDECYT 1230315 and ANID PIA/BASAL FB0002.

References

1. Abdelbar, A., Abdelshahid, S., Wunsch, D.: Fuzzy PSO: a generalization of particle swarm optimization. In: Proceedings. 2005 IEEE International Joint Conference on Neural Networks, vol. 2, pp. 1086–1091 (2005). https://doi.org/10.1109/IJCNN.2005.1556004
2. Alden, W.: Genetic algorithms for real parameter optimization. Found. Genetic Algorithms, vol. 1, pp. 205–218. Elsevier (1991). https://doi.org/10.1016/B978-0-08-050684-5.50016-1
3. Azuma, S.I., Kure, T., Sugie, T.: Structural bistability analysis of flower-shaped and chain-shaped Boolean networks. IEEE/ACM Trans. Comput. Biol. Bioinf. **17**(6), 2098–2106 (2020). https://doi.org/10.1109/TCBB.2019.2917196
4. Beneš, N., Brim, L., Kadlecaj, J., Pastva, S., Šafránek, D.: Exploring attractor bifurcations in Boolean networks. BMC Bioinformatics **23**(1) (2022). https://doi.org/10.1186/s12859-022-04708-9
5. Bertacchini, F., Scuro, C., Pantano, P., Bilotta, E.: Modelling brain dynamics by Boolean networks. Sci. Rep. (2022). https://doi.org/10.1038/s41598-022-20979-x
6. Bloomingdale, P., Nguyen, V.A., Niu, J., Mager, D.E.: Boolean network modeling in systems pharmacology. J. Pharmacokinet. Pharmacodyn. **45**(1), 159–180 (2018). https://doi.org/10.1007/s10928-017-9567-4

7. Chavarría, R., Cristina, S., Belén, G.M., Gregorio, R.: Modelling biological systems: a new algorithm for the inference of Boolean networks. Mathematics **9**(4), 373 (2021). https://doi.org/10.3390/math9040373
8. Cui, S., Jun, P.: CABEAN: a software for the control of asynchronous Boolean networks. Bioinformatics **37**(6), 879–881 (2020). https://doi.org/10.1093/bioinformatics/btaa752
9. Deal, I., Macauley, M., Davies, R.: Boolean models of the transport, synthesis, and metabolism of tryptophan in Escherichia coli. Bull. Math. Biol. (2023). https://doi.org/10.1007/s11538-023-01122-x
10. Demongeot, J., Goles, E., Morvan, M., Noual, M., Sené, S.: Attraction basins as gauges of robustness against boundary conditions in biological complex systems. PLoS ONE (2010). https://doi.org/10.1371/journal.pone.0011793
11. Eiben, A., Schoenauer, M.: Evolutionary computing. Inf. Process. Lett. **82**(1), 1–6 (2002). https://doi.org/10.1016/s0020-0190(02)00204-1
12. Goles, E., Montalva, M., Ruz, G.A.: Deconstruction and dynamical robustness of regulatory networks: application to the yeast cell cycle networks. Bull. Math. Biol. (2012). https://doi.org/10.1007/s11538-012-9794-1
13. Guebila, M.B., et al.: GRAND: a database of gene regulatory network models across human conditions. Nucleic Acids Res. (2021). https://doi.org/10.1093/nar/gkab778
14. Haotong, Q., Ruihao, G., Xianglong, L., Xiao, B., Jingkuan, S., Nicu, S.: Binary neural networks: a survey. Pattern Recogn. **105**, 107281 (2020). https://doi.org/10.1016/j.patcog.2020.107281
15. Hari, K., Harlapur, P., Gopalan, A., Ullanat, V., Duddu, A.S., Jolly, M.K.: Emergent properties of coupled bistable switches. J. Biosci. **47**(4) (2022). https://doi.org/10.1007/s12038-022-00310-6
16. Hopfensitz, M., et al.: Multiscale binarization of gene expression data for reconstructing Boolean networks. IEEE/ACM Trans. Comput. Biol. Bioinf. (2012). https://doi.org/10.1109/tcbb.2011.62
17. Kim, K.Y., Wang, J.: Potential energy landscape and robustness of a gene regulatory network: toggle switch. PLoS Comput. Biol. (2007). https://doi.org/10.1371/journal.pcbi.0030060
18. Kobayashi, K., Hiraishi, K.: Optimization-based approaches to control of probabilistic Boolean networks. Algorithms (2017). https://doi.org/10.3390/a10010031
19. Kumar, M., Husain, M., Upreti, N., Gupta, D.: Genetic algorithm: review and application. SSRN Electron. J. (2010). https://doi.org/10.2139/ssrn.3529843
20. Lin, J.C.W., Yang, L., Fournier-Viger, P., Hong, T.P., Voznak, M.: A binary PSO approach to mine high-utility itemsets. Soft Comput. (2016). https://doi.org/10.1007/s00500-016-2106-1
21. Liu, F., et al.: A system pharmacology Boolean network model for the TLR4-mediated inflammatory response in early sepsis. J. Pharmacokinet. Pharmacodyn. **49**(6), 645–655 (2022). https://doi.org/10.1007/s10928-022-09828-6
22. Ma, Z.: Probabilistic Boolean network modeling for fMRI study in Parkinson's disease. Ph.D. thesis, University of British Columbia (2008). https://doi.org/10.14288/1.0066945
23. Mendoza, L., Alvarez-Buylla, E.R.: Dynamics of the genetic regulatory network for Arabidopsis thaliana flower morphogenesis. J. Theor. Biol. (1998). https://doi.org/10.1006/jtbi.1998.0701
24. Mitchell, M., Holland, J., Forrest, S.: When will a genetic algorithm outperform hill climbing. In: Cowan, J., Tesauro, G., Alspector, J. (eds.) Advances in Neural Information Processing Systems, vol. 6. Morgan-Kaufmann (1993)

25. Nobile, M.S., Cazzaniga, P., Besozzi, D., Colombo, R., Mauri, G., Pasi, G.: Fuzzy self-tuning PSO: a settings-free algorithm for global optimization. Swarm Evol. Comput. (2018). https://doi.org/10.1016/j.swevo.2017.09.001
26. Ruz, G.A., Goles, E.: Learning binary threshold networks for gene regulatory network modeling. In: 2022 IEEE Conference on Computational Intelligence in Bioinformatics and Computational Biology (CIBCB), pp. 1–8 (2022). https://doi.org/10.1109/CIBCB55180.2022.9863056
27. Ruz, G.A., Goles, E.: Gene regulatory networks with binary weights. Biosystems **227–228**, 104902 (2023). https://doi.org/10.1016/j.biosystems.2023.104902
28. Ruz, G.A., Timmermann, T., Goles, E.: Reconstruction of a GRN model of salt stress response in Arabidopsis using genetic algorithms. In: 2015 IEEE Conference on Computational Intelligence in Bioinformatics and Computational Biology (CIBCB), pp. 1–8 (2015). https://doi.org/10.1109/CIBCB.2015.7300306
29. Sameon, F., Shamsuddin, S., Sallehuddin, R., Zainal, A.: Compact classification of optimized Boolean reasoning with particle swarm optimization. Intell. Data Anal. **6**, 915–931 (2012). https://doi.org/10.3233/ida-2012-00559
30. Solgi, R.: Genetic algorithm implementation in Python. https://github.com/rmsolgi/geneticalgorithm

Multilabel Classification of Intracranial Hemorrhages Using Deep Learning and Preprocessing Techniques on Non-contrast CT Images

Rodrigo Salas[1,2,3](✉), Juan Sebastian Castro[1], Marvin Querales[2,4], Carolina Saavedra[5], Claudia Prieto[3,6,7], and Steren Chabert[1,2,3]

[1] Biomedical Engineering School, Engineering Faculty, Universidad de Valparaíso, General Cruz 222, Valparaíso, Chile
juan.castro@postgrado.uv.cl, {rodrigo.salas,steren.chabert}@uv.cl
[2] Center of Interdisciplinary Biomedical and Engineering Research for Health - MEDING, Universidad de Valparaíso, Valparaíso, Chile
marvin.querales@uv.cl
[3] Millennium Institute for Intelligent Healthcare Engineering (iHealth), Santiago, Chile
[4] Medical Technology School, Medicine Faculty, Universidad de Valparaíso, Viña del Mar, Chile
[5] Departamento de Informática, Universidad Técnica Federico Santa María, Valparaíso, Chile
carolina.saavedra@usm.cl
[6] Engineering School, Pontificia Universidad Católica de Chile, Santiago, Chile
cdprieto@ing.puc.cl
[7] Department of Biomedical Engineering, King's College London, London, UK

Abstract. This study presents a comprehensive framework that integrates a deep learning model with advanced image preprocessing techniques to improve the multilabel classification of five types of intracranial hemorrhage—epidural, intraparenchymal, intraventricular, subarachnoid, and subdural—using non-contrast computed tomography (CT) images. The framework includes strategies to mitigate overfitting, data augmentation, and a custom loss function. It was rigorously evaluated on a dataset of over 25,000 non-contrast CT scans, each labeled by expert radiologists across six classes. The proposed model achieved 99% accuracy and 92.1% sensitivity in detecting intracranial hemorrhage, outperforming previously reported methods.

Keywords: Intracranial hemorrhage · Deep Learning · Multilabel Classification · Non-contrast Computed Tomography (CT) · Medical Image Processing

1 Introduction

Intracranial hemorrhage (ICH) occurs when bleeding inside the skull results from the rupture of blood vessels or capillaries. As a major cause of mortality worldwide, ICH represents a medical emergency, particularly in patients with neurological conditions, making rapid diagnosis essential. Computed tomography (CT) imaging is often preferred for its speed and cost-effectiveness compared to other diagnostic methods. However, the visualization of intracranial hemorrhages is influenced by surrounding tissue and the specific characteristics of the hemorrhage, making accurate identification a challenging task for medical professionals.

ICH can be classified based on its location in the skull, with intracerebral hemorrhage (IH) being the most common, followed by intraventricular hemorrhage (IVH), subarachnoid hemorrhage (SAH), subdural hemorrhage (SDH), and the less frequent epidural hemorrhage (EDH) [5,24]. Figure 1 shows examples of the different types of intracranial hemorrhage. SAH involves bleeding within the subarachnoid space, most commonly caused by trauma or the rupture of an intracranial aneurysm. IH, on the other hand, refers to bleeding within the brain parenchyma [24]. SDH occurs between the dura mater and the arachnoid, while EDH refers to bleeding between the dura mater and the skull. Early diagnosis of the type of hemorrhage can significantly reduce mortality, particularly in the less frequent cases of ICH.

Automatic diagnostic methods can be particularly valuable for rare types of hemorrhages. With the recent advancements in deep learning, computer vision algorithms have demonstrated exceptional performance, not only in image classification tasks but also in medical imaging applications (See [6–9,15,25,32,33]).

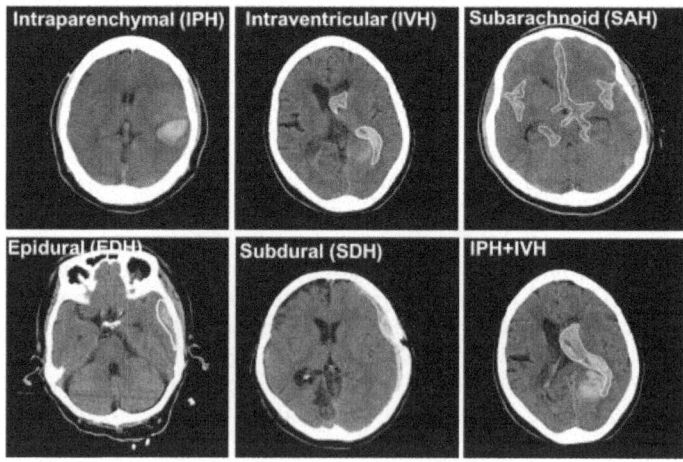

Fig. 1. Examples of the types of Intracranial hemorrhage in non-constrast CT images. Images obtained from the RSNA Intracranial Hemorrhage Detection and Classification Challenge [13]

Automatic and semi-automatic detection of intracranial hemorrhage (ICH) in non-contrast computed tomography (CT) images is a rapidly evolving area of research, bolstered by recent advancements in artificial intelligence and image processing [6]. For example, Bhadauria et al. [4] developed an integrated segmentation method for hemorrhage detection in brain CT images, combining Fuzzy C-Means (FCM) clustering with a region-based active contour approach.

The literature includes several studies that utilize deep learning for intracranial hemorrhage (ICH) classification. Many of these studies leverage convolutional neural network (CNN) models due to their high reliability and efficiency. For example, Chilamkurthy et al. [12] developed a model to detect and classify ICH types, cranial fractures, midline deviation, and mass effects. Their approach incorporated various deep learning networks, including ResNet18 [16] for hemorrhage classification and the UNet network [26] for image segmentation. Sage et al. [28] introduced a method for detecting five ICH subtypes in CT images using a double-branch CNN based on the ResNet-50 architecture. Watanabe et al. [34] assessed the impact of deep learning-based computer-assisted detection (CAD) on the performance of physicians at various levels, finding that CAD using the UNet network significantly enhanced diagnostic performance and reduced reading time for ICH detection. Additionally, Lee et al. [21] developed an interpretable deep learning system for detecting acute ICH and classifying five subtypes from unenhanced head CT scans. Their model employed four CNN architectures-VGG16, ResNet50, Inception-v3, and Inception-ResNet-v2-and demonstrated a high-performance system capable of handling small and imbalanced datasets.

However, there is a need for further research on incorporating a preprocessing stage for CT images before applying deep learning algorithms. Evidence from other imaging modalities suggests that preprocessing can enhance the classification performance of artificial intelligence models. Therefore, this study proposes an approach for the automatic detection of five types of intracranial hemorrhage (ICH) by combining image preprocessing techniques, convolutional neural networks (CNNs), and post-processing of metadata present in CT images to evaluate the effectiveness of the proposed method. The structure of this paper is as follows: Sect. 2 describes the proposed framework, Sect. 3 presents the results and discussion, and Sect. 4 summarizes the main conclusions.

2 Methods and Materials

2.1 Dataset Description

The dataset used in this study is sourced from the RSNA Intracranial Hemorrhage Detection and Classification Challenge. It comprises over 25,000 non-contrast brain CT studies collected by four research institutions: the Center for Artificial Intelligence & Imaging (AIMI) at Stanford University, Diagnostic Imaging and Learning Algorithms (DILA) at St. Michael's LKS-CHART, the Department of Radiology at Thomas Jefferson University, and Universidade Federal de São Paulo (Unifesp). Annotations for the dataset were provided by the American Society of Neuroradiology (ASNR), with contributions from over 60 volunteers who labeled the CT exams [13].

The dataset presents several challenges. First, intracerebral hemorrhage (ICH) cases make up slightly more than 14% of the total number of cases. Secondly, there is an imbalance problem [1] as certain types of bleeding, such as epidural hemorrhage (EDH) and intraventricular hemorrhage (IVH), being underrepresented in less than 4% of the images, accounting for only 0.4% and 3.5%, respectively. This phenomenon can cause traditional machine learning methods that try to minimize the overall error and maximize the accuracy to be unsuitable for these distributions. Therefore, accuracy is not a reliable performance indicator in datasets with imbalance problems, such as this one, especially when the classes with the lowest representation are the most critical. In general, there are two strategies for handling class imbalance in classification: (1) data-level approaches and (2) algorithm-level approaches (see, for example, [14,23] for more details). Data-level methods adjust the class imbalance ratio to create a more balanced distribution between classes. Algorithm-level techniques, on the other hand, include methods that directly learn from the imbalanced distribution, such as one-class learning, cost-sensitive learning, and ensemble methods.

2.2 Proposed Framework

The proposed framework is illustrated in Fig. 2. The central component is a sequence model that begins with preprocessing DICOM images using a custom batch training process. It then employs an EfficientNet B3 as the backbone, augmented with additional pooling and fully connected layers at the output.

Stage 1: CT Image Pre-processing. The Hounsfield unit (HU) is a quantitative measure of radio-density used by radiologists to interpret CT images. During the CT reconstruction process, the radiation absorption or attenuation coefficient within a tissue is used to generate a grayscale image. The Hounsfield unit is derived from a linear transformation of the baseline linear attenuation coefficient of the X-ray beam, with water defined as zero HU and air as -1000 HU. This linear transformation creates a Hounsfield scale that represents different radiodensities as varying gray levels [17]

The proposed image pre-processing pipeline includes:

1. **Hounsfield unit windowing.** In this study, three distinct Hounsfield unit (HU) windows were employed: a brain window (level = 40, width = 80), a subdural window (level = 80, width = 200) for blood visualization, and a bone window (level = 40, width = 380). These three windows were combined into a single image with three channels, as illustrated in Fig. 3.
2. **Crop blank space.** Minimum filters and masking kernels were applied to the CT images to reduce artifacts from the imaging process. Additionally, blank space removal was used to eliminate irrelevant information from the images, improving the focus on relevant features.
3. **Refine label, normalization, and resize.** A mask was applied to remove potential artifacts from the images, and additional steps including normalization and resizing were performed. The original slice size was 512×512

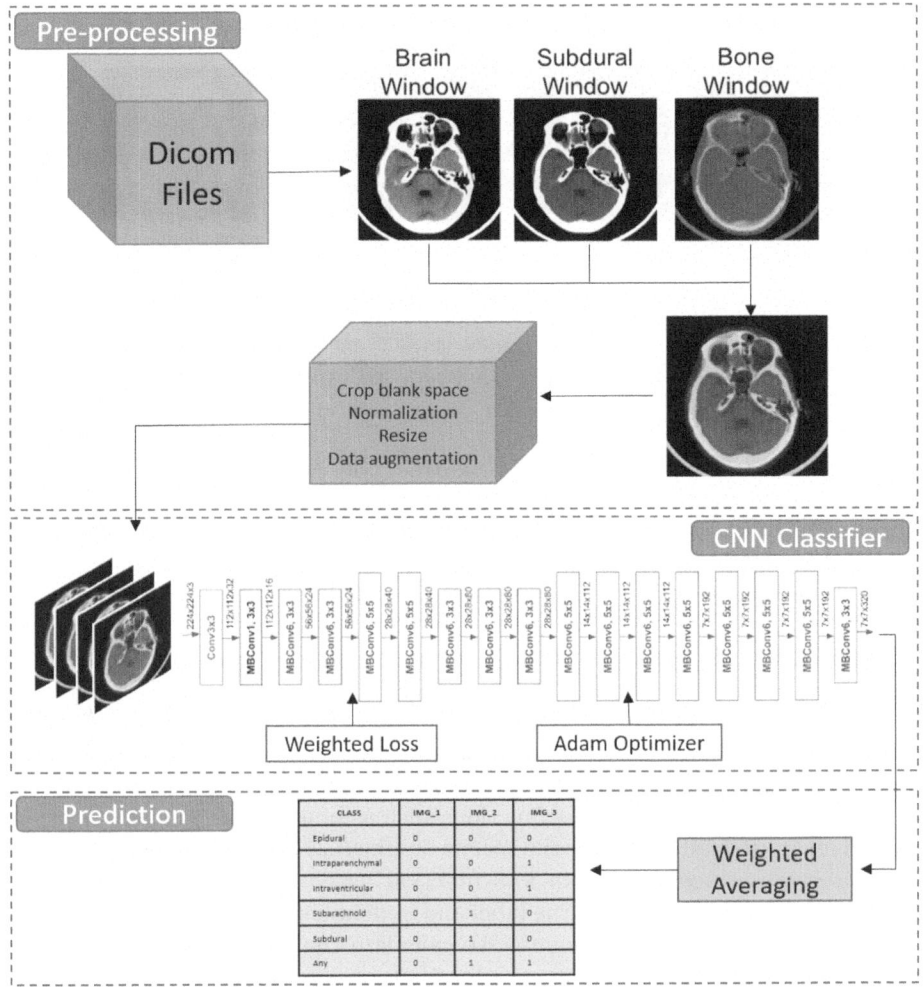

Fig. 2. Proposed deep learning framework with image preprocessing on non-contrast CT images for multilabel classification.

pixels. To reduce computational costs during training, the slices were resized to 224 × 224 pixels using bicubic interpolation [31]. Bicubic interpolation was chosen for its superior quality compared to traditional methods such as nearest neighbor or bilinear interpolation. Amanatiadis support this preference [2], who recommends prioritizing quality over performance in medical imaging applications.

4. **Data augmentation techniques.** To address label imbalance, several augmentation techniques were applied, including Random ShiftScaleRotate, Random Resize Crop, Horizontal Flip (Hflip), and Vertical Flip (Vflip).

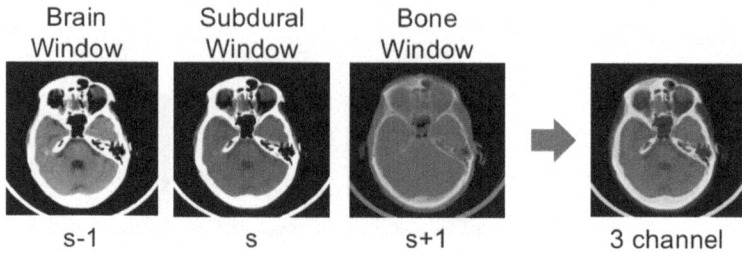

Fig. 3. Hounsdfield unit windowing for the Brain. From left to right: a brain window (level = 40, width = 80); a subdural window (level = 80, width = 200); a bone window (level = 40, width = 380); combined single image with three channels.

Stage 2: Convolutional Neural Network Classifier. This work proposes a deep learning framework based on Convolutional Neural Networks (CNNs), utilizing the EfficientNet family of models developed by Tan et al. [30]. EfficientNet has demonstrated superior accuracy and efficiency compared to conventional CNNs. It introduces a novel model scaling method that employs a simple yet highly effective compound coefficient to scale CNNs in a more structured manner. This approach optimizes model scaling, achieving high performance with significantly reduced parameters and faster inference times. Consequently, EfficientNet is well-suited for medical image classification tasks. Deep learning methods, such as those using EfficientNet, can automatically learn high-level features directly from training data, eliminating the need for manual feature engineering [20].

Traditional deep learning loss functions such as binary cross-entropy or mean squared error are inadequate for the specific application of hemorrhage detection. Therefore, we propose a custom loss function based on the log-loss equation (Eq. 1), where y represents the true label and y_i denotes the predicted probability for each label across N data points. We calculate the log-loss for each label in the dataset and incorporate class weights that can be adjusted throughout the experiments.

$$H_p(q) = -\frac{1}{N} \sum_{i=1}^{N} y_i \cdot log(p(y_i)) + (1 - y_i) \cdot log(1 - p(y_i)) \quad (1)$$

For the training process, we used a cross-validation subject-out scheme to ensure the independence of the training and test sets, as verified through the DICOM metadata. This approach guarantees that all slices from a given exam are assigned exclusively to either the training or the test dataset, but not to both.

Stage 3: Multilabel Prediction. In machine learning *single-label* refers to assigning one label l, from a set of mutually exclusive labels L, where $|L| >$

1. When $|L| = 2$, this classification problem is specifically known as binary classification.

In contrast, *multilabel* classification involves predicting a set of labels $Y \subseteq L$. In multilabel classification, an image can be simultaneously associated with multiple class labels, unlike multi-class classification, where each image is assigned to only one class. For instance, in the RSNA dataset, a CT image can be labeled with multiple categories. For example, an image labeled H might show both intraventricular hemorrhage (IVH) and epidural hemorrhage (EDH), making it positive for three labels: 'Any' (indicating the presence of hemorrhage), 'Intraventricular', and 'Epidural'.

2.3 Evaluation Metrics

The confusion matrix is used to evaluate the performance of classification models. It summarizes prediction results by comparing the predicted labels against the true labels. The parameters of the confusion matrix are: 1) True Positive (TP), the number of instances where the model correctly predicted the positive class; 2) False Negative (FN), the number of instances where the model incorrectly predicted the negative class when the actual class was positive; 3) False Positive (FP), The number of instances where the model incorrectly predicted the positive class when the actual class was negative; and, True Negative (TN): the number of instances where the model correctly predicted the negative class.

The following performance metrics can be obtained from the confusion matrix:

– **Accuracy:** The accuracy value refers to how close a measurement is to the true value.
$$Accuracy = \frac{FP + TN}{TP + TN + FP + FN} \quad (2)$$

– **Recall:** The measure of sensitivity or recall is the percentage of positive cases that were correctly labeled by the model.
$$recall = \frac{TP}{TP + FN} \quad (3)$$

– **Precision:** Precision is the percentage of correct classifications.
$$precision = \frac{TP}{TP + FP} \quad (4)$$

– **F1 score:** The F-Value score can be interpreted as a harmonic mean between precision and sensitivity.
$$F1\_score = 2 \left(\frac{precision \cdot recall}{precision + Recall} \right) \quad (5)$$

– **ROC curve.** The ROC (receiver-Operating-Characteristic) curve is a statistical tool used in classification analysis to assess the discriminant capacity of a diagnostic test. This tool compares two operational characteristics: the ratio of true positives (TP) and the ratio of false positives (FP), depending on the change in the threshold.

3 Results and Discussion

3.1 Multilabel Classification

The entire dataset was utilized to assess the multilabel classification performance of the proposed method, ensuring a balanced distribution between positive and non-positive cases. Data augmentation techniques were specifically applied to images with hemorrhages, which constitute less than 4% of the dataset, to enhance the representation of these less common classes.

The parameters used for the Deep Learning are the following. (1) The custom weighted loss function, which is highly emphasized for the 'Any' class, distinguishes between the binary classes of 'any hemorrhage' versus 'no hemorrhage'; (2) Sigmoid activation functions and average pooling were used; (3) Adam optimizer for the training phase; (4) The input images was $224 \times 224 \times 3$ pixels; (5) Batch size was 32, learning rate $5e^{-4}$, decay rate of 0.8; and (6) Five epochs were used in each fold in the cross-validation process.

An additional experiment was conducted to validate the suitability of the proposed model for the multilabel classification of ICH. In this experiment, the EfficientNet B3 CNN was trained and evaluated independently, without incorporating the preprocessing stage, weighted loss, or weighted averaging. Cross-validation was performed using five folds and five training epochs. The results from this experiment were directly compared to those obtained with the proposed model in the multilabel classification task.

As outlined in the previous section, the proposed method was evaluated through three experiments, starting with multilabel classification. The model was trained using TensorFlow on a GPU (AMD RX580 with 8 GB VRAM) for five epochs. Cross-validation with five folds was employed, and ROC curves were generated for each fold. Figure 4 presents the best (left) and worst (right) results, with classes defined as follows: class 0: 'Any', class 1: 'Epidural', class 2: 'Intraparenchymal', class 3: 'Intraventricular', class 4: 'Subarachnoid', and class 5: 'Subdural'.

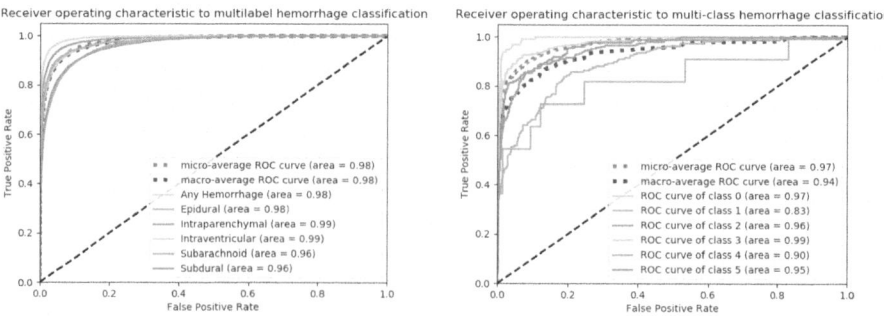

Fig. 4. (**left**) ROC curve of best result multilabel classification. (**right**) ROC curve of worst result multilabel classification.

As illustrated by the ROC curves, the results for the multilabel classification task are promising, with areas under the curve (AUC) ranging between 0.99 and 0.83. These AUC values indicate good performance of the proposed algorithm. All ROC curves are situated above the non-discrimination line, with the highest AUC of 0.99 achieved for class 3, which corresponds to intraventricular hemorrhage (IVH). This result is consistent with findings from Lee et al. [21], who reported AUC values exceeding 92%.

The lowest AUC of 0.83 was observed for class 1, which corresponds to epidural hemorrhage (EDH). This lower performance underscores the challenges associated with classifying EDH due to its distinctive morphological characteristics and its limited representation in the dataset. This difficulty is in line with observations by Sage and Badura [28], who noted similar low performance for this type of hemorrhage. Despite advancements in research, such as those by Lee et al. [21], achieving classification performance comparable to that of other hemorrhage subtypes remains a challenge.

In addition to individual class performance, the micro average and macro average metrics provide global indicators of the algorithm's overall effectiveness in multilabel classification. The AUC macro average, which is sensitive to low AUC values across different classes, typically yields a lower value compared to the AUC micro average. The micro average considers the class weights, resulting in a higher metric. In this study, the worst results from cross-validation produced an AUC macro average of 0.94 and an AUC micro average of 0.97, demonstrating superior performance in the ICH classification task.

3.2 Classification Without Pre-processing

Figure 5 displays the ROC curve for evaluating the EfficientNet B3 CNN in the multilabel classification task using the RSNA dataset, without incorporating preprocessing and the hyperparameters discussed previously. The curves for all evaluated classes are positioned on the discrimination line, indicating that the model struggled to classify the five types of intracranial hemorrhage (ICH) effectively. These results underscore the significant impact of preprocessing and hyperparameter tuning on the CNN's performance. Additionally, the custom loss function proposed for the model plays a crucial role in enhancing the network's training effectiveness.

Accuracy, F1 score, recall, the mean and standard deviation for each cross-validation fold were computed and are presented in Table 1. The results show that the accuracy for all types of intracranial hemorrhage (ICH) is above 0.91, with a standard deviation of less than 0.01. This indicates that the proposed model correctly identifies the label in at least 91% of cases, whether the image depicts a hemorrhage or a healthy brain.

The 'Any' class serves as a key indicator of the model's overall performance in ICH classification, as it denotes the presence of any type of ICH in the image. For the 'Any' class, the accuracy achieved is 0.91, recall is 0.92, and the F1 score is 0.76, reflecting good performance. The recall, which measures the model's ability

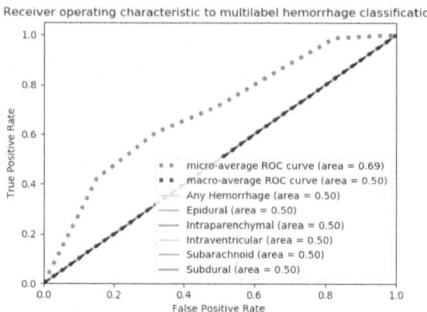

Fig. 5. ROC curve of multilabel classification without pre-processing in the model

to detect true positives, is 0.92, demonstrating a high capacity for identifying positive cases-an essential attribute for medical image classification.

These results align with those reported by Anupama et al. [3], who achieved a sensitivity of 94.01% and an accuracy of 95.73% with their GC-SDL model. However, our model outperforms the results reported by Lee et al. [22], who achieved a sensitivity of 82.5% in their best case.

Table 1. Cross-validation metrics of proposed method.

Label	Accuracy	F1 Score	Recall
Any	0.919 ± 0.010	0.768 ± 0.020	0.921 ± 0.014
Epidural	0.996 ± 0.000	0.379 ± 0.090	0.254 ± 0.084
Intraparenchymal	0.981 ± 0.000	0.804 ± 0.004	0.780 ± 0.032
Intraventricular	0.988 ± 0.000	0.832 ± 0.004	0.817 ± 0.019
Subarachnoid	0.968 ± 0.002	0.654 ± 0.014	0.620 ± 0.061
Subdural	0.958 ± 0.002	0.695 ± 0.006	0.752 ± 0.025

3.3 Hard-Class Classification

Classifying certain types of hemorrhages is particularly challenging due to their specific locations within the skull. In this study, these less common hemorrhage types make up less than 4% of the dataset and are represented by a limited number of diagnostic images. We trained the model specifically for these rare hemorrhage types to address this issue and applied data augmentation techniques to improve performance and address the class imbalance problem.

In the case of EDH, SAH, and SDH, the performance metrics shown in Table 1 are consistent with the observations from the ROC curves. For EDH, the accuracy remains stable, but the recall and F1 score are notably low, at 0.25 and

0.37, respectively. This suggests that the multi-class classification approach may not be well-suited for hemorrhage types that are difficult to classify and have lower incidence rates.

In contrast, the model demonstrated better performance for SDH and SAH, with recall scores of 0.75 and 0.62, respectively. While these scores are higher than those for EDH, they still indicate limited effectiveness for computer-aided diagnosis. This underscores the need for further refinement in detection capabilities for these less common hemorrhage types (Fig. 6).

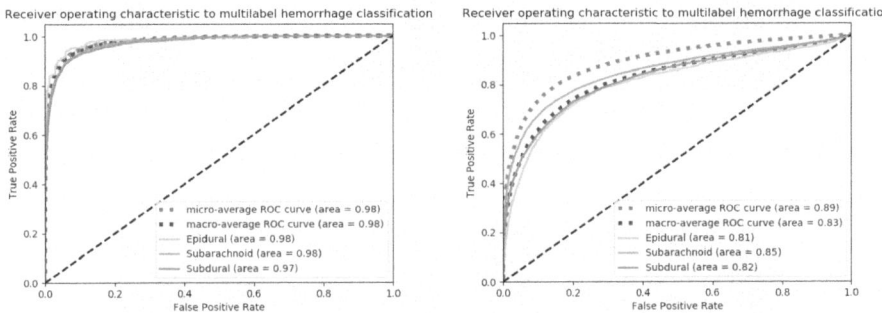

Fig. 6. (**left**) ROC curve of best result hard class classification. (**right**) ROC curve of worst result hard class classification

Figure 5 presents the ROC curves from the hard-class classification experiment, where cross-validation with four folds was employed to classify three specific types of hemorrhage: epidural, subarachnoid, and subdural. The model was tailored for this specific classification task. The performance results align with expectations, showing an AUC of 0.98 for the best case and 0.81 for the worst case. These results are consistent with those observed in the previous multi-class classification experiment involving all types of hemorrhage. In this context, label 1 corresponds to epidural hemorrhage (EDH), label 2 to subarachnoid hemorrhage (SAH), and label 3 to subdural hemorrhage (SDH). All three types classified in this experiment have a notably low incidence in the dataset compared to the hemorrhage types excluded from this study. Data augmentation techniques were applied to enhance performance and address the low incidence of these hemorrhage types (Figs. 7 and 8).

It is important to highlight that there is no significant difference between the ROC curves for each type of intracranial hemorrhage (ICH). This indicates that the model exhibits consistent performance across the three hemorrhage types included in the study. The behavior of the micro-average and macro-average AUC metrics is similar to that observed in the previous experiment. The differences between these two metrics are attributable to their calculation methods and potentially to the weighted averaging applied during the model's prediction stage.

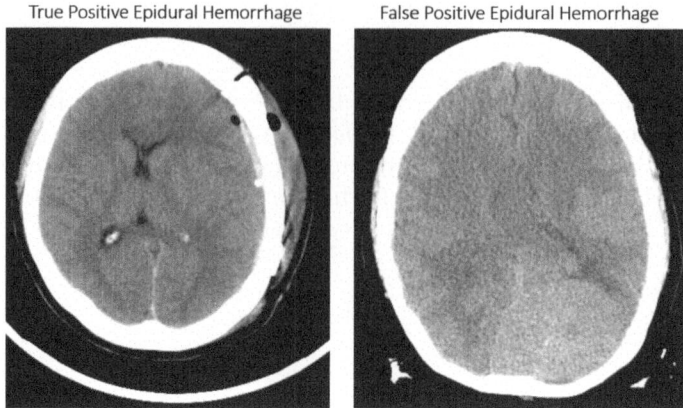

Fig. 7. True positive and false positive examples of epidural hemorrhage

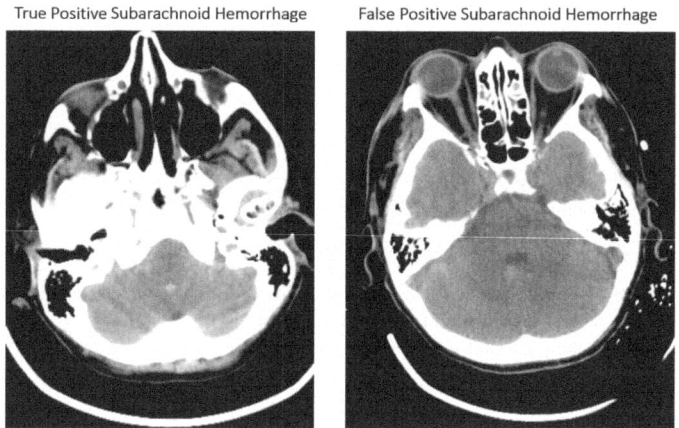

Fig. 8. True positive and false positive examples of subarachnoid hemorrhage

Table 2 summarizes the performance metrics for the hard-class classification experiment. The accuracy remains consistent with that observed in the multi-class classification experiment. However, the F1 score and recall have shown improvements. Specifically, recall for epidural hemorrhage (EDH) increased to 45.9%, and the F1 score rose to 56.2%, representing an enhancement of over 20% points compared to the previous experiment. For subarachnoid hemorrhage (SAH) and subdural hemorrhage (SDH), the improvements are smaller, reaching up to 10% points.

Table 3 compares the performance metrics of various deep learning models designed for intracranial hemorrhage (ICH) detection, as reported in the literature. The literature reports a range of performance metrics: accuracy varies between 80% and 97%, recall ranges from 81.8% to 97.1%, and the AUC con-

Table 2. Performance metrics of the proposed method for the hard-classification problem

Label	Accuracy	F1 Score	Recall
Epidural	0.996 ± 0.000	0.562 ± 0.060	0.459 ± 0.101
Subarachnoid	0.967 ± 0.004	0.731 ± 0.021	0.735 ± 0.013
Subdural	0.956 ± 0.004	0.742 ± 0.029	0.789 ± 0.029

sistently exceeds 90% across studies. Our model outperforms four of the six models reviewed in terms of all three metrics. Notably, the results from Lee et al. [21] and Chang, P [11], which also present multi-label deep learning models for detecting multiple types of ICH, show comparable performance to our model. These comparisons indicate that our proposed model aligns well with existing literature and, in some instances, achieves superior performance.

It is important to emphasize the distinct advantages of the proposed model compared to existing models in the literature [12,20]. Our model is specifically designed for multilabel classification, incorporating a custom loss function optimized for this purpose. Additionally, the preprocessing technique represents a significant advancement: it employs contrast windowing to create a three-channel image that distinctly represents the brain, meninges, and skull. This approach aligns with previous research, which has demonstrated that advanced preprocessing can significantly enhance the performance of classification and regression models [10,27,29].

Table 3. Comparison of the performance results for ICH classification in CT images.

Authors	Accuracy	Recall	AUC	Type of Task
[11]	97.5%	97.1%	98.0%	Any ICH vs non-hemorrhage
[12]	90.0%	NA	92%	High sensitivity operating point multi-label
[18]	97.2%	82.5%	NA	Segmentation ICH
[19]	94.7%	81.8%	90.0%	Single label best results
[21]	85.8%	92.4%	99.3%	High sensitivity operating point multi-label
Our Best Result	91.9%	92.1%	98.0%	Multi-label classification

The results indicate that the best performance was achieved for the 'Any' ICH classification, with a recall of 92.1%. However, performance was significantly lower for the less common and more challenging hemorrhage types. This reduced performance is attributed to the pathological characteristics of these hemorrhages and their lower prevalence, which results in limited image availability for training. Despite employing data augmentation techniques, the identification of EDH, SAH, and SDH did not improve significantly. For these less common ICH types, it may be beneficial to develop specialized deep learning models tailored to each specific class.

4 Conclusion

In this work, we present a novel deep learning model designed for the automatic multilabel detection of intracranial hemorrhage (ICH). The framework integrates image preprocessing, data augmentation, and custom loss functions to accurately identify both the presence of ICH and its five types in CT images. In the best-case scenario, the model achieved a sensitivity of 92.1% for detecting any hemorrhage. These results represent a significant advancement in computer-assisted automatic ICH detection—a crucial capability given the urgency of this medical emergency. Future work will focus on enhancing the model's ability to detect less common hemorrhage types and addressing the challenges associated with imbalanced datasets.

Acknowledgements. This research was funded by ANID FONDECYT project N° 1221938, 1231268, and 1210637, and ANID Millennium Science Initiative Program ICN2021_004.

Disclosure of Interests. The authors have no competing interests to declare that are relevant to the content of this article.

References

1. Ali, A., Shamsuddin, S.M., Ralescu, A.L.: Classification with class imbalance problem. Int. J. Adv. Soft Comput. Appl. **5**(3), 176–204 (2013)
2. Amanatiadis, A., Andreadis, I.: A survey on evaluation methods for image interpolation. Meas. Sci. Technol. **20**(10), 104015 (2009)
3. Anupama, C., Sivaram, M., Lydia, E.L., Gupta, D., Shankar, K.: Synergic deep learning model–based automated detection and classification of brain intracranial hemorrhage images in wearable networks. Pers. Ubiquit. Comput. 1–10 (2020)
4. Bhadauria, H., Singh, A., Dewal, M.: An integrated method for hemorrhage segmentation from brain CT imaging. Comput. Electr. Eng. **39**(5), 1527–1536 (2013)
5. Caceres, J.A., Goldstein, J.N.: Emergency medicine clinics. Intracranial Hemorrhage **30**(3), 771–794 (2012)
6. Castro, J.S., Chabert, S., Saavedra, C., Salas, R.F.: Convolutional neural networks for detection intracranial hemorrhage in CT images. In: CRoNe, pp. 37–43 (2019)
7. Cavieres, E., Tejos, C., Salas, R., Sotelo, J.: Automatic segmentation of brain tumor in multi-contrast magnetic resonance using deep neural network. In: 18th International Symposium on Medical Information Processing and Analysis, vol. 12567, pp. 81–89. SPIE (2023)
8. Chabert, S., et al.: Applying machine learning and image feature extraction techniques to the problem of cerebral aneurysm rupture. Res. Ideas Outcomes **3**, e11731 (2017)
9. Chabert, S., et al.: Hemodynamic response function description in patients with glioma. J. Neuroradiol. **51**(4), 101156 (2024)

10. Chabert, S., et al.: Impact of b-Value sampling scheme on brain IVIM parameter estimation in healthy subjects. Magnetic Resonance in Medical Sciences, pp. mp-2019 (2019)
11. Chang, P.D., et al.: Hybrid 3D/2D convolutional neural network for hemorrhage evaluation on head CT. Am. J. Neuroradiol. **39**(9), 1609–1616 (2018)
12. Chilamkurthy, S., et al.: Development and validation of deep learning algorithms for detection of critical findings in head CT scans. arXiv preprint arXiv:1803.05854 (2018)
13. Flanders, A.E., et al.: Construction of a machine learning dataset through collaboration: the RSNA 2019 brain CT hemorrhage challenge. Radiol. Artif. Intell. **2**(3), e190211 (2020)
14. Ganganwar, V.: An overview of classification algorithms for imbalanced datasets. Int. J. Emerg. Technol. Adv. Eng. **2**(4), 42–47 (2012)
15. Grewal, M., Srivastava, M.M., Kumar, P., Varadarajan, S.: RADnet: radiologist level accuracy using deep learning for hemorrhage detection in CT scans. In: 2018 IEEE 15th International Symposium on Biomedical Imaging (ISBI 2018), pp. 281–284. IEEE (2018)
16. He, K., Zhang, X., Ren, S., Sun, J.: Spatial pyramid pooling in deep convolutional networks for visual recognition. IEEE Trans. Pattern Anal. Mach. Intell. **37**(9), 1904–1916 (2015)
17. Hounsfield, G.N.: Computed medical imaging. Nobel lecture, December 8, 1979. J. Comput. Assist. Tomogr. **4**(5), 665–674 (1980)
18. Hssayeni, M.D., Croock, M.S., Salman, A.D., Al-khafaji, H.F., Yahya, Z.A., Ghoraani, B.: Intracranial hemorrhage segmentation using a deep convolutional model. Data **5**(1), 14 (2020)
19. Ker, J., Singh, S.P., Bai, Y., Rao, J., Lim, T., Wang, L.: Image thresholding improves 3-dimensional convolutional neural network diagnosis of different acute brain hemorrhages on computed tomography scans. Sensors **19**(9), 2167 (2019)
20. LeCun, Y., Bengio, Y., Hinton, G.: Deep learning. Nature **521**(7553), 436–444 (2015)
21. Lee, H., et al.: An explainable deep-learning algorithm for the detection of acute intracranial haemorrhage from small datasets. Nat. Biomed. Eng. **3**(3), 173–182 (2019)
22. Lee, J.Y., Kim, J.S., Kim, T.Y., Kim, Y.S.: Detection and classification of intracranial haemorrhage on CT images using a novel deep-learning algorithm. Sci. Rep. **10**(1), 1–7 (2020)
23. Li, Y., Sun, G., Zhu, Y.: Data imbalance problem in text classification. In: 2010 Third International Symposium on Information Processing, pp. 301–305. IEEE (2010)
24. Naidech, A.M.: Intracranial hemorrhage. Am. J. Respir. Crit. Care Med. **184**(9), 998–1006 (2011)
25. Oksuz, I., et al.: Deep learning-based detection and correction of cardiac MR motion artefacts during reconstruction for high-quality segmentation. IEEE Trans. Med. Imaging **39**(12), 4001–4010 (2020)
26. Ronneberger, O., Fischer, P., Brox, T.: U-Net: convolutional networks for biomedical image segmentation. In: Navab, N., Hornegger, J., Wells, W.M., Frangi, A.F. (eds.) MICCAI 2015. LNCS, vol. 9351, pp. 234–241. Springer, Cham (2015). https://doi.org/10.1007/978-3-319-24574-4_28
27. Saavedra, C., Salas, R., Bougrain, L.: Wavelet-based semblance methods to enhance the single-trial detection of event-related potentials for a BCI spelling system. Comput. Intell. Neurosci. **2019**, 8432953 (2019)

28. Sage, A., Badura, P.: Intracranial hemorrhage detection in head CT using double-branch convolutional neural network, support vector machine, and random forest. Appl. Sci. **10**(21), 7577 (2020)
29. Sotelo, J., Salas, R., Tejos, C., Chabert, S., Uribe, S.: Análisis cuantitativo de variables hemodinámicas de la aorta obtenidas de 4D flow. Revista chilena de radiología **18**(2), 62–67 (2012)
30. Tan, M., Le, Q.: EfficientNet: rethinking model scaling for convolutional neural networks. In: International Conference on Machine Learning, pp. 6105–6114. PMLR (2019)
31. Titus, J., Geroge, S.: A comparison study on different interpolation methods based on satellite images. Int. J. Eng. Res. Technol. **2**(6), 82–85 (2013)
32. Veloz, A., Orellana, A., Vielma, J., Salas, R., Chabert, S.: Brain tumors: how can images and segmentation techniques help; diagnostic techniques and surgical management of brain tumors (2011). ISBN: 978-953-307-589-1
33. Veloz, A., Chabert, S., Salas, R., Orellana, A., Vielma, J.: Fuzzy spatial growing for glioblastoma multiforme segmentation on brain magnetic resonance imaging. In: Rueda, L., Mery, D., Kittler, J. (eds.) CIARP 2007. LNCS, vol. 4756, pp. 861–870. Springer, Heidelberg (2007). https://doi.org/10.1007/978-3-540-76725-1_89
34. Watanabe, Y., et al.: Improvement of the diagnostic accuracy for intracranial haemorrhage using deep learning-based computer-assisted detection. Neuroradiology **63**(5), 713–720 (2021)

Segmentation of Brain Tumor Parts from Multi-spectral MRI Records Using Deep Learning and U-Net Architecture

Szabolcs Csaholczi[1,2], Ágnes Györfi[2,3], Levente Kovács[3], and László Szilágyi[2,3(✉)]

[1] Doctoral School of Applied Informatics and Applied Mathematics,
Óbuda University, Bécsi út 96/b, Budapest 1034, Hungary
[2] Computational Intelligence Research Group (CIRG), Sapientia University,
Tîrgu Mureş, Romania
{csaholczi.szabolcs,gyorfiagnes,lalo}@ms.sapientia.ro
[3] Physiological Controls Laboratory (PhysCon), Óbuda University, Budapest,
Hungary
{kovacs,laszlo.szilagyi}@uni-obuda.hu

Abstract. Brain tumor segmentation from MRI data is a vital task in medical image analysis that involves identifying and delineating regions of interest within brain images that correspond to tumor presence. This paper proposes a solution to the segmentation problem of tumor parts from multi-spectral MRI records, which combines a 2D convolution U-net architecture adapted to work with spatial features with a spatial histogram enhancement method that aims to improve the visibility of brain structures and lesions in the observed volume. The proposed method was trained and tested using the BraTS 2019 high-grade glioma data set, and evaluated using statistical accuracy benchmarks. The segmentation outcome is globally characterized by average Dice scores of 0.7368, 0.8005, 0.7912, and 0.8612, that were obtained in case of edema, enhanced core, tumor core, and whole tumor regions, respectively.

Keywords: brain tumor segmentation · magnetic resonance imaging · U-net · deep learning

1 Introduction

The segmentation of brain tumors from magnetic resonance imaging (MRI) data is a crucial task in medical imaging analysis. It involves identifying and delineating the boundaries of abnormal growths within the brain with the goal of aiding in diagnosis, treatment planning, and monitoring of tumor progression. Various imaging techniques, such as MRI, are used to capture detailed structural information that helps in detecting these tumors. Advanced algorithms and deep learning models are often employed to automate this segmentation process, leading to more efficient and accurate results. The continuous research in this field aims to improve the precision of tumor segmentation, ultimately contributing to better patient outcomes and personalized healthcare [1].

An accurate automatic brain tumor segmentation method can greatly aid medical staff in diagnosis, treatment planning, and post-intervention follow-up studies. However, segmenting brain tumors from MRI data presents significant challenges, including: the tumor's variable shape, location, and appearance, distortion and displacement of surrounding normal tissues, imperfections in the automatic registration process for aligning multi-spectral MRI data, the absence of a standardized intensity scale in MRI data, and the possible presence of intensity inhomogeneity.

Automatic brain tumor segmentation methods have been developed for several decades. Initially, the developed methods were strictly limited to 2D segmentation, and were using private data sets [2]. The algorithms evolved together with the computers, they have become more and more complex, they started dedicating extra efforts to handle noise phenomena (e.g. intensity non-uniformity [3,4]), and the volume of processed data gradually grew as well. Probably the most significant turning point in the field was caused by the Brain Tumor Segmentation (BraTS) Challenges [5,6], which provided standard data sets and standard conditions for objective comparison of the developed methods. Data availability caused a sudden intensification of the research in the field, and practically the whole arsenal of machine learning got involved in the process. Classical machine learning based methods needed sophisticated handcrafted features to provide acceptable quality in segmentation. The most relevant methods from this section include AdaBoost [7], random forest [8,9], ensemble of binary decision trees [10,11], and extremely randomized trees (ERT) [12].

Deep learning [13] opened the horizon for convolutional neural networks, which gradually conquered the field, and currently are dominating most medical imaging problems. Some remarkable deep learning based solutions involved sequential CNN networks [14], fully convolutional neural network [15], ensemble of various CNN structures [16], residual CNN networks [17], and U-net with 2D or 3D convolution [18]. Recent solutions (e.g., [19–22]) have incorporated specialized attention mechanisms and modules into CNN networks to obtain improved segmentation results. CNN-based methods learn significantly more parameters than traditional machine learning techniques, leading to improved segmentation performance. However, this comes at the cost of increased execution time and storage requirement, as well as the need for larger datasets.

This paper proposes a U-net based solution for the brain tumor segmentation problem, having the following items of originality: (1) it aims at distinguishing all lesion tissue types for which annotated data is available, including edema, enhancing tumor core, whole tumor core and whole tumor; (2) it combines the

U-net with a histogram enhancement method to provide better visibility of the tumor structures; (3) it employs 2D convolution to limit the computational load but uses neighbor slices for the segmentation of each slice, so that the algorithm can take advantage of spatial features.

The rest of this paper is structured as follows: Sect. 2 is dedicated to present the details of the proposed methodology. Section 3 gives a detailed report on the evaluation of the segmentation results provided by the proposed method. Section 4 discusses the results in the context of the state-of-the-art. Section 5 concludes this study.

2 Materials and Methods

2.1 Data

This study is based on the training data set of the MICCAI BraTS 2019 challenge, more exactly on its $N = 259$ high-grade glioma (HGG) records collected from various medical institutions and brought to a standard, easily accessible format. All records contain volumes of $155 \times 240 \times 240$ pixels, more exactly 155 transversal slices of 240×240 pixels each. Pixels have a standard size of 1 mm in each direction. All pixels have four observed feature values (T1, T2, T1c, and FLAIR), collected separately by the MRI device, and registered together thereafter using an automatic method described in [6]. All records were annotated by a team of human experts. The annotation of a pixel indicates to which tissue type the pixel belongs, and can be either normal, edema, necrotic tumor core or enhancing tumor core. The annotation can be used as expected value at training a machine learning method, and as ground truth at the evaluation of prediction outcome.

The skull was removed from the volumes, so the records contain approximately 1.5 million valid brain pixels, while all others belong to outer space. Each volume contains at least one contiguous tumor, consisting of annotated positive pixels, whose count ranges up to 250 thousand. These annotated positive pixels are further divided into edema, necrotic tumor core and enhancing tumor core. A major obstacle of reliable segmentation consists in the strong imbalance of different tissue classes.

MRI data is frequently contaminated by low frequency noise also known as intensity inhomogeneity or non-uniformity (INU) [3], which is caused by the turbulence of the magnetic fields in the MRI devices. However, in case of the BraTS 2019 data set, INU is neglectable. On the other hand, there is another phenomenon, which needs to be taken into account. The intensity scale of MRI records is not a standard one, so the histograms of various records coming from different MRI devices need to be aligned before feeding them to any machine learning algorithm. Based on previous experience [23] we decided to employ the so-called Contrast Limited Adaptive Histogram Equalization (CLAHE) algorithm [24], which adjusts pixel values according to the local intensity distribution, aiming to increase the contrast, or in other words, the local visibility of structures.

There are 259 HGG records in the BraTS 2019 dataset. Since our proposed methodology applies 2D convolution, individual slices are fed to the U-net architecture. Although the original slices are sized 240×240 pixels, the input size of the networks is 160×192 pixels, because all brains fit within a frame of this size.

2.2 Segmentation

The segmentation of the MRI volumes in our proposed method is performed slice by slice, using a U-net architecture that employs 2D convolution. U-Net was originally proposed by Ronneberger et al. [25] and it was initially deployed in a segmentation problem that worked on microscopy image. During the nine years of existence, U-Net has been used in any and every semantic segmentation tasks, due to its advantage of efficiently learning from reduced number of training samples [26]. Its name originates from the shape of the network architecture, consisting of a series of encoder and a series of decoder blocks, connected by a bridge in the middle. Our architecture contains four encoder and four decoder blocks, which was determined according to the size of the input images. The encoder branch acts as a feature extractor, because it learns an abstract representation for an image through multiple encoder blocks.

We adapted the U-Net structure to provide a solution to the volumetric segmentation problem. The network structure is exhibited in Fig. 1. The input data size is $12 \times 160 \times 192$, because when slice with index z is fed to the network, it receives slices $z - 1$ and $z + 1$ as well, all three slices with their four data channels. The output size of $5 \times 160 \times 192$ refer to the predicted probabilities of pixels from slice z to belong to five different classes.

As it was stated above, the network consists of four encoder blocks and four decoder block. Each encoder block contains two 2D convolution layers of kernel size 3×3 and padding set to same, which is followed by ReLU activation function and batch normalization. The last layer of each encoder block is a MaxPooling layer with 2×2 kernel, which halves the image size in two dimensions. This reduction also affects the number of trainable parameters and the computational cost as well. The number of filters (features) are doubled as we step from previous encoder to the next one. After the fourth encoder block there is a bridge which at its turn also contains two 2D convolution layers with kernel size 3×3, followed by ReLU activation and batch normalization.

The decoder branch receives through the bridge the abstract image representation generated by the encoder, and from the bridge and generates semantic segmentation masks. Each decoder block doubles the image size in both directions and halves the number of filters. Each decoder block starts with a transpose convolution layer of kernel size 2×2, followed by the concatenation of the transpose layer's output with the skip connection situated at the corresponding level. The skip connections have a double role: they help the decoder blocks to generate better semantic features, and also assures better gradient flow in backpropagation, consequently the network learns better representation. Two more 2D convolution layers come with kernel of size 3×3 and ReLU activation, each followed by batch normalization. In case of the last decoder block, the last

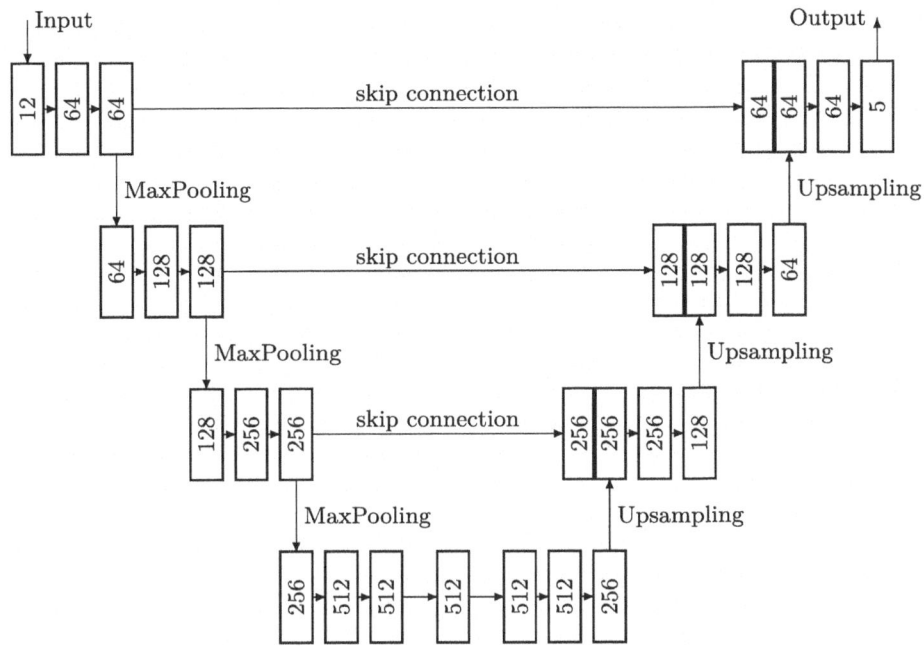

Fig. 1. The structure of the U-net network used for segmentation. Each rectangle stands for a Conv2D layer and the number in it represents the filter count. Each Conv2D layer is accompanied by batch normalization and dropout. Each encoding block ends with a MaxPooling operation that halves the image size in both directions. Each decoding block starts with an upsampling operation that serves in restoring the initial image size at the end of the processing chain. The four levels of the architecture works with images of sizes 160 × 192, 80 × 96, 40 × 48, and 20 × 24, respectively.

convolution layer has only 5 filters and it applies SoftMax activation to provide predicted probabilities. In all encoder and decoder blocks, dropout was used after the first batch normalization.

The network was trained using the sparse categorical cross-entropy loss function. The optimizer algorithm involved in training was the ADAM algorithm [27]. One hundred training epochs were sufficient to obtain high-quality results.

2.3 Evaluation

Statistical quality indicators are employed to evaluate the segmentation outcome. First of all, we need to clarify the pixel classes distinguished by the network. Pixel annotations given by the human experts, used as expected values at training and as ground truth as prediction, do not directly correspond to the classes we need to distinguish. The MICCAI training dataset gives us labels like edema, necrotic tumor core, enhancing core, and negatives. The latter also refers to non-brain pixels within the cubic volume. The MICCAI challenge demands the recognition of lesions like edema (ED), enhancing core (EC), tumor core (TC), and whole tumor (WT). TC includes pixels of necrotic tumor core and enhancing core. WT includes all lesion pixels. Non-brain pixels are considered those, which have zero intensity in all four data channels. So the proposed network is trained to distinguish five classes corresponding to the four different labels used by the annotation and the class of non-brain pixels. However, the predictions are interpreted according to the MICCAI problem formulation, concentrating on the correct recognition of ED, EC, TC, and WT classes.

Let us denote the set of positive label types by $\Psi = \{\text{ED}, \text{EC}, \text{TC}, \text{WT}\}$. For any MRI record with index $i = 1 \ldots N$, and for any tissue type $j \in \Psi$, let us define the following sets:

- Let $\Gamma_{i,j}$ be the set of pixels of volume i, whose annotation (ground truth) is compatible with tissue type j;
- Let $\Lambda_{i,j}$ be the set of pixels of volume i, that are assigned to a class compatible with tissue type j.

Now we can define the statistical accuracy indicators as follows:

- The recall or true positive rate corresponding to a given tissue type, extracted from record with index i, is defined as: $\text{TPR}_{i,j} = |\Lambda_{i,j} \cap \Gamma_{i,j}|/|\Gamma_{i,j}|$, where $|X|$ represents the cardinality of set X.
- The precision or positive predictive value corresponding to a given tissue type, extracted from record with index i, is defined as: $\text{PPV}_{i,j} = |\Lambda_{i,j} \cap \Gamma_{i,j}|/|\Lambda_{i,j}|$.
- The F1-score or Dice score corresponding to a given tissue type, extracted from record with index i, is defined as: $\text{DSC}_{i,j} = 2 \times |\Lambda_{i,j} \cap \Gamma_{i,j}|/(|\Lambda_{i,j}| + |\Gamma_{i,j}|) = 2 \times (\text{TPR}_{i,j} \times \text{PPV}_{i,j})/(\text{TPR}_{i,j} + \text{PPV}_{i,j})$.

The above statistical indicators describe the segmentation accuracy within individual MRI records. If we wish to establish a global accuracy descriptor for tissue type $j \in \Psi$, we may compute the average $\overline{\text{TPR}}_j$, $\overline{\text{PPV}}_j$, $\overline{\text{DSC}}_j$, with the formula

$$\overline{\text{DSC}}_j = \frac{1}{N} \sum_{i=1}^{N} \text{DSC}_{i,j} , \qquad (1)$$

or similarly in case of the other metrics. Further on, we may consider the whole collection of records a single set of pixels and we may extract overall $\widetilde{\text{TPR}}_j$, $\widetilde{\text{PPV}}_j$, and $\widetilde{\text{DSC}}_j$ values with formulas like

$$\widetilde{\text{DSC}}_j = 2 \times \frac{\left|\left(\bigcup_{i=1}^{N} \Lambda_{i,j}\right) \cap \left(\bigcup_{i=1}^{N} \Gamma_{i,j}\right)\right|}{\left|\bigcup_{i=1}^{N} \Lambda_{i,j}\right| + \left|\bigcup_{i=1}^{N} \Gamma_{i,j}\right|}. \qquad (2)$$

Accuracy (ACC) or rate of correct decisions is the number of pixels classified correctly divided by the total number of pixels. We may refer to the segmentation accuracy of a volume (ACC_i, $i = 1 \ldots N$), or the average or overall accuracy ($\overline{\text{ACC}}$ or $\widetilde{\text{ACC}}$), extracted with formulas similar to Eqs. (1) or (2), respectively.

Table 1. Average and overall values of various accuracy indicators

Tissue type	Value	TPR	PPV	DSC
Whole tumor	average	0.9078	0.8356	0.8612
	overall	0.9265	0.8538	0.8887
Enhancing core	average	0.8256	0.8082	0.8005
	overall	0.8386	0.8204	0.8294
Tumor core	average	0.7855	0.8371	0.7912
	overall	0.7685	0.8604	0.8119
Edema	average	0.7313	0.7805	0.7368
	overall	0.7703	0.8202	0.7945

Table 2. Expected Dice scores according to linear trends identified from Fig. 2

Tissue type	DSC at size $10\,\text{cm}^3$	At average size		DSC at size $100\,\text{cm}^3$
		size	DSC	
Whole tumor	0.7982	$95.1\,\text{cm}^3$	0.8612	0.8730
Enhancing core	0.7748	$22.6\,\text{cm}^3$	0.8005	0.9585
Tumor core	0.7786	$36.4\,\text{cm}^3$	0.7912	0.8214
Edema	0.6329	$58.7\,\text{cm}^3$	0.7368	0.8248

All the above presented indicators range between 0 and 1, higher values referring to better segmentation outcome.

3 Results

The proposed U-net architecture was deployed into a thorough evaluation process involving all 259 HGG glioma records from the BraTS 2019 training data set, using a seven-fold cross-validation scheme. The records were randomly divided into seven equal groups. Each group was used as evaluation data while the architecture was trained using the other six groups. Thus we obtained a segmentation result for all 259 records, which was predicted by U-nets that did not see the test volumes during training. The segmentation outcome was evaluated using the statistical benchmarks presented in Sect. 2.3.

Table 1 exhibits the global accuracy benchmark values, both the average and overall scores of the indicators. Overall values are usually higher than the mean by several percent, because of the imbalanced distribution of values on the two sides of the mean value. The mean accuracy value regardless to tissue type is ACC = 0.9786, meaning that one out of approximately 47 brain pixels get a mistaken label during segmentation.

Dice scores obtained for individual volumes, with respect to various tissue types are exhibited in several figures that follow. Figure 2 exhibits the Dice scores

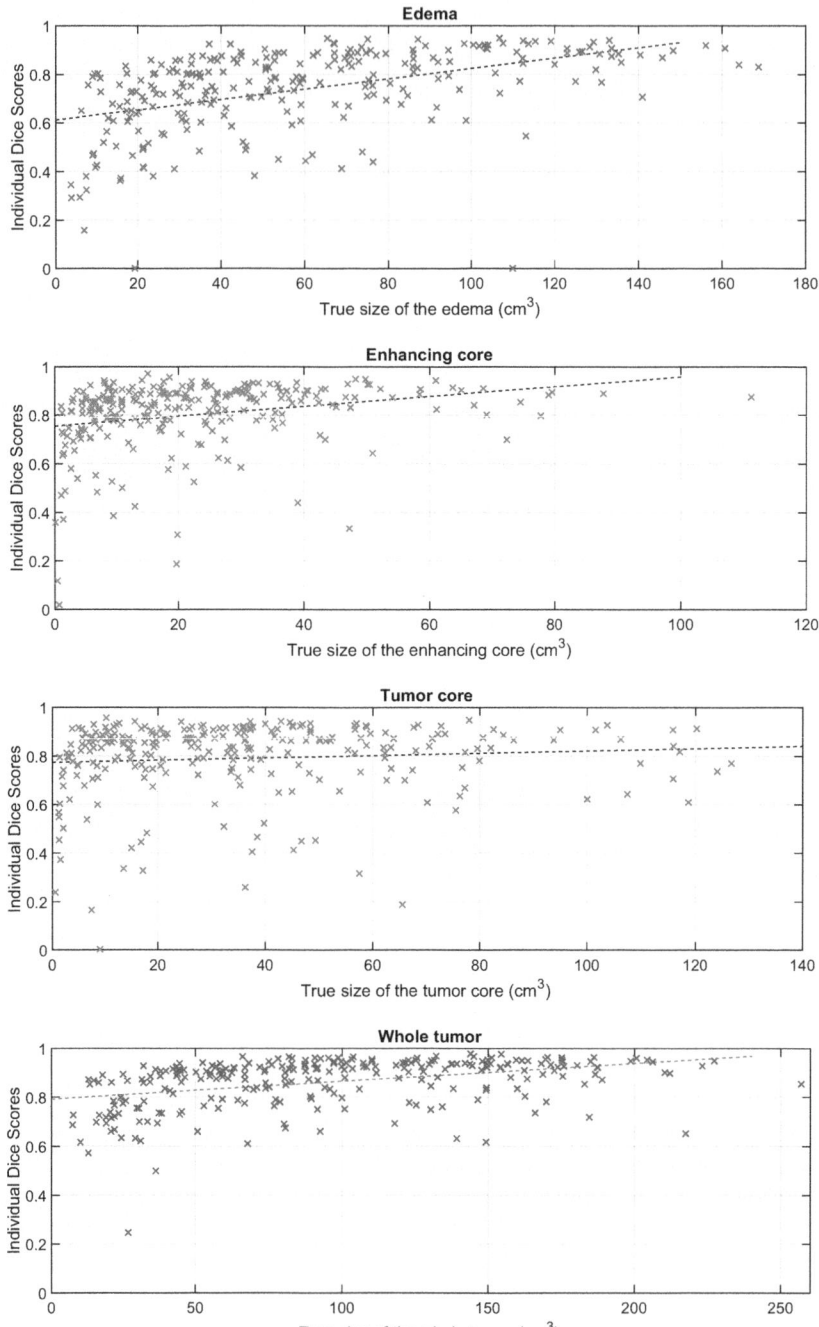

Fig. 2. Dice scores obtained for various tissue types from individual volumes, plotted against the true size of the respective tumor part.

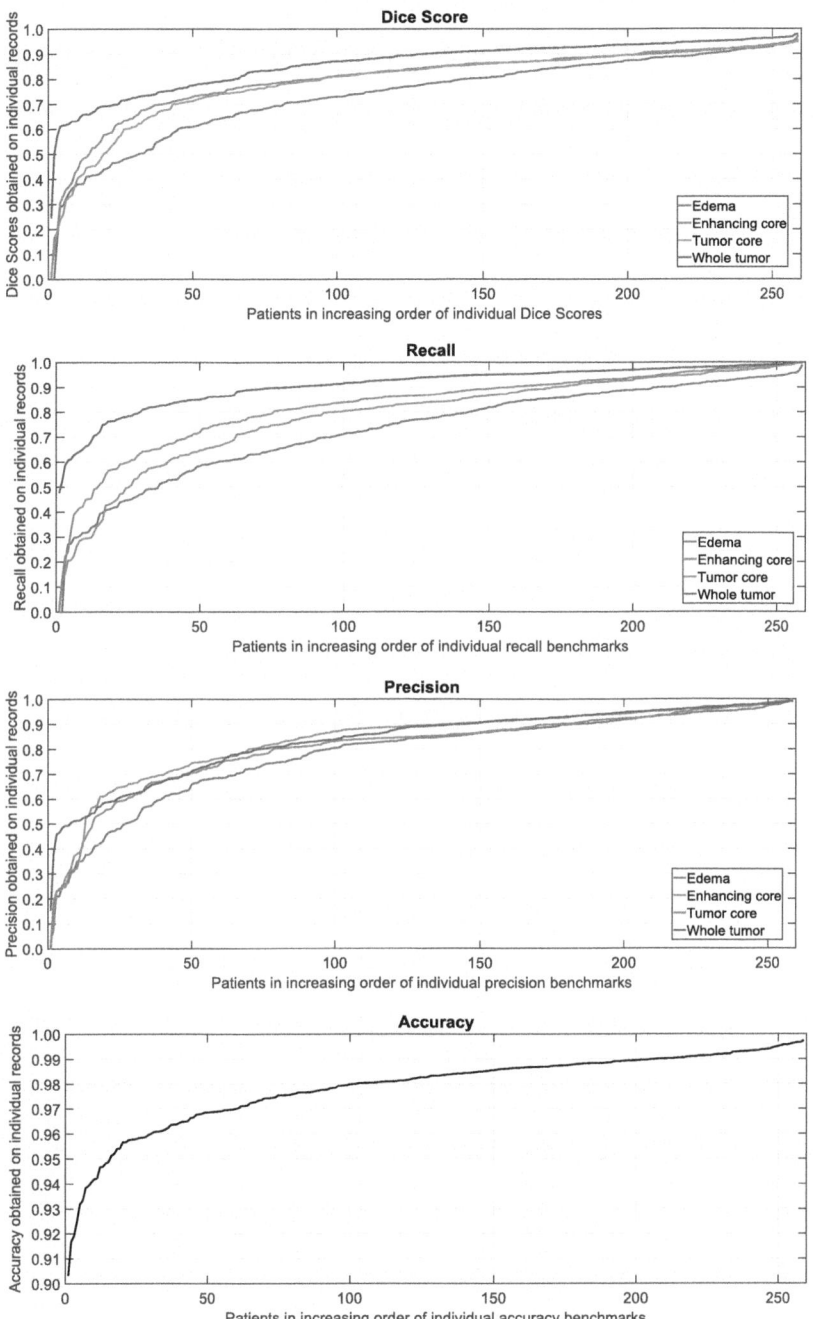

Fig. 3. Individual Dice scores, recall and precision benchmarks obtained for individual records and their tumor parts, and the segmentation accuracy of each volume, all plotted in increasing order of the given benchmark values.

plotted against the true size of the tumor parts whose segmentation accuracy they describe. The four graphs indicate values obtained for various tissue types, namely edema, tumor core, enhancing core and whole tumor. The true size in each case was established according to the ground truth. The linear trend of the obtained Dice scores was identified in each case using linear regression, and is represented with a dashed black line in each graph. Based on the linear trends, we established the expected value of Dices scores for some typical values of tumor sizes, and listed them in Table 2. The accuracy of whole tumor estimation can be substantially improved via post-processing, this has been shown several times in previous papers, e.g. [28]. In this study we did not apply post-processing.

Figure 3 shows the benchmark values obtained for each individual record from the 259 HGG volumes, each plotted in increasing order. The first three panels show Dice scores, recall and precision values for tissue types (ED, EC, TC, and WT) separately using different colors. The last panel contains a single curve that indicates the rate of correct decisions regarding all brain pixels of individual HGG records.

Figure 4 shows an example of segmentation outcome: 40 consecutive slices of a HGG volume are presented, indicating the correlation between ground truth and predicted labels for individual pixels, using color coding. Green shades from dark to light correspond to correctly identified necrotic core, edema, and enhancing core, respectively. Pink represents tumor pixels identified as tumor pixels belonging to the wrong tissue type. Red, blue, and gray indicate false negatives, false positives, and true negatives, respectively.

4 Discussion

In this paper we proposed a U-net based method for multi-class segmentation of brain tumors from multi-spectral MRI data. The proposed method performed a histogram standardization using a spatial CLAHE algorithm and then employed a U-net with 2D convolution to perform the segmentation including neighbor slices as well into the input data. The main hyperparameter of the proposed method is the number of filters in the first encoder block, which in Fig. 1 is indicated as 64. This value was chosen empirically, after a thorough evaluation process where multiples of 16 have been trialed up to 128. If the number of filters is too low, the architecture has less learnable parameters and thus it cannot adjust itself to very complex decision making. If the number of filters is too high, that increases the memory requirement and computational load, without improving the segmentation accuracy above a certain level. In our case, 64 filters in the first encoder block was found the best. This number directly affects the filter numbers in all other convolutional layers. Results exhibited in Sect. 3 were obtained with 64 filters. The great majority of the tests run with different filter numbers provided average Dice score of no more than one percent lower.

All the above presented results were obtained as the output of the U-net architecture, without applying any post processing. It is known from the literature that even a simple morphological operator can increase the average Dice

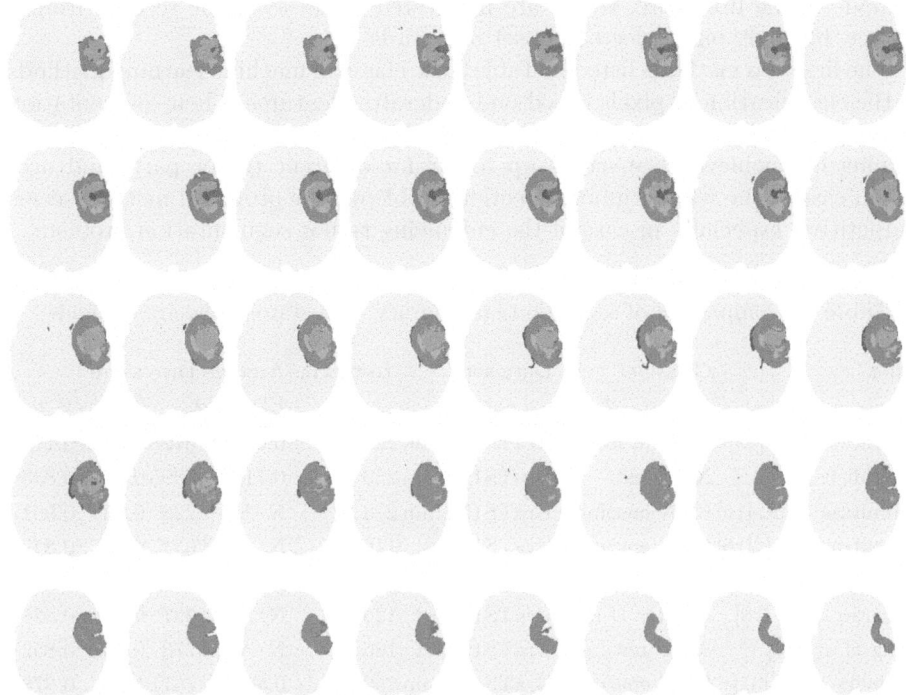

Fig. 4. Fourty consecutive slices from a HGG volume, with its segmentation outcome. Green shades from dark to light correspond to correctly identified necrotic core, edema, and enhancing core, respectively. Pink represents tumor pixels that received a different positive label, while red indicates false negatives – pixels that belong to any of the tumor parts and were labeled negative. False positives with any label appear in blue, while gray stands for true negatives. (Color figure online)

score of the whole tumor detection problem by up to 2% [23], and there exist more complex solutions for the problem involving conditional random fields [16] for example. However, these usually cannot improve the rate of correct decisions regarding tumor parts. Developing such a post-processing could be the topic of another paper.

Table 3 presents a collection of state-of-the-art methods published in recent years, as solutions given to the segmentation problem of the brain tumor parts. For each solution, the classifier method and test data set is shown, together with the obtained Dice scores as published originally by their authors. The average Dice scores are given for each study and for each tissue, wherever the value is available. Dice scores published with two decimals are not suitable for objective comparison as differences are low. Further on, Dice scores are not directly comparable from study to study, because they were not obtained under the same circumstances, using the same testing conditions. Further on, there are some methods in the literature, which are not tested on all available records from a dataset, but only on a selected subset of records.

The first two methods listed in Table 3 use classical machine learning methods for the classification of pixels based on handcrafted features. Their segmentation accuracy is easily overpassed by CNN based methods. Studies deploying deep learning can achieve Dice scores up to 0.8 for different tumor parts and over 0.85 in case of the whole tumor detection problem. Our proposed method seems competitive, especially in case of the enhancing tumor segmentation problem.

Table 3. Comparison of segmentation accuracy with state-of-the-art methods

Paper	Classifier	Data set	Test items	Average Dice score			
				ED	EC	TC	WT
Csaholczi et al. [29]	random forest	BraTS15 train	220	0.6566	0.6728	0.6554	0.7722
Unpublished	XGBoost	BraTS19 train	259	0.7112	0.7895	0.7795	0.8339
Kamnitsas et al. [16]	CNN ensemble	BraTS15 train	274	N/A	0.728	0.754	0.901
Kamnitsas et al. [16]	CNN ensemble	BraTS15 test	110	N/A	0.634	0.667	0.849
Ding et al. [17]	ResNet	BraTS15 test	93	N/A	0.63	0.71	0.86
Bhalerao et al. [18]	3D Res U-net	BraTS19 test	125	N/A	0.697	0.772	0.828
Wang et al. [30]	3D U-net	BraTS19 test	125	N/A	0.778	0.798	0.852
Lefkovits et al. [31]	CNN ensemble	BraTS19 train	259	0.8005	0.7671	N/A	0.878
Proposed	U-net	BraTS19 train	259	0.7368	0.8005	0.7912	0.8612

5 Conclusions

This paper attempted to establish an accurate and reliable solution for the segmentation problem of brain tumor parts. The proposed method combined a spatial CLAHE based preprocessing with a U-net that works with 2D convolution with multiple slices. The proposed method was tested on the whole HGG glioma data set of the BraTS 2019 challenge, and evaluated using statistical accuracy benchmarks. Average Dice scores of 0.7368, 0.8005, 0.7912, and 0.8612 were obtained in the segmentation of edema, enhanced core, tumor core, and whole tumor, respectively. These values can be still improved using an appropriate post-processing.

References

1. Mohan, G., Subashini, M.M.: MRI based medical image analysis: survey on brain tumor grade classification. Biomed. Signal Process. Control **39**, 139–161 (2018)
2. Gordillo, N., Montseny, E., Sobrevilla, P.: State of the art survey on MRI brain tumor segmentation. Magn. Reson. Imaging **31**, 1426–1438 (2013)

3. Vovk, U., Pernuš, F., Likar, B.: A review of methods for correction of intensity inhomogeneity in MRI. IEEE Trans. Med. Imaging **26**, 405–421 (2007)
4. Szilágyi, L., Szilágyi, S.M., Benyó, B.: Efficient inhomogeneity compensation using fuzzy c-means clustering models. Comput. Meth. Prog. Biomed. **108**, 80–89 (2012)
5. Bakas, S., et al.: Identifying the best machine learning algorithms for brain tumor segmentation, progression assessment, and overall survival prediction in the BRATS challenge. arXiv: 1181.02629v3, 23 April 2019
6. Menze, B.H., et al.: The multimodal brain tumor image segmentation benchmark (BRATS). IEEE Trans. Med. Imaging **34**, 1993–2024 (2015)
7. Islam, A., Reza, S.M.S., Iftekharuddin, K.M.: Multifractal texture estimation for detection and segmentation of brain tumors. IEEE Trans. Biomed. Eng. **60**, 3204–3215 (2013)
8. Tustison, N.J., et al.: Optimal symmetric multimodal templates and concatenated random forests for supervised brain tumor segmentation (simplified) with ANTsR. Neuroinformatics **13**, 209–225 (2015)
9. Soltaninejad, M., et al.: Supervised learning based multimodal MRI brain tumour segmentation using texture features from supervoxels. Comput. Meth. Prog. Biomed. **157**, 69–84 (2018)
10. Kapás, Z., Lefkovits, L., Szilágyi, L.: Automatic detection and segmentation of brain tumor using random forest approach. In: Torra, V., Narukawa, Y., Navarro-Arribas, G., Yañez, C. (eds.) MDAI 2016. LNCS (LNAI), vol. 9880, pp. 301–312. Springer, Cham (2016). https://doi.org/10.1007/978-3-319-45656-0_25
11. Szilágyi, L., Iclănzan, D., Kapás, Z., Szabó, Z., Győrfi, Á., Lefkovits, L.: Low and high grade glioma segmentation in multispectral brain MRI data. Acta Univ. Sapientia, Informatica **10**(1), 110–132 (2018)
12. Pinto, A., Pereira, S., Rasteiro, D., Silva, C.A.: Hierarchical brain tumour segmentation using extremely randomized trees. Pattern Recogn. **82**, 105–117 (2018)
13. LeCun, Y., Bengio, Y., Hinton, G.: Deep learning. Nature **521**, 436–444 (2015)
14. Pereira, S., Pinto, A., Alves, V., Silva, C.A.: Brain tumor segmentation using convolutional neural networks in MRI images. IEEE Trans. Med. Imaging **35**, 1240–1251 (2016)
15. Zhao, X.M., Wu, Y.H., Song, G.D., Li, Z.Y., Zhang, Y.Z., Fan, Y.: A deep learning model integrating FCNNs and CRFs for brain tumor segmentation. Med. Image Anal. **43**, 98–111 (2018)
16. Kamnitsas, K., et al.: Efficient multi-scale 3D CNN with fully connected CRF for accurate brain lesion segmentation. Med. Image Anal. **36**, 61–78 (2017)
17. Ding, Y., Li, C., Yang, Q.Q., Qin, Z., Qin, Z.G.: How to improve the deep residual network to segment multi-modal brain tumor images. IEEE Access **7**, 152821–152831 (2019)
18. Bhalerao, M., Thakur, S.: Brain tumor segmentation based on 3D residual U-Net. In: Crimi, A., Bakas, S. (eds.) BrainLes 2019. LNCS, vol. 11993, pp. 218–225. Springer, Cham (2020). https://doi.org/10.1007/978-3-030-46643-5_21
19. Zhu, Z.Q., He, X.Y., Qi, G.Q., Li, Y.Y., Cong, B.S., Liu, Y.: Brain tumor segmentation based on the fusion of deep semantics and edge information in multimodal MRI. Inf. Fusion **91**, 376–387 (2023)
20. Cao, Y., Zhou, W.F., Zang, M., An, D.L., Feng, Y., Yu, B.: MBANet: a 3D convolutional neural network with multi-branch attention for brain tumor segmentation from MRI images. Biomed. Signal Process. Control **80**, 104296 (2023)
21. Chang, Y.K., Zheng, Z.Z., Sun, Y.W., Zhao, M.M., Lu, Y., Zhang, Y.: DPAFNet: a residual dual-path attention-fusion convolutional neural network for multimodal brain tumor segmentation. Biomed. Signal Process. Control **79**, 104037 (2023)

22. Liu, H.X., Huo, G.Q., Li, Q., Guan, X., Tseng, M.L.: Multiscale lightweight 3D segmentation algorithm with attention mechanism: brain tumor image segmentation. Expert Syst. Appl. **214**, 119166 (2023)
23. Győrfi, Á., Szilágyi, L., Kovács, L.: A fully automatic procedure for brain tumor segmentation from multi-spectral MRI records using ensemble learning and atlas-based data enhancement. Appl. Sci. **11**, 564 (2021)
24. Stimper, V., Bauer, S., Emstorfer, R., Schölkopf, B., Xian, R.P.: Multidimensional contrast limited adaptive histogram equalization. IEEE Access **7**, 165437–165447 (2019)
25. Ronneberger, O., Fischer, P., Brox, T.: U-Net: convolutional networks for biomedical image segmentation. In: Navab, N., Hornegger, J., Wells, W.M., Frangi, A.F. (eds.) MICCAI 2015. LNCS, vol. 9351, pp. 234–241. Springer, Cham (2015). https://doi.org/10.1007/978-3-319-24574-4_28
26. Siddique, N., Paheding, S., Elkin, C.P., Devabhaktuni, V.: U-Net and its variants for medical image segmentation: a review of theory and applications. IEEE Access **9**, 82031–82057 (2021)
27. Kingma, D.P., Ba, J.: Adam: a method for stochastic optimization. arXiv: 1412.6980, 22 December 2014
28. Csaholczi, S., Iclănzan, D., Kovács, L., Szilágyi, L.: Brain tumor segmentation from multi-spectral MR image data using random forest classifier. In: Yang, H., Pasupa, K., Leung, A.C.-S., Kwok, J.T., Chan, J.H., King, I. (eds.) ICONIP 2020. LNCS, vol. 12532, pp. 174–184. Springer, Cham (2020). https://doi.org/10.1007/978-3-030-63830-6_15
29. Csaholczi, S., Kovács, L., Szilágyi, L.: Automatic segmentation of brain tumor parts from MRI data using a random forest classifier. In: Proceedings of 19th IEEE World Symposium on Applied Machine Intelligence and Informatics (SAMI 2021), pp. 471–476 (2021)
30. Wang, F., Jiang, R., Zheng, L., Meng, C., Biswal, B.: 3D U-net based brain tumor segmentation and survival days prediction. In: Crimi, A., Bakas, S. (eds.) BrainLes 2019. LNCS, vol. 11992, pp. 131–141. Springer, Cham (2020). https://doi.org/10.1007/978-3-030-46640-4_13
31. Lefkovits, S., Lefkovits, L., Szilágyi, L.: HGG and LGG brain tumor segmentation in multi-modal MRI using pretrained convolutional neural networks of Amazon Sagemaker. Appl. Sci. **12**, 3620 (2022)

Exploiting the *Segment Anything Model* (SAM) for Lung Segmentation in Chest X-ray Images

Gabriel Bellon de Carvalho and Jurandy Almeida(✉)

Department of Computing, Federal University of São Carlos UFSCar, 18052-780 Sorocaba, SP, Brazil
gabrielbellon@estudante.ufscar.br, jurandy.almeida@ufscar.br

Abstract. *Segment Anything Model* (SAM), a new AI model from Meta AI released in April 2023, is an ambitious tool designed to identify and separate individual objects within a given image through semantic interpretation. The advanced capabilities of SAM are the result of its training with millions of images and masks, and a few days after its release, several researchers began testing the model on medical images to evaluate its performance in this domain. With this perspective in focus—i.e., optimizing work in the healthcare field – this work proposes the use of this new technology to evaluate and study chest X-ray images. The approach adopted for this work, with the aim of improving the model's performance for lung segmentation, involved a transfer learning process, specifically the fine-tuning technique. After applying this adjustment, a substantial improvement was observed in the evaluation metrics used to assess SAM's performance compared to the masks provided by the datasets. The results obtained by the model after the adjustments were satisfactory and similar to cutting-edge neural networks, such as U-Net.

Keywords: Deep Learning · Segment-Anything Model (SAM) · Medical Image Analysis · Lung Segmentation · Chest X-Ray Images

1 Introduction

Segment Anything Model (SAM) [6] is a tool that, since its release in April 2023, has proven to be very promising in the task of image segmentation. Its approach involves using a variety of input prompts to identify different objects in images, such as points and bounding boxes. To predict the masks, SAM uses three components: (*i*) an image encoder, (*ii*) a prompt encoder, and (*iii*) a mask decoder. Additionally, the model can automatically segment anything in an image and generate multiple valid masks for ambiguous inputs, which is innovative in the field.

Given this and the immense amount of data used in its training—11 million images and over 1 billion masks [6]—many researchers have recognized the potential of this technology in the medical field and have begun to investigate its effectiveness in this area. However, despite having this large volume of data in its training, there are no medical images among the domains in which SAM was trained, which makes its generalization ability moderate when it comes to this area [4,8].

This study aims to advance the application of SAM in the field of medical image analysis, especially, for lung segmentation in chest X-ray images. Understanding the effectiveness of SAM in this domain is of paramount importance in the development of new technologies for the diagnosis, treatment and follow-up of lung diseases. We finetune SAM on two collections of chest X-ray images, known as the Montgomery and Shenzhen datasets [5]. This well-established practice aims to leverage the benefits of representations previously learned on a larger database to optimize the training of a network on a smaller dataset. Our exploration also involved testing SAM across such datasets using various input prompts, like bounding boxes and individual points. The obtained results show that our finetuned SAM can perform similar to state-of-the-art approaches for lung segmentation, like U-Net [9].

2 Related Work

Several studies have conducted a comprehensive evaluation of SAM on a variety of medical image segmentation tasks [4,8], demonstrating that the model achieved satisfactory segmentation results, especially on targets with well-defined boundaries. However, it is evident that SAM has certain limitations due to the lack of contour in the regions of the images in question, making it difficult to identify certain patterns such as the shapes of organs and tissues [8].

Among such studies, it is worth mentioning the work of Ma and Wang [7]. They introduce *MedSAM*, which was developed on an unprecedented set of over 1 million medical image-mask pairs. Also, they evaluated the fine-tuning of the model, the same technique adopted in this work for chest X-ray images, and the obtained results demonstrate the great potential of SAM in medicine. In most of the 86 tasks evaluated, *MedSAM* ranked first, surpassing the performance of specialized models, like U-Net [9] and DeepLabV3+ [2].

In spited of all the advances, SAM achieved a maximum F1-Score of 60% for lung segmentation of chest X-ray images using points as input, indicating poor performance, as highlighted in the work of He et al. [3]. This study also revealed that the model's performance with bounding box prompts was even worse.

3 Segment Anything Model (SAM)

Referred to as SAM, the *Segment Anything Model* was trained on a dataset containing over a billion masks [6], presenting itself as an innovative proposal in the field of segmentation. The primary purpose of this model is to generate a valid mask for any input image, relying on two essential functionalities.

In the first of these, *panoptic segmentation*, the goal is to automatically generate various masks for the objects identified in an image. The result of its execution is a list of masks, where each mask is a dictionary with descriptor fields. It is important to note that there are several adjustable parameters in the automatic mask generation, allowing control over interesting factors such as the density of sampling points and limits for removing duplicate masks. An example of using this functionality can be seen in Fig. 1.

(a) original image   (b) panoptic segmentzation

Fig. 1. Example of a given image and its corresponding masks after a panoptic segmentation. Adapted from Google Images.

In addition to this, another very interesting functionality for the purpose of this work is *instance segmentation*, which uses points or bounding boxes to select an area of interest in the image and extract related masks from it, as demonstrated in Fig. 2. After selecting the bounding boxes, simply pass them as a parameter for the predict function to make the prediction.

(a) original image   (b) instance segmentation

Fig. 2. Example of a given image with areas of interest selected by bounding boxes and its corresponding masks after an instance segmentation. Adapted from Google Images.

3.1 Model Development

In the development of the model, three stages were used to create the masks that make up the database used during training. First, labels were manually and interactively placed on each image mask in a browser, and after that, the network was trained 6 times.

In the second stage, called semi-automatic, the masks were delivered to the involved professionals, who had the function of checking if any object was not considered in the mask production process by the network. Again, the masks were fed into the network, which was trained 5 more times.

In the final stage, the production of masks was carried out automatically without any human intervention. At this stage, 32×32 points were evenly distributed throughout the image to produce masks covering the entire data extension. After completing this process, the masks underwent filtering and post-processing. In the filtering, one step was to remove masks covering 95% or more of the image, as it does not make sense to keep a mask that identifies the image as a whole. On the other hand, in post-processing, masks with an area smaller than 100 pixels were removed [6].

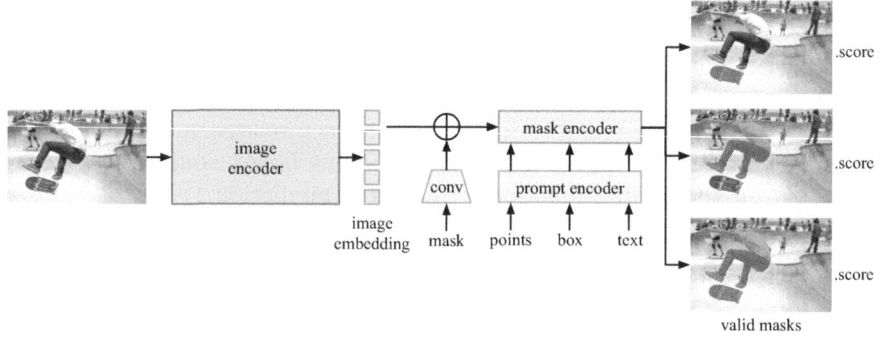

Fig. 3. Organization of the Segment-Anything Model. Adapted from [6].

3.2 Components of SAM

SAM uses three components to perform mask prediction [6], as depicted in Fig. 3.

Image Encoder. This component is responsible for processing the input image. For this, a pre-trained Vision Transformer (ViT) was used. The encoder is executed once per image and can be applied before the input is fed to decoder.

Prompt Encoder. This deals with the prompts provided to the model for the segmentation process. There are two sets of prompts considered: sparse (such as points and bounding boxes) and dense (masks). For representation purposes, sparse prompts use positional encodings and learned embeddings, while dense ones are processed by convolutions and summed with the image representation.

Mask Decoder. This component is responsible for generating the segmentation mask based on the encoded input information. This is done by modifying a transformer decoder block and layers that perform dynamic mask prediction.

4 Materials and Methods

4.1 Datasets

Since the procedure of identifying parts in an image will not be performed in this work, the proposal is to use publicly recognized datasets that have been reviewed by qualified medical professionals.

It is important to highlight the availability of actual masks provided by healthcare professionals in both datasets. The absence of these masks would hinder proper evaluation and comparison of the model's results, compromising the integrity of the analysis.

To work with these datasets, preprocessing of both the images and provided masks was necessary. Initially, to enable the use of SAM in predicting images, as well as fine-tuning and evaluating the model's accuracy, standardization of resolution to 256×256 was required.

Montgomery Dataset [5]. Comprising 138 images, this dataset was acquired from the Maryland Department of Health and Human Services and published by the United States National Library of Medicine. It consists of chest X-ray images whose corresponding masks for the left and right lungs are stored separately and, for this reason, a pre-processing step was necessary to merge them.

Shenzhen Dataset [5]. This dataset contains 566 chest X-ray images obtained from the Shenzhen Hospital, also published by the United States National Library of Medicine.

4.2 Technical Approach

The approach used to improve SAM's performance, as previously mentioned, was fine-tuning. After conducting tests with different prompts for the adjustment, including bounding boxes, points extracted from average images, and a combination of both, it was found that the most effective approach was using sets of points obtained from the average images of each dataset. This is due to their better performance.

It is important to highlight that, to avoid any bias or interference in the validation and test sets, only the samples designated for training were used in constructing such images and points. This ensures an objective evaluation of the model's predictions. Figure 4 presents an example of a mean image obtained from the Montgomery dataset.

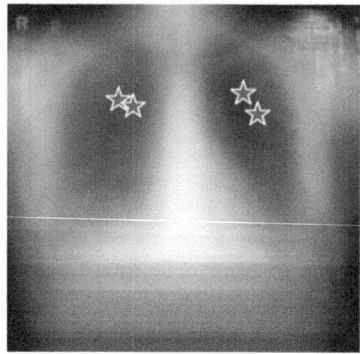

Fig. 4. Example of a mean image obtained on the Montgomery dataset.

For the bounding boxes needed for training, a function was used to extract the boxes from the image masks and apply perturbations to their coordinates to improve the model's robustness and generalization.

During the fine-tuning process, it was necessary to ensure that gradients were calculated exclusively in the mask decoder, to avoid undesired changes in the other components of SAM. This approach aims to preserve the representations learned in the upper layers of the neural network, which may contain valuable information for the segmentation process.

4.3 Evaluation Metrics

Intersection over Union (IoU). In the Intersection over Union (IoU) metric, the similarity between the mask produced by the model (M_p) and the provided ground truth mask (M_v) is evaluated. This is calculated as the ratio between the intersection and the union of the two masks:

$$\text{IoU}(M_v, M_p) = \frac{|M_v \cap M_p|}{|M_v \cup M_p|}, \quad (1)$$

where $M_v \cap M_p$ and $M_v \cup M_p$ are, respectively, the intersection and union between sets M_v and M_p, and $|\cdot|$ is their cardinality.

F1-Score. The F1-Score, with a purpose similar to the previous metric, is an evaluation measure that combines precision and recall into a single value, given by the following expression:

$$\text{F1-Score}(M_v, M_p) = \frac{2 \cdot |M_v \cap M_p|}{|M_v| + |M_p|}. \tag{2}$$

4.4 Training Procedure

Before starting the experiments, the datasets were divided into validation, training, and test sets. This division is a common practice in machine learning to evaluate and adjust the training and applied techniques. In this work, 20% of the data was set aside for testing, 60% for training, and 20% for validation. The validation set was used to adjust some hyperparameters during the process, such as the segmentation threshold and the loss function.

For comparison with the results obtained with the U-Net network [9], a 5-fold cross-validation was performed in the same manner as the work of Brioso [1]. Cross-validation is a technique to assess the generalization capability of the model on a given dataset. In short, cross-validation splits the data into several subsets called folds. After this procedure, the model is trained, tested, and validated with different portions of the data until all parts have been used as test sets. At the end of the process, the evaluation metrics, such as F1-Score and IoU, are aggregated from all the training and testing iterations to provide a more robust and less biased estimate of performance.

4.5 Loss Function and Optimizer

Regarding the loss function, the "*DiceFocalLoss*" from the Monai library[1] was chosen, which computes both Dice and Focal losses and returns a weighted sum of them. This approach demonstrated superior performance compared to other available loss functions for image segmentation, such as "*DiceCELoss*", which combines Dice and Cross Entropy losses. As for the optimizer, Adam was chosen, a widely-used method that combines the benefits of RMSProp and SGD Momentum to converge quickly to the global minimum.

5 Experiments and Results

This section presents the experiments conducted to evaluate and optimize SAM's performance. Initially, the results of the model before fine-tuning are discussed, followed by an analysis of the learning curve. Subsequently, we explore the influence of the segmentation threshold on the quality of the masks generated by SAM and detail the hyperparameter tuning process that optimized the model's performance. Finally, experiments about the model's adaptability and its comparison with other baselines are discussed.

---
[1] https://monai.io/ (As of July, 2024).

5.1 Initial Evaluation of SAM

Before fine-tuning SAM for lung segmentation in chest X-ray images, an initial evaluation was conducted to measure the performance of the neural network in its original form [6]. This step is important to establish a baseline reference for the model with respect to the Montgomery and Shenzhen datasets [5]. For this, the datasets were divided into five subsets to follow the 5-fold cross-validation procedure. Tables 1 and 2 present the mean and standard deviation of the evaluation metrics calculated on the resulting subsets obtained from the Montgomery and Shenzhen datasets, respectively.

Table 1. Initial evaluation of SAM on the Montgomery dataset.

Prompt	F1-Score	IoU
Bounding Box	0.718 ± 0.033	0.586 ± 0.033
Points	0.860 ± 0.013	0.774 ± 0.018
Both	0.848 ± 0.006	0.746 ± 0.007

Table 2. Initial evaluation of SAM on the Shenzhen dataset.

Prompt	F1-Score	IoU
Bounding Box	0.782 ± 0.009	0.661 ± 0.013
Points	0.726 ± 0.021	0.593 ± 0.026
Both	0.863 ± 0.005	0.765 ± 0.008

5.2 Learning Curve

To achieve satisfactory results, various experiments were conducted during the model's training procedure. Initially, to determine the appropriate number of epochs for fine-tuning SAM, the learning curve was used. This curve represents the average loss obtained by the neural network over the training epochs and is an essential tool for evaluating its progress and performance. The learning curve allows analysis, through the decrease in loss over time, of whether the model is optimizing over the epochs or stagnating, the latter indicating that the training can be stopped.

From the aforesaid tests, it was found that none of the datasets and none of the different prompts evaluated showed a significant improvement in loss after the hundredth epoch. Figure 5 shows an example of one of the learning curves obtained on the Shenzhen dataset.

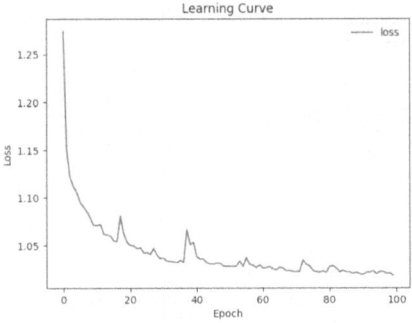

Fig. 5. Example of a learning curve obtained on the Shenzhen dataset.

5.3 Segmentation Threshold

Another critical factor influencing SAM's performance after fine-tuning is the threshold used to convert the segmented mask into a binary mask after model's prediction. This is because the mask returned by the model has continuous values between 0 and 1, representing the probability of a given pixel belonging to the region of interest. The choice of threshold to binarize this mask directly affects the results, as an inappropriate selection of this value can lead to inaccurate or incomplete segmentation masks. Low values (e.g., below 0.5) may include many pixels that do not belong to the mask, while high values (e.g., above 0.7) may exclude many pixels that belong to the mask. For this reason, in this study, threshold values from 0.50 to 0.70 were evaluated on the validation set.

Table 3. F1-Score obtained for different thresholds in the Montgomery dataset.

Threshold	Bounding Box	Points	Both
0.50	**0.828**	0.898	**0.894**
0.55	0.812	0.932	0.893
0.60	0.743	0.957	0.879
0.65	0.596	**0.960**	0.835
0.70	0.410	0.938	0.759

The results obtained for the Montgomery and Shenzhen datasets are presented in Tables 3 and 4, respectively. The best F1-Score for each dataset and prompt is highlighted in bold. This information helped identify the most suitable threshold value for each dataset and input, thereby improving the overall accuracy and quality of predictions.

Table 4. F1-Score obtained for different thresholds in the Shenzhen dataset.

Threshold	Bounding Box	Points	Both
0.50	**0.837**	0.880	**0.846**
0.55	0.127	**0.918**	0.782
0.60	0.003	0.870	0.417
0.65	0.000	0.829	0.105
0.70	0.000	0.789	0.005

5.4 Hyperparameter Tuning

Hyperparameter tuning is a crucial step in the SAM fine-tuning process as it determines the values of learning rate and weight decay, which directly impact the model's performance. Briefly, the learning rate controls the size of steps that the optimization algorithm takes during the learning process. A too high value can lead to instability and prevent or hinder algorithm convergence, whereas a too low value can result in slow or stagnant training. On the other hand, weight decay controls the magnitude of the model's weights by adding a penalty to prevent them from becoming too large, which can help prevent overfitting.

In this study, a grid of predefined values was used: $[1\times10^{-5}, 1\times10^{-4}, 1\times10^{-3}]$ for the learning rate and $[0, 1\times10^{-1}, 1\times10^{-3}]$ for the weight decay. After the search, a learning rate of 1×10^{-5} and a weight decay of 0 were adopted.

5.5 Model Adaptability

To verify the model's adaptability, two experiments were conducted in which SAM was trained on one dataset and tested on another. This procedure allows for evaluating the model's generalization capability in different contexts, which is important for ensuring utility and effectiveness in real-world situations where data may vary. It is worth noting that, unlike the other tests, this was performed only with the points obtained from the average image.

First, fine-tuning was performed on the Shenzhen dataset and then applied to the Montgomery dataset. The results of this experiment can be seen in Table 5. For a comparison of the results, the F1-Score and IoU values for the Montgomery dataset trained with its own images were also included.

Table 5. Adaptability of the model trained on the Shenzhen dataset and tested on the Montgomery dataset.

Metric	Shenzhen → Montgomery	Montgomery → Montgomery
F1-Score	0.924	0.943
IoU	0.860	0.897

Similarly, another experiment was conducted where the neural network was initially fine-tuned on the Montgomery dataset and then applied to the Shenzhen dataset. Interestingly, the evaluated results were better compared to the fine-tuning performed directly on the Shenzhen dataset.

Table 6. Adaptability of the model trained on the Montgomery dataset and tested on the Shenzhen dataset.

Metric	Montgomery → Shenzhen	Shenzhen → Shenzhen
F1-Score	0.933	0.915
IoU	0.875	0.845

5.6 Best Models

Tables 7 and 8 present the results of the best models for the Montgomery and Shenzhen datasets, respectively. They were trained using the best hyperparameters found in the previously conducted experiments and with different prompts. The results were obtained based on 5-fold cross-validation, as described in Sect. 4.4. Additionally, to measure the variability of the results with respect to the mean, the standard deviation is also indicated.

Table 7. Detailed results on the Montgomery dataset.

Prompt	F1-Score	IoU
Bounding Box	0.818 ± 0.040	0.707 ± 0.045
Points	0.943 ± 0.007	0.897 ± 0.012
Both	0.876 ± 0.015	0.787 ± 0.023

Table 8. Detailed results on the Shenzhen dataset.

Prompt	F1-Score	IoU
Bounding Box	0.797 ± 0.014	0.667 ± 0.024
Points	0.915 ± 0.011	0.845 ± 0.018
Both	0.845 ± 0.012	0.735 ± 0.018

5.7 Comparison with U-Net

A comparison with the best results reported by Brioso [1] for by U-Net network [9] is also performed. The values used for comparison with U-Net represent

Table 9. SAM compared to U-Net (F1-Score).

Dataset	SAM	U-Net [1]
Montgomery	0.943 ± 0.007	0.973 ± 0.014
Shenzhen	0.915 ± 0.011	0.941 ± 0.047

the best performance achieved by SAM on the test set. These results correspond to the models trained with points extracted from the average images.

Above, in Table 9, F1-Score metric values are compared between SAM and U-Net, as presented in [1]. The results indicate that the U-Net network, known for its effectiveness in this task, achieved superior performance on both the Montgomery and Shenzhen datasets. Unlike the findings of He et al. [3], despite a slightly lower F1-Score, SAM demonstrated satisfactory results similar to those of U-Net, indicating its feasibility and potential for future applications.

5.8 Predicted Masks

Finally, for illustration purposes, Fig. 6 shows the best and worst predictions performed by SAM on the Shenzhen dataset. Similarly, in Fig. 7, the best and worst predictions obtained on the Montgomery dataset can be seen.

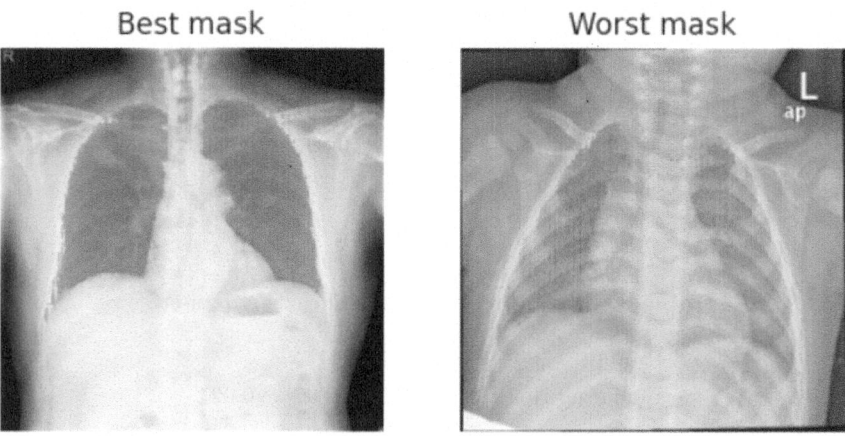

Fig. 6. Best and worst masks obtained on the Shenzhen dataset, respectively.

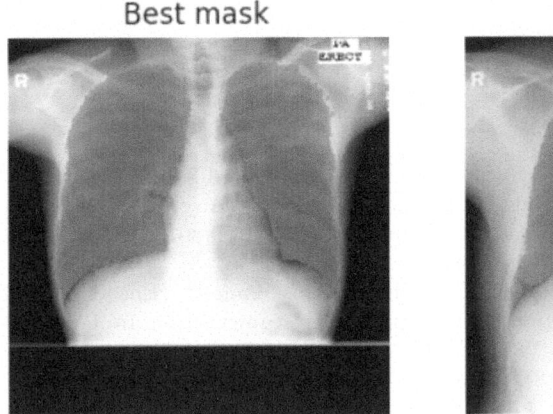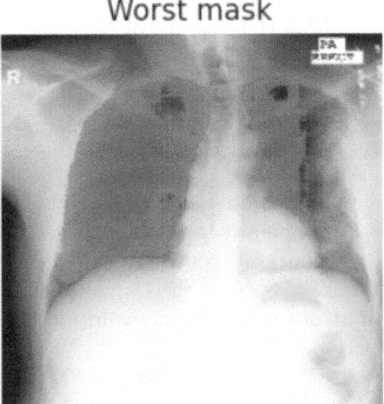

Fig. 7. Best and worst masks obtained on the Montgomery dataset, respectively.

6 Conclusions

In light of the promising neural network proposed by Meta AI, this work sought ways to fine-tune SAM and adapt it for lung segmentation in chest X-ray images provided by the Shenzhen and Montgomery datasets [5]. Its main objective is to contribute to the automation and accuracy of medical diagnosis, as well as to provide insights and opportunities regarding this new technology.

Throughout this study, various strategies to fine-tuning SAM were investigated, including the use of different prompts, such as bounding boxes, points selected from the average image, and a combination of both. Additionally, different numbers of training epochs and loss functions were explored, with the latter being a crucial choice for satisfactory results.

The experiments conducted during the mask prediction phase allowed for evaluating the impact of the threshold used in the binarization process, that is, converting the masks from the soft mask format to the hard mask format. This analysis provided valuable information on the model's sensitivity to different thresholds and their significant influence on mask quality. Furthermore, tests conducted to verify the model's adaptability showed its utility in real-world situations where datasets exhibit variations.

Considering the complexity of the challenging task of segmenting lungs in chest X-ray images, future work can explore different transfer learning techniques not covered in this study to further improve the model's performance.

Acknowledgements. This research was supported by São Paulo Research Foundation - FAPESP (grant 2023/17577-0) and Brazilian National Council for Scientific and Technological Development - CNPq (grants 315220/2023-6, 420442/2023-5, and 146570/2023-5).

Disclosure of Interests. The authors have no competing interests to declare that are relevant to the content of this work.

References

1. Brioso, E.R.C.Q.: Anatomical Segmentation in Automated Chest Radiography Screening. Ph.D. thesis, Faculdade de Engenharia da Universidade do Porto, Porto, Portugal (July 2022), https://hdl.handle.net/10216/143015, outras ciências da engenharia e tecnologias, openAccess
2. Chen, L.-C., Zhu, Y., Papandreou, G., Schroff, F., Adam, H.: Encoder-decoder with atrous separable convolution for semantic image segmentation. In: Ferrari, V., Hebert, M., Sminchisescu, C., Weiss, Y. (eds.) ECCV 2018. LNCS, vol. 11211, pp. 833–851. Springer, Cham (2018). https://doi.org/10.1007/978-3-030-01234-2_49
3. He, S., Bao, R., Li, J., Grant, P.E., Ou, Y.: Accuracy of segment-anything model (SAM) in medical image segmentation tasks. arXiv preprint arXiv:2304.09324 (2023)
4. Huang, Y., et al.: Segment anything model for medical images? Med. Image Anal. **92**, 103061 (2024). https://doi.org/10.1016/J.MEDIA.2023.103061
5. Jaeger, S., Candemir, S., Antani, S., Wáng, Y.X.J., Lu, P.X., Thoma, G.: Two public chest X-ray datasets for computer-aided screening of pulmonary diseases. Quant. Imaging Med. Surg. **4**(6), 475–477 (2014). https://doi.org/10.3978/j.issn.2223-4292.2014.11.20
6. Kirillov, A., et al.: Segment anything. arXiv preprint arXiv:2304.02643 (2023)
7. Ma, J., He, Y., Li, F., Han, L., You, C., Wang, B.: Segment anything in medical images. Nat. Commun. **15**(1), 654 (2024). https://doi.org/10.1038/s41467-024-44824-z
8. Mazurowski, M.A., Dong, H., Gu, H., Yang, J., Konz, N., Zhang, Y.: Segment anything model for medical image analysis: an experimental study. Medical Image Anal. **89**, 102918 (2023). https://doi.org/10.1016/J.MEDIA.2023.102918
9. Ronneberger, O., Fischer, P., Brox, T.: U-net: convolutional networks for biomedical image segmentation. In: Navab, N., Hornegger, J., III, W.M.W., Frangi, A.F. (eds.) MICCAI 2015, Part III. LNCS, vol. 9351, pp. 234–241. Springer, Heidelberg (2015). https://doi.org/10.1007/978-3-319-24574-4_28

Predicting Next Phases of Multi-Stage Network Attacks: A Comparative Study of Statistical and Deep-Learning Models

Antonia Severín[1], Claudio Canales[1], Romina Torres[1(✉)], César Roudergue[2], and Rodrigo Salas[3]

[1] Engineering Faculty, Universidad Adolfo Ibáñez, Viña del Mar, Chile
{aseverin,clcanales}@alumnos.uai.cl, romina.torres.t@uai.cl
[2] Engineering Faculty, Universidad Andres Bello, Viña del Mar, Chile
c.rouderguefuentes@uandresbello.edu
[3] Biomedical Engineering School, Engineering Faculty, Universidad de Valparaíso, Valparaíso, Chile
rodrigo.salas@uv.cl

Abstract. Multi-Stage Network Attacks (MSNAs) are complex, coordinated sequences of malicious activities that can unfold over extended periods-lasting hours, days, or even months. Detecting and mitigating these attacks is challenging due to their prolonged nature, and the cost of defense increases significantly depending on the stage at which the attack is detected. Organizations often face multiple concurrent MSNAs, and limited resources necessitate a strategic approach to prioritize threats, particularly those closest to their final stages. This study investigates existing methodologies for predicting the next phase of an already detected MSNA attack. We evaluate three distinct models—Hidden Markov Models (HMM), Random Forest (RF), and Long Short-Term Memory (LSTM) networks—using two well-known datasets, DARPA and CTF22, to analyze attack sequences and intrusion detection system (IDS) alert data. Our comparative analysis of the models' predictive performance, based on the F1 score, shows that HMM performed best (67.5%) on the DARPA dataset, while RF excelled on the CTF dataset (75.1%). These findings provide valuable insights for prioritizing responses to critical network threats and improving the strategic allocation of defensive resources.

Keywords: Multi-stage Network Attack · Cybersecurity · Hidden Markov Models · Random Forest · Machine Learning · Long-Short Term Memory · Deep Learning

1 Introduction

In the context of computer networks, and in line with ISO/EIC 27000 standards [1], a cyber-attack is defined as an "attempt to destroy, expose, alter, disable, steal, gain access to, or make unauthorized use of an asset". A Multi-Stage Network Attack (MSNA) is a coordinated series of intrusion activities carried out by one or more attackers, all aimed at achieving a specific intrusion goal [2]. These stages follow a sequence commonly referred to as the "Cyber Kill Chain" (CKC), which outlines the steps an attacker takes to execute a successful cyberattack [3].

The CKC model delineates the progressive phases of an attack, beginning with reconnaissance (R), where the attacker gathers information on potential targets, followed by weaponization (W) preparation, delivery (D), exploitation (E) attacks, installation (I), command and control (C2), and finally, actions on objectives (final goal). In this final stage, the attacker achieves their end goals, such as data exfiltration (confidentiality), data corruption (integrity), or system compromise (availability).

The traditional Cyber Kill Chain (CKC) model includes phases difficult to measure within an attack, such as weapon creation, which often occurs outside the target network. This complexity makes the standard CKC less practical for certain types of assessments. To address this, a simplified version of the CKC model was proposed [4], reducing the process to four core stages: reconnaissance, access, execution, and persistence.

Figure 1 presents a selection of the diverse pathways an attack might undertake. Notably, the diagram in the first row, second column, shows that an attack can loop back to the reconnaissance phase. This non-linear progression allows attackers to refine their tactics based on previously gathered intelligence. The figure also emphasizes a key phase where, following successful reconnaissance, the development of scripts or exploits targeting known vulnerabilities is completed. The attack then proceeds to the delivery phase, which may involve sending a malicious link via email, setting the stage for the subsequent installation of the weapon. This diagram underscores cyberattack's complexity and evolving nature, highlighting the importance of understanding each potential route to strengthen defensive strategies.

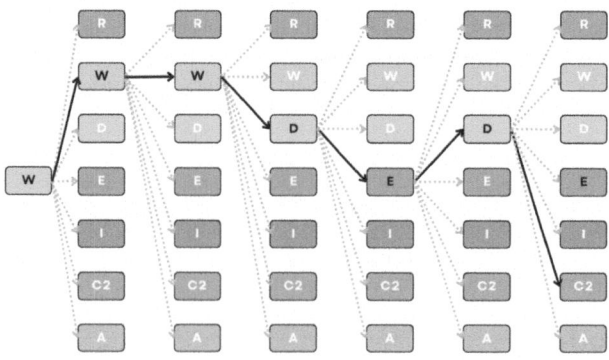

Fig. 1. Multi-stage Network Attack. The diagram shows the evolution of the attack between an attacker and a victim.

On one hand, an organization may face hundreds of cyberattacks daily, each potentially representing a step in a Multi-Stage Network Attack (MSNA) that can take hours to months to reach its objective. These attacks are frequent and intricate, involving multiple stages, which makes correlating detection and response efforts increasingly challenging.

On the other hand, organizations are further strained by resource constraints, limiting their ability to defend against sophisticated threats. Financial, technical, and human resource limitations can expose critical vulnerabilities, challenging effective defense. The situation worsens when multiple attacks occur simultaneously, forcing organizations to prioritize their limited resources. This prioritization is crucial for mounting an effective defense and minimizing the impact on operations and the security of sensitive information.

In general, traditional models like Hidden Markov Models (HMM) [5,6] have shown potential for predicting the progression of a Multi-Stage Network Attack (MSNA) [7]. Still, they face limitations in adaptability and are prone to overfitting. While Long Short-Term Memory (LSTM) models have consistently outperformed traditional methods across various domains, there is a notable lack of research, models, and records in cybersecurity literature that focus specifically on using LSTM to predict the next stage of an attack. Therefore, in this work, we explore various approaches to address this gap and contribute to developing more accurate predictive models in cybersecurity.

2 State of the Art

Over the years, the models used to predict the next step or stage of a MSNA have evolved significantly. Initially, these models relied on attack graphs [8], which offered a structured method for visualizing potential attack paths. As the field advanced, more sophisticated models emerged, incorporating data mining techniques to identify patterns and predict attacks [9].

Martinez et al. [10] present a study aimed at optimizing Network Intrusion Detection Systems (NIDS) through the use of feature selection techniques and machine learning models. Their comparative analysis of Random Forest and XGBoost highlights the potential for identifying a relevant subset of features that can efficiently detect network intrusions in real-time environments, ensuring high detection performance and operational efficiency without significant loss in accuracy. On the other hand, Torres et al. [11] introduce a novel nested cascade model designed to address the challenges of imbalanced datasets in cyberattack classification. The model improves the classification of underrepresented attacks by employing binary classifiers across multiple tiers, each focusing on a specific attack type. This approach demonstrates promise in better handling the dynamic nature of underrepresented cyberattacks.

Recently, Chadza et al. [7] utilized Hidden Markov Models (HMMs) to predict the next step and stage of an attack based on Intrusion Detection System (IDS) alerts. Their model, evaluated on the DARPA LLDOS 1.0 dataset [12], achieved an average accuracy of 60% across different attack stages. Similarly, Farhadi et al. [13] reported a precision of 82% in predicting the next attack stage using the same dataset. These studies highlight the potential of HMMs as dynamic tools for modeling and analyzing the progression of cyberattacks.

Holgado et al. [14] also leveraged the DARPA LLDOS 1.0 dataset [12], categorizing alerts into five DARPA-defined stages. Their model assembled the most likely sequences of events, allowing it to predict the next step and estimate the completion percentage toward the attacker's anticipated goal. The authors highlighted the model's efficiency, with an average inference processing time of 0.14 milliseconds, underscoring its capability for real-time threat assessment.

Kholidy et al. [15] applied HMM to the alerts from the DARPA 2000 LLDOS 1.0 dataset, categorizing them into specific states and warning categories. Their model successfully issued warning alarms up to 3,882 s before the execution of the LLDOS 1.0 attack and 2,377 s before the detection phase, within a total scenario duration of 11,836 s.

Li et al. [16] addressed the limitations of traditional multi-stage network attack detection by integrating temporal sequences into a Hidden Markov Model (HMM), which was then transformed into a real-time updatable Bayesian network. When tested on the DARPA 2000 LLDOS 1.0 dataset, their enhanced HMM-PI-UCM model significantly outperformed a standard HMM, achieving 29.13% accuracy in predicting the attacker's intent-over six times higher than the standard HMM's accuracy of 3.27%.

Cyberattacks often occur in correlated sequences, making architectures like LSTM well-suited for modeling them. Fan et al. [17] introduced the ALEAP model, an Attention-based LSTM with Event Embedding, as a deep-learning approach to address this challenge. The model consists of three layers: a preprocessing layer for event extraction and database creation, an event embedding layer that classifies events into four stages of an attack to capture correlations and context, and an attention-based LSTM layer that identifies relevant hidden events within the data sequence. Each event is characterized by four features: 'Subject' (the entity performing the attack, such as an IP address), 'Object' (the victim of the attack, like a file or system service), 'Action' (the specific behavior executed, such as file manipulation), and 'Other' (additional relevant details, like detecting a malicious file). The model was tested on four private datasets constructed from accurate user security system data, achieving an accuracy of approximately 72%.

3 Material and Methods

In this section, we describe the datasets used, outline the process of creating and preprocessing the dataframes, and provide a detailed explanation of how machine learning models were applied to predict the next stage of an attack.

The proposed methodology is illustrated in Fig. 2. The process begins with monitoring network traffic, which generates alerts that are then processed and categorized into different stages according to the attack's progression.

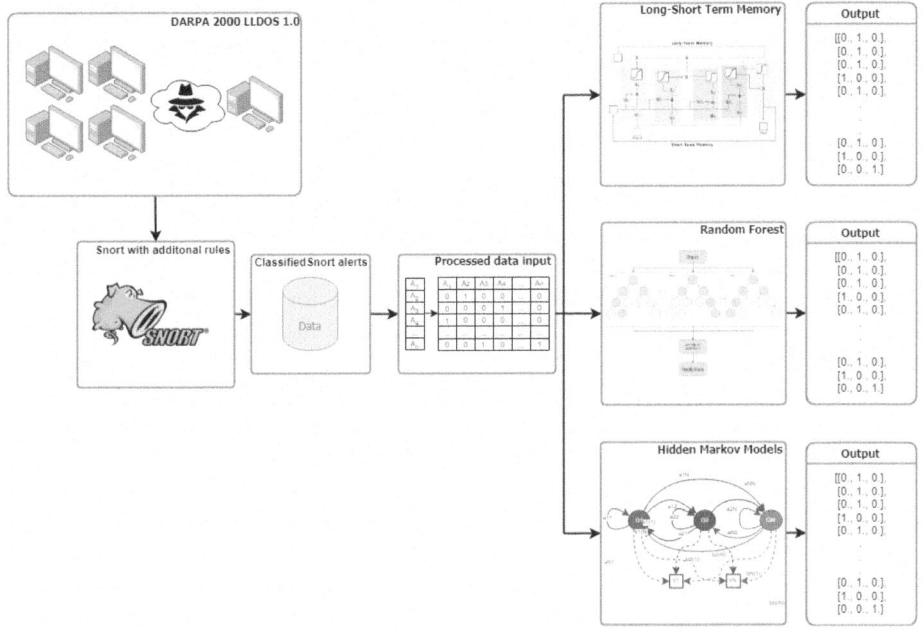

Fig. 2. Proposed methodology for predicting the next phases of a Multi-Stage Network Attack.

3.1 DARPA 2000 LLDOS 1.0 Dataset

The dataset used in this study is the DARPA 2000 LLDOS 1.0 dataset[1] [12], developed by the Defense Advanced Research Projects Agency (DARPA) under the United States Department of Defense (DoD). It captures an intrusion detection evaluation scenario involving a Distributed Denial of Service (DDoS) attack. The scenario unfolds across multiple network and audit sessions, organized into five distinct stages, where an adversary scans the network, exploits vulnerabilities, installs malicious software, and ultimately launches a DDoS attack.

The attack scenario progresses as follows:

- **Stage 1**: IP Sweep of Air Force Base (AFB) from a remote site.
- **Stage 2**: Probe of live IPs to detect the sadmind daemon running on Solaris hosts.
- **Stage 3**: Breaking into hosts using the sadmind vulnerability, both successful and unsuccessful.
- **Stage 4**: Installation of the mstream DDoS software on compromised hosts.
- **Stage 5**: Launching the DDoS attack from the compromised hosts.

The dataset is provided in PCAP format and primarily contains TCP traffic. The attack stages are mapped to a simplified Cyber Kill Chain (CKC) model,

---

[1] https://archive.ll.mit.edu/ideval/data/2000/LLS_DDOS_1.0.html.

following the method proposed by [18]. In this model, each stage corresponds to a specific phase of the attack: reconnaissance, access and exploitation, execution, persistence, and the final attack.

3.2 CTF 22 Dataset

DEFCON, the world's largest cybersecurity conference, features a 'Capture the Flag' (CTF) competition[2], where participants showcase their skills in cybersecurity offense and defense. Expert hackers legally demonstrate their abilities in a public forum in this setting. Each team is responsible for defending its own 'flag'- a data file stored on its server-while simultaneously attempting to compromise other teams' flags.

At the end of DEFCON, all files and data generated during the competition are made publicly available. The DEFCON 22 dataset contains data packets captured during the 22nd edition of the event, held in 2014. This dataset offers detailed insights into various phases of real-world attacks executed during the competition. For this analysis, we utilized the network captures from the winning team, Plaid Parliament of Pwning (PPP), with a compressed size of 4.47 GB. These captures offer a comprehensive view of the strategies and techniques employed by the team during the competition. The dataset was downloaded using the provided torrent file, which supports download recovery in case of failure.

3.3 Data Preprocessing

The raw data from both datasets was processed through the Snort Intrusion Detection System (IDS) [19] to generate security alerts. These alerts include various fields related to network traffic, such as timestamps, IP addresses, protocols, and alert messages. Python with the Pandas toolbox was used to construct a dataframe for storing and organizing the information generated by Snort.

To categorize the alerts into their respective attack stages, we applied the method proposed by [18], which maps stages according to a simplified Cyber Kill Chain (CKC) model. The initial dataframe consisted of 1,242 rows and 28 variables for DARPA dataset and 168,300 rows and 28 variables for the CTF dataset. The next step involved preparing the dataset for machine learning models, including Hidden Markov Models (HMM), Random Forests, and Long Short-Term Memory (LSTM) networks, by ensuring the data was properly formatted for model training.

During the initial data cleaning phase, six variables (`sig_generator`, `tcplen`, `icmptype`, `icmpcode`, `icmpid`, and `icmpseq`) were removed due to being empty. After cleaning, 521 source-destination pairs remained, representing individual attack instances. A train-test split was then conducted, resulting in 1,116 rows available for the DARPA dataset and 140,973 rows for the CTF dataset for training the machine learning models.

---

[2] https://defcon.org/html/defcon-22/dc-22-index.html.

Table 1. Summary of Alert Fields

Field	Description
timestamp	Time when the alert was generated
sig_generator	Component of Snort that generated the alert
sig_id	Signature ID from Snort
sig_rev	Signature revision number
msg	Alert message
proto	Protocol of the triggering traffic
src	Source IP of the traffic
srcport	Source port of the traffic
dst	Destination IP
dstport	Destination port
ethsrc	Source Ethernet address
ethdst	Destination Ethernet address
ethlen	Length of the Ethernet frame
tcpflags	TCP flags set in the triggering traffic
tcpseq	TCP sequence number
tcpack	TCP acknowledgment number
tcplen	TCP payload length
tcpwindow	TCP window size
ttl	IP packet time-to-live
tos	IP packet type-of-service
id	IP packet identification number
dgmlen	IP datagram length
iplen	IP packet length
icmptype	ICMP packet type
icmpcode	ICMP packet code
icmpid	ICMP packet ID
icmpseq	ICMP sequence number
Stage	Stage assigned using the proposal of [18]

After data cleaning, transformation steps were applied to prepare the dataset for model training. One-hot encoding was used for categorical variables such as the `msg` (alert message) and `proto` (protocol) fields, generating new variables. Additionally, the source and destination IP addresses, originally in string format, were converted to integers by splitting each IP into its four octets and multiplying each octet by the corresponding power of 256. Additionally, hexadecimal values were identified and converted to numerical values to ensure compatibility with the models. Following these transformations, the dataset expanded from 28 to 46 columns. The final processed dataset was then used to train the machine-learning models (Table 1).

3.4 Statistical and Machine Learning Methods

In this study, we perform a comparative analysis of three types of statistical and machine learning models that differ in their capacity to retain past information. The HMM and LSTM models are distinguished by their memory capabilities, while the Random Forest (RF) model is memoryless. Despite being a memoryless model, the Random Forest is well-suited for this study due to its ability to identify the most relevant features in a multiclass scenario accurately.

Hidden Markov Model: A Hidden Markov Model (HMM) is a probabilistic model that describes the evolution of a system over time. It is termed "hidden" because the system's underlying states are not directly observable; instead, we observe outputs or emissions related to these hidden states. In the context of network security, these emissions could be Intrusion Detection System (IDS) alerts, which provide insights into the system's current state.

An HMM is defined by a 6-tuple of parameters (Q, V, A, B, π, O) [7], according to the following definition:

1. Set of system states, Q. This finite set Q has N states represented as:
$$Q = \{q_1, q_2, \ldots, q_N\} \tag{1}$$

2. Set of distinct observations, V. Composed of M symbols given by:
$$V = \{v_1, v_2, \ldots, v_M\} \tag{2}$$

3. Observation sequence, O. It has a variable length, T, and represents a series of observations whose elements belong to the set of distinct symbols V. The set is denoted as
$$O = \{o_1, o_2, \ldots, o_T\} \tag{3}$$

4. State transition matrix, A. This is an $N \times N$ matrix whose row sum is unity.
$$A = \begin{bmatrix} a_{11} & a_{12} & \cdots & a_{1N} \\ a_{21} & a_{22} & \cdots & a_{2N} \\ \vdots & \vdots & \ddots & \vdots \\ a_{N1} & a_{N2} & \cdots & a_{NN} \end{bmatrix} \tag{4}$$

5. Observation probability matrix, B. An $N \times M$ matrix represents the probabilities of emitting each observation symbol from each state.
$$B = \begin{bmatrix} b_{11} & b_{12} & \cdots & b_{1M} \\ b_{21} & b_{22} & \cdots & b_{2M} \\ \vdots & \vdots & \ddots & \vdots \\ b_{N1} & b_{N2} & \cdots & b_{NM} \end{bmatrix} \tag{5}$$

6. Initial state probability vector, π: A vector of length N representing the initial probabilities for each state.

$$\pi = [\pi_1, \pi_2, \ldots, \pi_N] \tag{6}$$

Decision Trees and Random Forest: The *Decision Tree* is a machine-learning model with an architecture that resembles a tree structure [20]. Each internal node represents a test or decision based on an input variable, and the test outcomes are branches leading to other nodes. The algorithm recursively partitions the input space, terminating the process at leaf nodes, which provide the final classification or decision.

The *Random Forest* (RF) model is an ensemble of randomized decision trees, constructed using the Bootstrap Aggregating technique (Bagging), where each training set is generated from random samples with replacement [21]. In this study, default parameters were used, including the number of trees in the forest (n_estimators): 100, the function to measure the quality of a split: Gini, the strategy for choosing splits (splitter): best, the maximum tree depth: None, the minimum number of samples required to split an internal node (min_samples_split): 2, and the minimum number of samples required at a leaf node (min_samples_leaf): 1.

Long Short-Term Memory (LSTM): Long Short-Term Memory (LSTM) is a type of recurrent neural network designed to model sequential data [22]. LSTMs have an internal memory mechanism that enables them to retain information over long periods. Each LSTM memory block consists of a cell and three gates-Input Gate, Forget Gate, and Output Gate-which regulate the flow of information through the network [23]. Each gate plays a distinct role in managing the information flow, as briefly described below:

1. **Input gate:** It controls the addition of new information to the cell's memory. It uses a sigmoid function to determine which information should be retained, while a hyperbolic tangent function generates a candidate memory vector to be added to the cell.

$$i_i = \sigma(W_i \cdot x_t + U_i \cdot h_{t-1} + b_i) \tag{7}$$

2. **Forget gate:** It regulates how much old information is retained in the cell's memory. It employs a sigmoid function to decide which information should be kept and which should be discarded.

$$f_t = \sigma(W_f \cdot x_t + U_f \cdot h_{t-1} + b_f) \tag{8}$$

3. **Output gate:** It regulates the information passed from the cell's memory to the network's output. It uses a sigmoid function to determine which information should be transmitted and a hyperbolic tangent function to generate the network's final output.

$$o_i = \sigma(W_o \cdot x_t + U_o \cdot h_{t-1} + b_o) \tag{9}$$

$$c_t = f_t \odot c_{t-1} + i_t \odot \tanh(W_c \cdot x_t + U_c \cdot h_{t-1} + b_c) \quad (10)$$

$$h_t = o_t \odot \tanh(c_t) \quad (11)$$

Here, for each presented equation, W_i, W_f, W_o, U_i, U_f, and $U_o \in R_{dxd}$ are the weighted matrices, and b_i, b_f, and $b_o \in R_d$ represent the LSTM bias, with i, f, and o representing the input gate, forget gate, and output gate, respectively. σ represents the sigmoid function, $\odot$ represents element-wise multiplication, x_t represents the security event type embedding, and the hidden layer vector is h_t [17].

4 Results

In this section, we conducted an experiment to predict the most probable next phase of an MSNA attack. A comparative study was performed using three models: Hidden Markov Model (HMM), Random Forest (RF), and Long Short-Term Memory (LSTM). These models were applied to two datasets-the DARPA dataset and the CTF 22 PPP dataset-to assess their performance across different contexts.

For all experiments, Python version 3.10[3] was used, along with common data processing libraries: pandas 2.2.2[4], numpy 1.26.4[5], and scikit-learn 1.5.1[6]. We worked with the hmmlearn[7] library to import the Hidden Markov Model (HMM). These libraries were employed to compute model performance metrics. Additionally, all experiments leveraged the processing power of an Nvidia RTX 2060 graphics card to optimize execution efficiency.

The data transformation process for the HMM model involves arranging the observations into chronologically ordered lists, with each stage representing the model's states. Each observation is uniquely linked to the corresponding *sig_id* of each alert message. These observations are then input into the model, which is trained using the Baum-Welch algorithm with the following parameters: 4 hidden states, 13 categorical features, and a maximum of 500 iterations. The random seed is initialized using the Mersenne Twister algorithm to ensure reproducibility. This training process generates a matrix of transition probabilities between observations and corresponding stages. The stage with the highest transition probability is the predicted next stage. It is important to note that for HMM, only a single $src - dst$ pair is worked on at a time due to the nature of the model.

The data transformation process for both the RF and LSTM models is identical. Using the same dataset, we applied One Hot Encoding to categorical variables of the message (msg), creating a new column to represent the next stage of each alert in the sequence. The source (src) and destiny (dst) pairs were treated as part of a single MSNA attack. In this way, each alert is assigned the next

---
[3] https://www.python.org/downloads/release/python-3100/.
[4] https://pandas.pydata.org/docs/index.html.
[5] https://numpy.org/devdocs/release/1.26.4-notes.html.
[6] https://scikit-learn.org/stable/whats_new/v1.5.html.
[7] https://hmmlearn.readthedocs.io/en/latest/.

stage corresponding to the alert with the subsequent timestamp in the dataset. Let $S = \{s_1, s_2, \ldots, s_n\}$ represent a sequence for a specific MSNA, where each element s_i in the sequence consists of three features: the alert identifier (sig_id), source (src), destination (dst), and a One-Hot-Encoding representation of the messages ($msg = [msg_{i,0} \ldots msg_{i,m}]$), with m being the number of distinct alert messages.

The LSTM model was implemented using the TensorFlow and Keras 2.10 libraries[8]. The architecture consists of an input layer, followed by two LSTM layers with 16 and 8 units, respectively. The first LSTM layer is configured to return the full sequence. The output layer is a Dense layer with 4 units-corresponding to the four recorded stages-and uses the 'softmax' activation function. The model is optimized using the Adam optimizer with a learning rate of 0.0001 and *categorical_crossentropy* as the loss function. Each experiment was executed ten times to obtain averaged performance metrics.

4.1 Discussion

In this context, the F1 Score is the most critical metric to prioritize. While minimizing false negatives is crucial-since missing an attack could lead to severe consequences-avoiding excessive false positives is equally important. Halting production or suspending operations due to false alarms can be extremely costly for companies or organizations. The F1 Score balances precision and recall, ensuring the model detects attacks accurately without overwhelming the system with unnecessary disruptions. By accounting for both false positives and false negatives, this metric offers a more balanced and practical performance measure, making it ideal for cybersecurity applications.

Table 2 presents the results of the models in predicting the next stage of the multi-attack sequence for both datasets. The models rely solely on the current state to predict the subsequent step. In this scenario, the model performances differ from previous cases, with Random Forest emerging as the best performer, achieving an average F1 score of 59.6% and 75.1% for the DARPA and CTF22 datasets, respectively. LSTM obtained F1 scores of 59.3% and 70.6% for the DARPA and CTF22 datasets, respectively. Finally, the HMM model shows significant variability, with an F1 score of 67.5% on the DARPA dataset but only 19.2% on the CTF22 dataset.

These results align with our expectations. Although the sequences from the CTF dataset are incomplete for the HMM model, they are longer than those from the DARPA dataset, and HMM tends to struggle with larger data sequences. In contrast, the RF and LSTM models performed well, particularly since the dataset was treated as a complete set without separating the vectors by victim-attacker pairs. This setup favored RF, as expected. The situation might have been different if the data had been segmented into blocks based on individual instances, where the memory capabilities of the LSTM model would have played

---

[8] https://www.tensorflow.org/.

Table 2. Performance comparison of models predicting the next stage in multi-stage network attacks.

Model	DARPA				CTF 22			
	Precision	Accuracy	F1 Score	Recall	Precision	Accuracy	F1 Score	Recall
HMM	67.5%	48.9%	67.5%	56.5%	17.8%	26.4%	19.2%	26.4%
RF	59.6%	61.0%	59.6%	52.1%	79.5%	76.4%	75.1%	76.4%
LSTM	59.3%	56.5%	59.3%	50.8%	71.3%	76.0%	70.6%	76.0%

a more prominent role. However, this approach was not feasible due to the limitations of the DARPA dataset, which contains only a single MSNA instance. In comparison, the CTF22 dataset offers over three hundred MSNA instances, enabling a more comprehensive analysis.

5 Conclusions and Future Work

In the first case with the DARPA dataset, HMM outperformed the other methods, achieving an $F1$ score of 67.5%, compared to 59.6% for RF and 59.3% for LSTM. In contrast, in the second case using the CTF22 dataset, Random Forest outperformed the other models with an $F1$ score of 75.1%, compared to 70.6% for LSTM and 19.2% for HMM. These results highlight the need for LSTM and RF to capture longer temporal sequences rather than focusing solely on the most recent step. Improving LSTM's ability to predict subsequent stages of multi-stage network attacks is part of our ongoing work.

Furthermore, testing these models on larger datasets with more instances per attacker and longer attack chains is likely beneficial. A more comprehensive evaluation may provide deeper insights into their performance in more complex scenarios. Expanding the dataset size, the length, and the number of attack sequences could also help fine-tune the models, particularly the LSTM, allowing it better to capture the temporal dynamics of multi-stage network attacks.

Future research should focus on a more granular analysis of how these models perform at different stages of an attack and whether their predictive accuracy improves or declines across these stages. It is also crucial to examine the models' dependency on the volume of instance examples and assess how the results are affected when the dataset is segmented by attacker-victim pairs or aggregated across all instances of an attacker, regardless of the victim. These inquiries will be essential for improving the precision of predictive models and developing robust strategies to prioritize and mitigate the most critical cyber threats.

For future work, we aim to refine our approach by recognizing that attackers often employ repetitive and patterned strategies against their victims. To help organizations counter these threats, we propose developing individual models tailored to each attacker. One possible solution involves creating a mixture of expert models, where each expert is dedicated to a known attacker profile, complemented by a gating network that attributes ongoing attacks to the most likely perpetrator. We anticipate significant improvements in predictive accuracy by implementing specialized models, such as HMM or LSTM, for each unique sequence of multi-stage attacks.

Additionally, we plan to enhance our models by incorporating attention mechanisms into the LSTM architecture to improve its ability to identify critical patterns within the data. We will also explore advanced architectures, such as transformers, which have effectively handled complex sequential tasks. Our ultimate goal is to integrate these models into a cohesive ensemble, thereby boosting the accuracy and reliability of predictions for proactive cyber defense strategies.

Finally, expanding the current dataset will validate our findings across diverse environments and attack types. This approach will improve our prediction accuracy and ensure the models remain relevant and applicable in today's rapidly evolving cybersecurity landscape.

Acknowledgements. This research was funded by ANID FONDECYT project N° 11221155 "Framework to detect, monitor and analyze the evolution of multistage attacks during runtime". Researchers are also partially supported by National Center for Artificial Intelligence CENIA Basal ANID FB210017, and Millennium Science Initiative Program ANID ICN2021_004.

Disclosure of Interests. The authors have no competing interests to declare that are relevant to the content of this article.

References

1. ISO/IEC. ISO/EIC 27000:2018 Information technology - Security techniques - Information security management systems - Overview and vocabulary. ISO, 5 edn. (2018)
2. Levshun, D., Kotenko, I.: A survey on artificial intelligence techniques for security event correlation: models, challenges, and opportunities. Artif. Intell. Rev., 1–44 (2023)
3. Hutchins, E.M., Cloppert, M.J., Amin, R.M., et al.: Intelligence-driven computer network defense informed by analysis of adversary campaigns and intrusion kill chains. In: Leading Issues in Information Warfare & Security Research, vol. 1, no. 1, p. 80 (2011)
4. Singh, S., Silakari, S.: A survey of cyber attack detection systems. Int. J. Comput. Sci. Netw. Secur. **9**(5), 1–10 (2009)
5. Baum, L.E., Petrie, T., Soules, G., Weiss, N.: A maximization technique occurring in the statistical analysis of probabilistic functions of Markov chains. Ann. Math. Stat. **41**(1), 164–171 (1970)
6. Rabiner, L.R.: A tutorial on hidden Markov models and selected applications in speech recognition. Proc. IEEE **77**(2), 257–286 (1989)
7. Chadza, T., Kyriakopoulos, K.G., Lambotharan, S.: Analysis of hidden Markov model learning algorithms for the detection and prediction of multi-stage network attacks. Futur. Gener. Comput. Syst. **108**, 636–649 (2020)
8. Hughes, T., Sheyner, O.: Attack scenario graphs for computer network threat analysis and prediction. Complexity **9**(2), 15–18 (2003)

9. Li, Z.-T., Lei, J., Wang, L., Li, D.: A data mining approach to generating network attack graph for intrusion prediction. In: Fourth International Conference on Fuzzy Systems and Knowledge Discovery (FSKD 2007), vol. 4, pp. 307–311. IEEE (2007)
10. Martinez, V., Salas, R., Tessini, O., Torres, R.: Machine learning techniques for behavioral feature selection in network intrusion detection systems. IET (2021)
11. Torres, R., Solis, M.A., Martínez, V., Salas, R.: A nested-cascade machine learning based model for intrusion detection systems. In: 2023 IEEE CHILEAN Conference on Electrical, Electronics Engineering, Information and Communication Technologies (CHILECON), pp. 1–6 (2023)
12. Data web page of DARPA 2000 LLDOS 1.0. Accessed 15 Mar 2024
13. Farhadi, H., AmirHaeri, M., Khansari, M.: Alert correlation and prediction using data mining and HMM. ISeCure **3**(2) (2011)
14. Holgado, P., Villagrá, V.A., Vazquez, L.: Real-time multistep attack prediction based on hidden Markov models. IEEE Trans. Dependable Secure Comput. **17**(1), 134–147 (2017)
15. Kholidy, H.A., Erradi, A., Abdelwahed, S., Azab, A.: A finite state hidden Markov model for predicting multistage attacks in cloud systems. In: 2014 IEEE 12th International Conference on Dependable, Autonomic and Secure Computing, pp. 14–19. IEEE (2014)
16. Li, T., Liu, Y., Liu, Y., Xiao, Y., Nguyen, N.A.: Attack plan recognition using hidden Markov and probabilistic inference. Comput. Secur. **97**, 101974 (2020)
17. Fan, S., et al.: ALEAP: attention-based LSTM with event embedding for attack projection. In: 2019 IEEE 38th International Performance Computing and Communications Conference (IPCCC), pp. 1–8. IEEE (2019)
18. Cano, M., Torres, R.: Caracterización de ataques multietapa en ejercicios capture the flag. Revista Política y Estrategia **141**, 133–151 (2023)
19. Roesch, M.: Network intrusion detection & prevention system (1998)
20. Breiman, L., Friedman, J., Stone, C., Olshen, R.: Classification and Regression Trees. Taylor & Francis (1984)
21. Breiman, L.: Random forests. Mach. Learn. **45**, 5–32 (2001)
22. Schmidhuber, J., Hochreiter, S., et al.: Long short-term memory. Neural Comput. **9**(8), 1735–1780 (1997)
23. Aggarwal, C.C.: Neural Networks and Deep Learning, vol. 10, no. 978, p. 3. Springer, Cham (2018). https://doi.org/10.1007/978-3-319-94463-0

Improving Suicide Ideation Screening with Machine Learning and Questionnaire Optimization Through Feature Analysis

Ignacio Martínez[1(✉)], César Astudillo[1], and Daniel Núñez[2]

[1] Department of Computer Science, Universidad de Talca, Curicó, Chile
imartinez17@alumnos.utalca.cl
[2] Psychology Faculty, Universidad de Talca, Talca, Chile

Abstract. This study explores data science and machine learning techniques to predict suicidal ideation in young individuals, aiming to identify the most effective subset of questions from a comprehensive questionnaire. We benchmarked various machine learning models, including ElasticNet, Ridge Regression, Support Vector Regressor (SVR), Gradient Boosting Regressor, Random Forest Regressor, and XGBoost Regressor. Ridge Regression emerged as the most suitable model, achieving a Mean Squared Error (MSE) of 146.77, Mean Absolute Error (MAE) of 8.68, and an R^2 of 0.57. Utilizing SHAP (SHapley Additive exPlanations) analysis, we identified the 20 most influential questions from the dataset. This refined approach not only enhances the efficiency of the questionnaire but also aids targeted intervention strategies by facilitating the early detection of suicidal ideation. The findings provide mental health professionals with a streamlined and effective tool to assess and address suicidal tendencies among students and young people, thereby improving preventative measures and outcomes.

Keywords: Machine Learning · Suicidal Ideation · Ridge Regression · SHAP Analysis · Questionnaire Optimization · Feature Analysis · Early Detection

1 Introduction

Suicide remains a significant public health concern, ranking among the top 10 causes of death globally among individuals aged 10–64 years, with alarming increases in incidence rates observed from 2000 to 2021 [1]. Beyond the tragic loss of life, suicide profoundly impacts families, friends, and communities worldwide [3]. The prevalence of suicidal thoughts and attempts underscores the critical need for effective monitoring and prevention strategies [2,3].

In recent years, machine learning has emerged as a promising approach to enhance the prediction of suicide risk and ideation. Walsh et al. [4] demonstrated the superiority of machine learning algorithms over traditional clinical assessments in predicting suicide attempts. A systematic review by Lai et al. [5]

underscored the potential and challenges of employing machine learning in this domain. Additionally, Park et al. [6] emphasized the significance of feature selection and data quality in improving predictive accuracy across different machine learning models.

Recent studies have explored diverse applications and methodologies in suicide risk prediction. Kumar and Dixit [9] tailored predictive models specifically for older adults, while Lee et al. [10] validated these techniques within a Korean demographic, highlighting cultural nuances in suicide prevention strategies. Furthermore, research by Fekih-Romdhane et al. [11] examined comorbidities associated with suicide risk, and Chattopadhyay et al. [12] integrated machine learning into public health frameworks for enhanced prevention efforts.

Recent advancements also include the use of natural language processing by Shoaib et al. [13] for early detection of suicide ideation and the development of predictive models focusing on depression and suicide risk by Zhang et al. [14]. Moreover, Park et al. [16] conducted comparative effectiveness studies on various machine learning techniques, while Al-Garadi et al. [17] explored social media-based approaches for predicting suicide.

Despite these advancements, there remains a critical need to refine and optimize predictive models for specific demographic groups and to identify the most relevant predictors for accurate screening. This study aims to address these gaps by focusing on predicting suicidal ideation among young people [7,8].

A systematic review of suicide ideation prediction using machine learning was conducted by Heckler (2022) [5]. Among the papers included in this comprehensive study, only one, Zhang et al. (2015) [15], reports regression metrics. In contrast, all other studies within the review approach suicide ideation as a classification problem [15] specifically attempts to detect suicide ideation by analyzing data from microblog users, utilizing regression techniques to predict outcomes. In our study, the focus is on estimating the suicide ideation score, and therefore, it is approached as a regression problem. Our data consider a group of young people responding to surveys, which provides a unique dataset for this regression-based approach. This highlights a distinct methodological perspective, diverging from the predominant classification-focused approaches in the field.

This study contributes to the growing body of research aimed at improving suicide risk assessment through machine learning, offering tailored solutions to address the unique challenges faced by young individuals. By refining predictive models and identifying key risk factors, we strive to strengthen preventive measures and promote better mental health outcomes among adolescents and young adults.

2 Data and Methods

2.1 Data

This study is using 1539 students and young people from the age of 13–20 who completed a comprehensive set of questionnaires consisting in a total of 116 questions. The questionnaires were meticulously crafted to assess various psychological and emotional dimensions potentially associated with suicidal ideation.

The questionnaires selected for this study are part of a project funded by the Chilean Agency of Research and Development (FONDECYT Regular 1210093). They aim to measure prevalent psychopathological aspects and emotional regulation strategies linked to suicidal ideation in adolescents. The measurement is conducted within the framework of a Randomized Controlled Trial (RCT) to evaluate the reduction of suicidal ideation and explore associated mechanisms.

The questionnaires cover a wide range of dimensions including depression, anxiety, hopelessness, and emotional regulation strategies relevant to understanding suicidal thoughts among adolescents.

The comprehensive set of 116 questions covers a broad spectrum of psychological and emotional factors. Some key areas include: Psychopathological Symptoms, Cognitive Patterns, Emotional Regulation, Behavioral Indicators. The psychopathological symptoms are measured through questions related to symptoms of depression, anxiety, and other mental health issues. The cognitive patterns, correspond to items that assess thoughts of hopelessness, self-worth, and future outlook. The emotional regulation is measured through questions that evaluate how adolescents manage and respond to their emotions. Finally, the behavioral indicators are queries about past behaviors and experiences that may indicate risk factors for suicidal ideation. These questionnaires were developed based on established psychological scales and have been validated by a specialist. These questionnaires provide a comprehensive assessment of the factors contributing to suicidal ideation in adolescents.

Finally these questionnaires was answer by students and young people aged 13–20 and were recruited from various schools and educational institutions from Chile.

2.2 Method

The aim of this paper is to predict the suicidal ideation score from questionnaire responses. As the score is a continuous variable, this constitutes a regression problem. In essence, regression is a supervised learning technique used to predict a target variable based on input features. Here, the target variable is the suicidal ideation score, and the input features are the students' questionnaire responses.

We compared several regression models including: ElasticNet [19], Ridge Regression [20], Support Vector Regressor (SVR) [21], Gradient Boosting Regressor [23], Random Forest Regressor [24], and XGBoost Regressor [25]. Each model was chosen based on its unique strengths and recent success in similar applications.

- **ElasticNet:** Combines L1 and L2 penalties to achieve both variable selection and regularization, making it suitable for datasets with many correlated features. L1 and L2 regularization are techniques used to prevent overfitting by adding penalties to the loss function based on the coefficient weights of the model [19].

- **Ridge Regression:** Includes an L2 regularization term to mitigate multicollinearity and prevent overfitting, particularly effective when predictor variables are highly correlated [18,20].
- **Support Vector Regressor (SVR):** Utilizes Support Vector Machines (SVM) principles for regression tasks, fitting a hyperplane within a specified margin of tolerance. It is robust to outliers and performs well in high-dimensional spaces [21,22].
- **Gradient Boosting Regressor:** Builds models sequentially, each correcting errors of its predecessor, and is effective in capturing complex patterns in data [23].
- **Random Forest Regressor:** Constructs multiple decision trees and aggregates their predictions to enhance accuracy and robustness, making it suitable for large datasets with high dimensionality [24].
- **XGBoost Regressor:** An optimized gradient boosting framework that excels in speed and performance, offering state-of-the-art results in various data science challenges [25].

These models were selected for their diverse strengths and complementary capabilities in handling different types of data and prediction challenges. The goal was to ensure a comprehensive evaluation and to identify the model that best fits the specific characteristics of our dataset.

The performance of each model was assessed using the following metrics:

- **Mean Squared Error (MSE):** Measures the average squared difference between predicted and actual values, providing a measure of predictive accuracy.

$$MSE = \frac{1}{n} \sum_{i=1}^{n} (y_i - \hat{y}_i)^2 \qquad (1)$$

- **Mean Absolute Error (MAE):** Provides the average absolute difference between predicted and actual values, offering insights into the magnitude of errors in the predictions.

$$MAE = \frac{1}{n} \sum_{i=1}^{n} |y_i - \hat{y}_i| \qquad (2)$$

- **Coefficient of Determination (R^2):** Indicates the proportion of the variance in the dependent variable that is predictable from the independent variables, assessing how well the model fits the data.

$$R^2 = 1 - \frac{\sum_{i=1}^{n} (y_i - \hat{y}_i)^2}{\sum_{i=1}^{n} (y_i - \bar{y})^2} \qquad (3)$$

The performance of these models was compared using Mean Squared Error (MSE), Mean Absolute Error (MAE), and the coefficient of determination (R^2), shown in (1), (2), and (3), respectively. MSE measures the average squared difference between the predicted and actual values, penalizing larger errors more severely. MAE provides the average absolute difference between predicted and actual values, giving a sense of the average prediction error. R^2 indicates the proportion of the variance in the dependent variable that is predictable from the independent variables, providing a measure of how well the predictions fit the actual data. Where y_i is the actual value, $\hat{y}_i$ is the predicted value, $\bar{y}$ is the mean of the actual values, and n is the number of observations. These metrics are essential for assessing the accuracy and fit of predictive models [26,28,29].

Hyperparameter optimization was performed using Bayesian Search, a method that efficiently explores the hyperparameter space to find the optimal set of parameters for each model. Bayesian Search was chosen over traditional grid search due to its ability to converge faster and more efficiently by using prior knowledge to guide the search process [27].

This approach helped in fine-tuning the models to achieve the best possible performance for predicting suicidal ideation.

2.3 Result and Analysis

The performance of various regression models was evaluated to predict suicidal ideation based on questionnaire responses. Table 1 summarizes the key performance metrics of each model.

Table 1. Model Performance Comparison

Model	MSE	MAE	R^2
ElasticNet	343.78	14.03	0.00
Ridge Regression	146.77	**8.68**	**0.57**
SVR	161.99	9.04	0.53
Gradient Boosting Regressor	176.63	9.04	0.49
Random Forest Regressor	146.18	**8.09**	**0.57**
XGBoost Regressor	165.75	8.53	0.52

Random Forest Regressor and Ridge Regression emerged as the top performers, demonstrating the lowest Mean Squared Error (MSE) and Mean Absolute Error (MAE), as well as the highest coefficient of determination (R^2). This suggests that these models are most effective for predicting suicidal ideation scores from questionnaire responses [18,20,24].

Support Vector Regressor (SVR), Gradient Boosting Regressor, and XGBoost Regressor also showed competitive performance but were generally inferior to Random Forest Regressor and Ridge Regression [21–23,25]. Conversely, ElasticNet exhibited the poorest performance with the highest MSE and MAE, and a negative R^2, indicating that it is not well-suited for this dataset [19].

The Mean Absolute Error (MAE) of 8.68 for the Ridge Regression model indicates that, on average, predictions are within ±8.68 units of the actual values. This level of uncertainty suggests that Ridge Regression provides reliable estimates of suicidal ideation scores, making it a valuable tool for early detection and intervention strategies.

3 Feature Selection

The selection of key questions from the questionnaire is driven by the need for interpretability and actionable insights in predicting suicidal ideation. Beyond accuracy, understanding which specific factors or questions contribute most significantly to predictions is crucial for informing targeted interventions and preventive measures in clinical practice.

Feature importance analysis reveals the questions that have the greatest impact on the model's predictions. By identifying these influential questions, we can optimize the questionnaire to enhance efficiency without compromising predictive power. This approach is vital in clinical settings where efficient risk assessment can lead to timely interventions [7,8].

We utilized SHAP (SHapley Additive exPlanations) analysis, a method rooted in cooperative game theory, to quantify the contribution of each question to the model's predictions. This method ensures consistency and additivity in feature importance metrics, providing a clear picture of each question's impact [31].

After selecting Ridge Regression as the preferred model, using SHAP analysis identified most influential questions. The following table summarizes only the 20 most important features scores from the Ridge model, highlighting the top questions and their impact on predicting suicidal ideation (Table 2).

The SHAP analysis pinpointed these questions as critical, shedding light on the factors most strongly associated with suicidal ideation. This refined set of questions not only streamlines the questionnaire but also provides actionable insights for mental health professionals to tailor interventions effectively.

The identified questions primarily focus on symptoms and emotional regulation strategies, aligning with established literature on suicidal behavior [30].

- Symptoms: These questions probe aspects such as depression, hopelessness, anxiety, psychotic experiences, and family history of suicide.
- Emotional Regulation Strategies: Questions related to problem-solving skills and coping mechanisms in identifying situations.

According to the validation provided by the mental health specialist, the above-mentioned results are consistent with literature showing that depression, hopelessness, psychotic experiences, and anxiety are symptoms associated with suicidal ideation. Similarly, emotional regulation strategies are associated with both phenomena in this population, although their specific contribution is not yet clear. They seem to be more context-dependent rather than inherently beneficial or harmful.

Table 2. Question Impact on Suicide Ideation Prediction

No.	Question	Impact
1	Have you felt bad about yourself or felt that you are a failure or have let yourself or your family down?	3.99
2	I believe my problems can be solved	3.42
3	The future seems vague and uncertain to me	3.20
4	Have you felt down, depressed, irritable, or hopeless?	3.13
5	All I can see ahead of me is more unpleasant than pleasant things	2.95
6	I feel fearful when I have important problems	2.84
7	Feeling so agitated that you cannot sit still?	2.82
8	I can't make things change, and there's no reason to believe I can in the future	2.58
9	My future seems dark to me	2.50
10	Have you felt as if a double had taken the place of a member of your family, friend, or acquaintance?	2.46
11	Have you had thoughts so intense that you worried others might hear them?	2.27
12	Have you seen objects, people, or animals that others cannot see?	2.15
13	I never get what I want, so it's absurd to desire anything	1.74
14	Previous treatment	1.73
15	Identifying situations that make my symptoms worse	1.65
16	Have you had trouble concentrating on activities like schoolwork, reading, or watching TV?	1.58
17	Previous family treatment	1.55
18	Family suicide	1.49
19	Have you moved or spoken so slowly that other people could have noticed? Or the opposite, have you been so restless that you were moving a lot more than usual?	1.44
20	Have you had little appetite, lost weight, or overeaten?	1.30

3.1 Prediction Model with Feature Selection

To evaluate the impact of using different numbers of questions, a systematic selection of relevant questions was conducted within the range of 15 to 30. For each scenario, the most effective questions were identified using SHAP analysis. Subsequently, a Ridge regression model was trained with the same training data and parameters optimized as described in Sect. 2.2, and validated with the test set. The results, presented in Table 3, include MSE, MAE, and R2 values for the test set, corresponding to the number of relevant questions specified in the first column.

Table 3. Benchmark With Different Number of Questions

N	MSE	MAE	R^2
30	150.92	8.82	0.56
29	150.73	8.75	0.56
28	151.44	8.78	0.56
27	150.96	8.75	0.56
26	151.34	8.76	0.56
25	151.77	8.80	0.56
24	150.15	8.76	0.56
23	150.95	8.80	0.56
22	150.50	8.78	0.56
21	150.87	8.79	0.56
20	**151.68**	**8.81**	**0.56**
19	152.65	8.84	0.56
18	152.19	8.82	0.56
17	155.12	8.89	0.55
16	155.39	8.88	0.55
15	155.60	8.89	0.55

From this benchmark it was decided to use the 20 most important questions, since the model with 20 questions has a similar performance to the model with 30 questions, but with fewer questions, which makes the questionnaire more easier to administer.

When comparing the results from Table 1, where Ridge regression had an MAE of ±8.68, the new model with reduced features shows an MAE of ±8.81. This marginal decrease in predictive accuracy indicates the effectiveness and robustness of the feature selection process, despite the reduction in the number of features.

4 Discussion and Conclusion

This study aimed to predict suicidal ideation among adolescents and young adults aged 13–20 using machine learning techniques while optimizing a questionnaire through advanced feature selection. We evaluated multiple regression models and found that both Ridge Regression and RandomForestRegressor demonstrated strong performance in predicting suicidal ideation based on questionnaire responses. RandomForestRegressor showed slightly better predictive accuracy, but Ridge Regression's interpretability makes it a practical choice, especially in settings where understanding model decisions is crucial.

The integration of SHAP analysis allowed us to identify a refined set of fewer questions that significantly influence the predictive model's outcomes. This streamlined questionnaire improves the ease of application for mental health professionals, facilitating quicker identification of individuals at risk. While reducing the number of questions, the questionnaire maintains a strong level of predictive power, making it a practical and efficient tool in clinical settings. This trade-off between questionnaire length and predictive accuracy enhances usability without compromising its effectiveness in detecting psychological factors associated with suicidal thoughts.

Our findings contribute significantly to the field of mental health assessment among young populations. By leveraging machine learning and rigorous feature selection techniques, we have developed an effective approach to early detection of suicidal ideation. This not only improves screening processes but also supports targeted intervention strategies tailored to individual risk profiles.

In conclusion, the combination of Ridge Regression and RandomForestRegressor, supported by SHAP analysis, provides a robust framework for predicting and understanding suicidal ideation among adolescents and young adults. Future research should focus on broader validation across diverse demographics and incorporate longitudinal data to enhance predictive accuracy and generalizability. By advancing our understanding and application of these insights, we can foster more effective mental health interventions and ultimately reduce the prevalence of suicidal ideation in vulnerable populations.

Acknowledgement. Chilean Agency of Research and Development (Fondecyt Regular 1210093).

References

1. Trinh, E., Ivey-Stephenson, A.Z., Ballesteros, M.F., Idaikkadar, N., Wang, J., Stone, D.M.: CDC Guidance for Community Assessment and Investigation of Suspected Suicide Clusters, United States, 2024, MMWR Suppl. **73**(Suppl-2), 8–16 (2024). Centers for Disease Control and Prevention (CDC)
2. National Institute of Mental Health. Suicide Prevention (2021). https://www.nimh.nih.gov/health/topics/suicide-prevention
3. World Health Organization. Preventing Suicide: A Global Imperative. ISBN: 978 92 4 156477 9. World Health Organization (2024)
4. Walsh, C.G., Sharman, J., Kaminsky, R.S.W.: Predicting suicide attempts using machine learning (2017). https://doi.org/10.1177/2167702617691560
5. Heckler, W.F., de Carvalho, J.V., Barbosa, J.L.V.: Machine learning for suicidal ideation identification: a systematic literature review. Comput. Hum. Behav. **128**, 107095 (2022). https://doi.org/10.1016/j.chb.2021.107095
6. Park, S., Choi, E., Park, J., Lee, K.: Comparative analysis of machine learning models for suicide ideation prediction (2021). https://doi.org/10.1016/j.invent.2021.100424
7. Obermeyer, Z., Emanuel, E.J.: Predicting the future-big data, machine learning, and clinical medicine. New England J. Med. **375**(13), 1216–1219 (2016). https://doi.org/10.1056/NEJMp1606181
8. Shatte, A., Hutchinson, D.M., Teague, S.J.: Machine learning in mental health: a scoping review of methods and applications. Psychol. Med. **49**(9), 1426–1448 (2019). https://doi.org/10.1017/S0033291719000151

9. Kumar, A., Dixit, A.: Suicide ideation prediction in older adults using machine learning. Techniques (2020). https://doi.org/10.1080/09720502.2020.1721674
10. Lee, H.J., Lee, S.J., Park, S.M.: Application of machine learning techniques for suicide ideation prediction in Korean. Population (2019). https://doi.org/10.30773/pi.2019.0270
11. Fekih-Romdhane, F., Ouanes, H., Ben Rejeb, M., Cheour, F.: Comorbidities and suicide risk prediction using machine learning (2023). https://doi.org/10.1016/j.ajp.2023.103725
12. Chattopadhyay, S., Panda, R.D., Verma, S.B.: Applications of machine learning in suicide ideation detection (2020). https://doi.org/10.1093/comjnl/bxz120
13. Shoaib, M., Ali, A., Shah, S.A., Haq, I.: Early detection of suicide ideation using natural language processing (2023). https://doi.org/10.1016/j.artmed.2022.102395
14. Zhang, X., Zhu, Y., Gao, Q., Zhao, L.: Depression and suicide risk prediction using machine learning models (2024). https://doi.org/10.3390/jcm9030658
15. Zhang, L., Huang, X., Liu, T., Li, A., Chen, Z., Zhu, T.: Using linguistic features to estimate suicide probability of Chinese microblog users, pp. 549–559 (2015). https://doi.org/10.1007/978-3-319-15554-8_45,2015
16. Park, S., Lee, K., Kim, J., Choi, E.: Machine learning techniques for predicting suicide ideation: a comparative study (2024). https://doi.org/10.21123/bsj.2020.17.4.1328
17. Al-Garadi, M.A., Varathan, A., Lim, D.C.: Twitter-based suicide prediction using machine learning (2022). https://doi.org/10.3390/jpm12040516
18. Hosmer, Jr. D.W., Lemeshow, S., Sturdivant, R.X.: Applied Logistic Regression, Wiley (2013)
19. Zou, H., Hastie, T.: Regularization and variable selection via the elastic net. J. Roy. Stat. Soc. Ser. B Stat. Methodol. **67**(2), 301–320 (2005). https://doi.org/10.1111/j.1467-9868.2005.00503.x
20. Hoerl, A.E., Kennard, R.W.: Ridge regression: biased estimation for nonorthogonal problems. Technometrics **12**(1), 55–67 (1970). https://doi.org/10.1080/00401706.1970.10488634
21. Drucker, H., Burges, C., Kaufman, L., Smola, A., Vapnik, V.: Support vector regression machines. In: Advances in Neural Information Processing Systems (NeurIPS), vol. 28, pp. 779–784 (1997)
22. Cortes, C., Vapnik, V.: Support-vector networks. Mach. Learn. **20**, 273–297 (1995). https://doi.org/10.1007/BF00994018
23. Friedman, J.H.: Greedy Function approximation: a gradient boosting machine. Ann. Stat. (2001). https://doi.org/10.1214/aos/1013203451
24. Breiman, L.: Random forests. Mach. Learn. **45**, 5–32 (2001). https://doi.org/10.1023/A:1010933404324
25. Chen, T., Guestrin, C.: XGBoost: a scalable tree boosting system. In: Proceedings of the 22nd ACM SIGKDD International Conference on Knowledge Discovery and Data Mining (2016). https://doi.org/10.1145/2939672.2939785
26. Montgomery, D.C., Peck, E.A., Vining, G.G.: Introduction to Linear Regression Analysis. Wiley (2012)
27. Snoek, J., Larochelle, H., Adams, R.P.: Practical Bayesian optimization of machine learning algorithms. In: Advances in Neural Information Processing Systems (NeurIPS) (2012). https://doi.org/10.48550/arXiv.1206.2944
28. Willmott, J., Matsuura, K.: Advantages of the mean absolute error (MAE) over the root mean square error (RMSE) in assessing average model performance. Climate Res. **30**, 79–82 (2005). https://doi.org/10.3354/cr030079

29. Chai, T., Draxler, R.R.: Root mean square error (RMSE) or mean absolute error (MAE)? - Arguments against avoiding RMSE in the literature. Geosci. Model Dev. **7**(3), 1247–1250 (2014). https://doi.org/10.5194/gmd-7-1247-2014
30. Franklin, J.C., Ribeiro, K.R., Fox, J.D.: Risk factors for suicidal thoughts and behaviors: a meta-analysis of 50 years of research. Psychol. Bull. **143**(2), 187–232 (2017). https://doi.org/10.1037/bul0000084
31. Lundberg, S.M., Lee, S.-I.: A unified approach to interpreting model predictions In: Guyon I. et al. (eds.) Advances in Neural Information Processing Systems, vol. 30, Curran Associates, Inc. (2017). https://doi.org/10.48550/arXiv.1705.07874

Aquila Optimizer for Hyperparameter Metaheuristic Optimization in ELM

Philip Vasquez-Iglesias[1](✉), David Zabala-Blanco[1], Amelia E. Pizarro[1], Juan Fuentes-Concha[1], and Paulo Gonzalez[2]

[1] Facultad de Ciencias de la Ingeniería, Universidad Católica del Maule, Talca, Chile
{fvasquez,dzabala}@ucm.cl, juan.fuentes.01@alumnos.ucm.cl
[2] Facultad de Economía y Negocios, Universidad de Talca, Talca, Chile
paulo.gonzalezg@utalca.cl

Abstract. This paper introduces the adaptation of the Aquila Optimizer (AO) metaheuristic to optimize the hyperparameters of the Extreme Learning Machine (ELM). The AO algorithm is a metaheuristic based on swarm intelligence that optimizes the objective function by simulating the hunting behavior of aquilas. The ELM belongs to the family of single hidden layer feed-forward network algorithms, where the hidden layer weights are randomly initialized and whose training is based on the Moore-Penrose pseudoinverse. It is known for faster convergence than traditional methods, providing promising performance with minimal programmer intervention.

The proposed method focuses on optimizing the hidden neurons of the ELM by maximizing the most popular performance metrics, namely Accuracy and G-Mean. This method offers an alternative to the classic grid search method by avoiding the need to go through all possible combinations in search of the optimal value. We evaluated three typical datasets and found that our proposal achieves an average efficacy of 99% compared to the global maximum found by the grid search, reducing the search time to an average of 20%. In other words, our method can achieve performance close to the global maximum in a fraction of the time required by the brute-force methodology.

Keywords: Aquila Optimizer · Heuristic Parametrization · Extreme Learning Machine · Soft Computing

1 Introduction

Optimization involves finding the best solution by evaluating a given objective function at one or more values within a specified domain. Different optimization methods, such as classical methods, are based on the derivative information like gradient descent [2]. On the other hand, metaheuristic methods do not rely on derivative information; instead, they use heuristics to find the best solutions.

Evolutionary optimization methods are a branch of Soft Computing that take inspiration from nature to mimic behaviors that have allowed many species to survive. These methods are motivated by the behavior of ant colonies, bee colonies, and fish schools. They offer the following advantages over traditional optimization methods: (1) Do not require differentiability, and (2) Tend to be more resistant to getting stuck in local optimal solutions in complex, multimodal functions. Some well-known algorithms in this category include Genetic Algorithms (GA) [6], Particle Swarm Optimization (PSO) [9], Ant Colony Optimization [3], and Artificial Bee Colony (ABC) [8].

In metaheuristic optimization, the parameterization is one problem that can be addressed. The following steps are required to obtain it: (1) Identifying an objective function (OF), typically a performance function that determines whether to maximize or minimize; and (2) Determining the necessary constraints, which usually involve different parameters and their valid domains. In [5], the Anisotropic Diffusion filter of Perona and Malik [12] was parameterized using Karaboga's Artificial Bee Colony algorithm [8] by adjusting the filter parameters (numOfIteration, scaleParameter, and timeStep) and performance functions to measure the quality of an image (Signal to Noise Ratio (SNR) and Mean Squared Error (MSE)). In the ongoing research, the AO algorithm is employed to optimize the hyperparameter corresponding to the number of neurons in the hidden layer of a neural network model known as an Extreme Learning Machine on various data sets.

The ELM is widely used to solve various classification problems. It is becoming an exciting alternative to health-related problems due to its good performance without the need to invest time in training. In this paper, we present a method capable of achieving a performance close to the best value found by an exhaustive search, limiting the run instances of the ELM performed by AO to approximately 25% of the ELM executions done by the exhaustive search.

This paper is divided into five sections. Section 1 introduces metaheuristic optimization. Section 2 presents the base knowledge of the AO and ELM algorithms to understand the terminology used. Section 3 describes the proposed method for using the AO metaheuristic to optimize ELM networks. Section 4 presents the results achieved by ELM across several datasets using both exhaustive search and the AO algorithm as hyperparameter optimization methods, and their performance is discussed. Finally, Sect. 5 summarizes the conclusions of the proposed work.

2 Theoretical Framework

2.1 Aquila Optimizer Algorithm

The Aquila Optimizer (AO) algorithm, introduced in [1], is based on the hunting behavior of aquilas. The algorithm begins with a random distribution of agents (aquilas) in the search space, and the quality of each location is assessed using an objective function. The obtained information is stored and used by the rest of the agents to generate new solutions in an iterative process. The best solution discovered is periodically reported until the final iteration of the algorithm, which depends on a predetermined maximum number of iterations.

The hunting options modeled in the AO include (1) High vertical flight to select the search space (Expanded Exploration), (2) Short glide attack, allowing divergent exploration of the search space (Narrow Exploration), (3) Slow descent attack to exploit the convergent search space (Expanded Exploitation), and (4) Ground attack to capture prey (Narrow Exploitation). It is important to note that options 1 and 2 are exploration or coarse search strategies, while alternatives 3 and 4 are exploitation or refined search strategies.

In the iterative process, each agent will be relocated using one of four hunting options based on the progress criterion, which depends on the percentage of iterations completed. Let us say T is the total number of iterations for each agent in each cycle t. If $t \leq \frac{2}{3} \times T$, the new position for the agent will be determined using an exploration criterion from either Eq. 1 or 2, chosen randomly following a uniform distribution. If this situation is not met, one-third of the total number of cycles remains. Therefore, the new position for the agent will be determined using an exploitation criterion from either Eq. 7 or 8, also randomly selected.

In Eq. 1, the first kind of attack, the Expanded Exploration, is modeled.

$$X_{\text{New}}(t+1) = X_{\text{Best}}(t) \times \left(1 - \frac{t}{T}\right) + (X_{\text{Mean}}(t) - X_{\text{Best}}(t) \times \text{rand}(\cdot)), \quad (1)$$

such that $X_{\text{New}}(t+1)$ corresponds to the solution of the next iteration to the current t, $X_{\text{Best}}(t)$ is the best position found in the current iteration, $X_{\text{Mean}}(t)$ is the average among all positions of the current iteration and $rand(\cdot)$ comes to be a random number between 0 and 1.

In Eq. 2, the second kind of attack, Narrow Exploration, is modeled.

$$X_{\text{New}}(t+1) = X_{\text{Best}}(t) \times Levy(D) + X_{\text{R}}(t) + (y - x) \times rand(\cdot), \quad (2)$$

such that, D is dimension space, $X_{\text{R}}(t)$ is a random solution taken between 1 and N at the ith iteration, and $Levy(D)$ is the Levy flight distribution function, calculated using Eq. 3.

$$Levy(D) = s \times \frac{u \times \sigma}{|v|^{\frac{1}{\beta}}}, \quad (3)$$

such that s is a constant value equal to 0.01, u and v are random numbers between 0 and 1, β is a constant value equal to 1.5, and σ is a number calculated using Eq. 4.

$$\sigma = \left(\frac{\Gamma(1+\beta) \times \sin\left(\frac{\pi\beta}{2}\right)}{\Gamma\left(\frac{1+\beta}{2}\right) \times \beta \times 2^{\left(\frac{\beta-1}{2}\right)}}\right). \quad (4)$$

In Eq. 2, y and x are used to present the spiral in the search, calculated using Eqs. 5 and 6, respectively.

$$y = (r_1 + U \times D_1) \times \cos\left(-\omega \times D_1 + \frac{3 \times \pi}{2}\right) \tag{5}$$

$$x = (r_1 + U \times D_1) \times \sin\left(-\omega \times D_1 + \frac{3 \times \pi}{2}\right), \tag{6}$$

such that r_1 is a value equal to several iterations, U is a constant value equal to 0.00565, D_1 is an array with numbers from 1 to the length of the search space (Dim), and ω is a constant value equal to 0.005.

In Eq. 7, the third kind of attack, Expanded Exploitation, is modeled.

$$X_{\text{New}}(t+1) = (X_{\text{Best}}(t) - X_{\text{Mean}}(t)) \times \alpha - rand(\cdot) + ((UB - LB) \times rand(\cdot) + LB) \times \delta, \tag{7}$$

such that, α and δ are the exploitation adjustment parameters fixed in the original paper to 0.1, LB and UB denotes to the lower and upper bound, respectively.

In Eq. 8, the fourth kind of attack, Narrowed exploitation, is modeled.

$$X_{\text{New}}(t+1) = QF(t) \times (X_{\text{Best}}(t) - (G_1 \times X(t) \times rand(\cdot)) - G_2 \times Levy(D) + rand(\cdot) \times G_1, \tag{8}$$

such that $QF(t)$ is the quality function used to equilibrium the search strategies in the current iteration, which is calculated using Eq. 11. G_1 denotes the motions of the Aquila that are used to track the prey, is calculated by Eq. 11. G_2 is a decreasing value from 2 to 0, denote the flight slope of the Aquila that is used to follow the prey during the slope from the first location to t^{th} location, which is calculated using Eq. 11.

$$QF(t) = t^{\frac{2 \times rand(\cdot) - 1}{(1-T)^2}}, \tag{9}$$

$$G_1 = 2 \times rand(\cdot) - 1, \tag{10}$$

$$G_2 = 2 \times \left(1 - \frac{t}{T}\right), \tag{11}$$

The Expanded Exploration (1) and Narrow Exploration (2) equations allow a coarse search in the space exploration, while the Expanded Exploitation (7) and Narrowed Exploitation (8) equations allow a refined search in the space exploitation.

The component that makes AO attractive is its robust mathematical foundation and low computational cost relative to other swarm intelligence algorithms. In [1], an extensive performance comparison is reported between AO and other metaheuristic algorithms applied to the optimization of fundamental problems and 23 classical mathematical functions of recurrent use in the literature, classifiable in different categories and difficulties (unimodal, multimodal and fixed-dimensional multimodal). The study reports a favorable result for AO. In [4], an improvement of the Kennedy PSO algorithm is proposed [9], where very similar results are reported with AO, leaving behind other algorithms such as the original PSO and the ABC of Karaboga [8].

2.2 Extreme Learning Machine

The Extreme Learning Machine (ELM) is a type of neural network characterized by a hidden layer with randomly initialized weights [7]. It is part of the family of random weights neural networks, eliminating the need to adjust the weights of the hidden layer model iteratively. As a result, the training time for neural networks is significantly reduced compared to the gradient descent method. It is worth noting that other similar methods, such as Random Vector Functional Link (RVFL) [10] and those discussed in [13], have been proposed with slight differences, mainly related to residual connections.

Figure 1 provides a visual representation of the ELM algorithm's basic structure. Here, $\mathbf{X}_{d \times N}$ denotes the dataset, a collection of N samples each with d features. The diagram also includes $\tilde{N}$ as the number of neurons in the hidden layer, m the number of neurons in the output layer, $\mathbf{b}$ the biases of the hidden layer neurons (which are randomly generated), $\mathbf{H}$ the output of the hidden layer, β the weights of the hidden layer, $\mathbf{T}$ the expected output, and $\mathbf{W}$ the weights between the input and the hidden layer (also randomly generated).

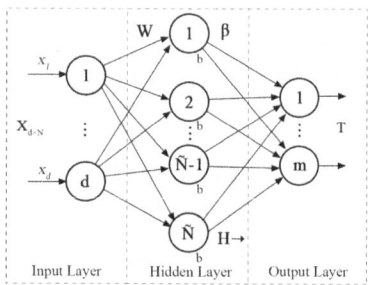

Fig. 1. ELM model.

The training of the network consists of the search for the least squares solution $\hat{\beta}$ of the linear system given by $\mathbf{H}\beta = \mathbf{T}$. If the amount of training data N and the number of neurons $\tilde{N}$ are equal, the solution will be exact with $\mathbf{H}$ square matrix and invertible, fitting the training data without error. In practice, these values differ, so $\mathbf{H}$ will correspond to a rectangular matrix and, therefore, no inverse. This results in the above system of equations containing multiple solutions. According to [7], one of these solutions is given by:

$$\hat{\beta} = \mathbf{H}^{\dagger}\mathbf{T}, \tag{12}$$

where $\mathbf{H}^{\dagger}$ corresponds to the Moore-Penrose pseudoinverse. This solution is optimal because it minimizes the network's error and norm weights [7].

3 Proposed Method

The proposed method consists of adapting the AO algorithm to optimize the hyperparameters of a neural network. In this work, AO is utilized to find the optimum number of neurons in an ELM by maximizing an objective function that reflects its performance.

It is necessary to modify the equations representing the aquila attacks presented in the previous chapter (1, 2, 7 and 8) by incorporating a rounding process because the solutions are indexes that allow access to the values of the vectors containing the valid domain for each dimension.

The cardinality of the entire space search is denoted by EB and is calculated as shown in Eq. 13.

$$EB = \prod_{i=1}^{Dim} \#EB_i, \tag{13}$$

such that $\#EB_i$ is the cardinality of the ith Space Search, where each ith cardinality is calculated based on the domain of the ith dimension.

In order to define the restriction of the Space Search's domain of the number of neurons, its cardinality is represented by $EB_{N\underline{o}Neural}$ and is computed according to the Eq. 14, where $N\_samples$ is the number of samples in the dataset, and $STEP$ is the step size. P is a percentage equal to 80% of the number of samples present in the used dataset, meaning that a dataset with 1000 samples is processed by an ELM with a domain in the interval $[1, 800]$.

$$EB_{N\underline{o}Neural} = \left\lceil \frac{P \times N\_samples}{STEP} \right\rceil. \tag{14}$$

In the AO algorithm, two parameters can affect the obtained results: the number of aquilas and the number of iterations. The number of aquilas is typically set to 10 based on previous literature. The number of iterations is determined using the following Eq. 15, where $Perc$ is a user-defined value between 1 and 100 representing the approximate percentage of time for an exhaustive search in $EB_{N\underline{o}Neural}$, and N is the number of aquilas set as a parameter in the method.

$$T = \left\lceil \frac{Perc \times EB}{100 \times N} \right\rceil \tag{15}$$

From the clearance of expression 15, it is possible to determine the total number of trips made by all aquilas within the same configuration, resulting in Eq. 16. This number is equivalent to the total number of ELM runs performed by AO, while the number of ELM runs in each iteration of AO is equal to N.

$$Total\_Travels = T \times N = \left\lceil \frac{Perc \times EB}{100} \right\rceil. \tag{16}$$

To use an objective function that allows numerically measuring the neural network's performance in different situations and to apply optimization methods such as the proposed one, two metrics (accuracy and G-Mean) are used together with the average, both of which are fundamental in evaluating the classification methods.

Accuracy measures the overall prediction rate of the classifiers, quantifying the accuracy with which the models can classify the samples as a whole. It is from the raw comparison between the value obtained from the network and the desired value. We obtained the accuracy from the Eq. 17.

$$\text{Accuracy} = \sqrt{\frac{1}{N}\sum_{i=1}^{N}(t_i - \bar{t}_i)^2}, \qquad (17)$$

where N represents the number of samples, t_i and $\bar{t}_i$ denote the actual and predicted values, respectively.

The G-Mean is a measure that focuses on evaluating the specific prediction rate of classifiers, especially in the case of unbalanced classes. From the product, we obtain the accuracy of each class. The G-Mean considers the success of each class, including the minority ones, equally important, making it a more representative metric. According to [11], we use the Eq. 18 to obtain the G-Mean.

$$\text{G-Mean} = \sqrt[L]{\beta_1 \beta_2 ... \beta_L}, \qquad (18)$$

where L corresponds to the number of classes and $beta_i$ the accuracy in the ith class.

In addition to the previously described metrics, this work considers pertinent the optimization of the average of both from a linear weighting for each, as shown in Eq. 19.

$$\text{Ponderated} = \frac{Accuracy + \text{G-Mean}}{2}. \qquad (19)$$

Once the aforementioned adaptations have been made, the AO algorithm executes in each iteration a number of ELMs equal to the number of agents established. It receives as parameters the training and testing sets and the value of the hyperparameter associated with the position in which the agent is located. From this configuration, the neural network is trained, evaluates its performance and returns as a result a number between 0 and 1 belonging to one of the previously described metrics (G-Mean, Accuracy or Ponderated). AO uses this value as a quality criterion of the index content and will be used to generate new solutions in the following iterations from the Eqs. 1, 2, 7 or 8, as appropriate. During the executions, the information on the content of the index that reports the highest performance is stored. This value, which represents the best solution found by the AO algorithm, will be the final output of the algorithm.

4 Discussion and Results

We use the UCI Heart Disease Dataset, Diabetes Dataset, and UCI Cardiotocography 3-Class Dataset because all of them are related to health problems and have different characteristics and levels of difficulty. The first one is a balanced binary-class problem, the second is a unbalanced binary class problem, and the third is a unbalanced 3-class problem. In each case, we determine the number of iterations of AO using T (Eq. 15), with approximately 25% run percentage

($Perc$) and ten aquilas ($N = 10$). In order to compare with the proposed method, we conducted an exhaustive search with a cardinality of $EB = EB_{N\underline{o}Neural}$ and $STEP$ equal to 1. The number of ELM executions performed by the exhaustive search is equal to EB, while the number of ELM executions performed by AO is calculated by multiplying $T \times N$.

The metrics used to assess the performance of the ELM are *Accuracy*, G-Mean, and *Ponderated*, calculated using the formulas 17, 18, and 19, as described in the previous section. In all experiments, the datasets were standardized to values between -1 and 1 to prevent any single feature from dominating the model, and a stratified k-fold cross-validation was performed to prevent overfitting. Each of the three metrics was used independently as an objective function for the AO experiments. In each experiment, the results of all three metrics were reported, regardless of which one was used as the objective function. Three graphics are presented for each dataset, Figs. 2, 4, and 6, by showing the result of the exhaustive search. Figures 3(a), 5(a), and 7(a) depict the results of AO adopting each metric as an objective function. The full dot represents the point where the superior value of the objective function was reached, while the empty dot represents the best value of the metrics when they were not the objective function; the segmented lines demonstrate the best value found in the exhaustive search for each metric as a reference case. Finally, Figs. 3(b), 5(b) and 7(b) show the AO results together, where each metric is identified with a different color and each objective function with a particular line type, by highlighting the best value found in each curve via the following markers: dot, cross or asterisk.

4.1 UCI Heart Disease Dataset Results

This dataset contains 1025 samples characterized by 12 inputs and a single output (binary classification). The class represents either the absence or presence of disease, having 499 and 526 samples, respectively. Namely, the classification problem is balanced. For these experiments, $EB = 820$, $T = 21$, and $T \times N = 210$, inferring that the AO occupies 25.61% runs of the brute force ELM.

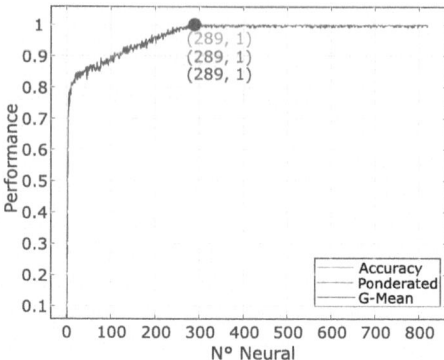

Fig. 2. Performance in terms of the ELM hidden neurons for the UCI Heart Disease Dataset and the exhaustive search approach.

In Fig. 2, it is evident that the optimal point is reached at 289 hidden neurons, achieving 100% in the three metrics used as the objective function. The results remain consistent beyond this number of neurons.

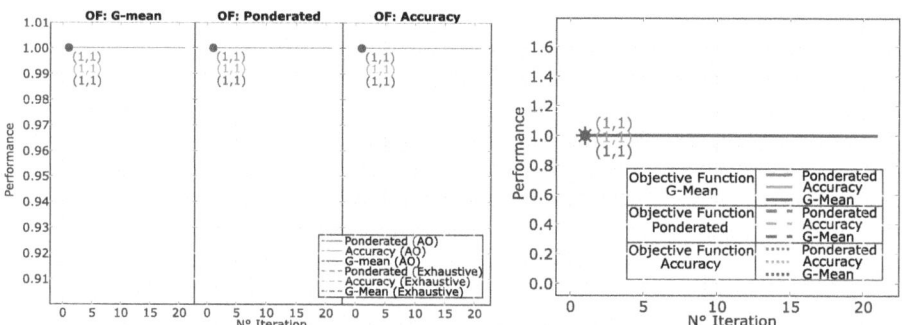

(a) Results for the AO by comparing with the exhaustive search values.

(b) Summary for the nine results among the three objective functions.

Fig. 3. Results and experiments summarization of AO for UCI Heart Disease Dataset.

Figures 3(a) and 3(b) show that AO finds the best solution in the first iteration. In this case, the robustness of AO is still not exploited because, in the first iteration, AO has performed only a random search. In any case, the UCI Heart Disease Dataset is an easy problem for ELM. If we assume that after 288 neurons, the performance of the metrics is always 1, each of the ten aquilas has a high possibility of 63.3875% to randomly select several neurons over 288. Consequently, AO has a possibility in the first iteration of approximately 99.97% to have at least one aquila with a neuron over 288, resulting in a performance of 100% in the metrics.

In general, the UCI Heart Disease Dataset results show that it is an easy problem for an ELM to tackle. Likewise, using AO for the ELM hyperparameter optimization allows for the easy finding of the optimal value due to the behavior of ELM in this dataset, independent of the objective function.

4.2 Diabetes Dataset Results

This dataset comprises a total of 768 samples, each with 9 attributes and a binary class. Class 0, indicating the absence of diabetes, is represented by 505 samples, while class 1, signifying its presence, is represented by 263 samples, making it an unbalanced problem that we aim to address.

For these experiments, $EB = 615$, $T = 16$, and $T = 160$, this implies that AO perform 26.02% of the total ELM runs done by the exhaustive search.

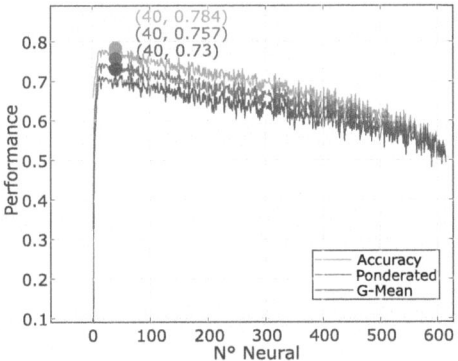

Fig. 4. Performance in terms of the ELM hidden neurons for the Diabetes Dataset and the exhaustive search approach.

Figure 4 shows a steady increase in the value of the metrics until the optimal value is reached using 40 neurons. After this point, the performance of each metric begins to decay.

Figure 5(a) reveals a behavior that consistently increases across all metrics, regardless of the objective function used. This behavior tends to approach the best value found in the exhaustive search and, in some cases, even surpasses it. Notably, the most superior results were achieved when the Ponderated metric was used as the objective function, with an Accuracy value that was only marginally lower than when the same Accuracy was used as the objective function.

Figure 5(b) shows that the results for the same metric across the different objective functions tend to cluster together, with all the results for *Accuracy* being at the top, followed by *Ponderated* and $G - Mean$.

In general, the use of Accuracy as the objective function does not always lead to a good G-Mean result. Conversely, using G-Mean as the objective function does not necessarily decrease the value of Accuracy. In this dataset, the

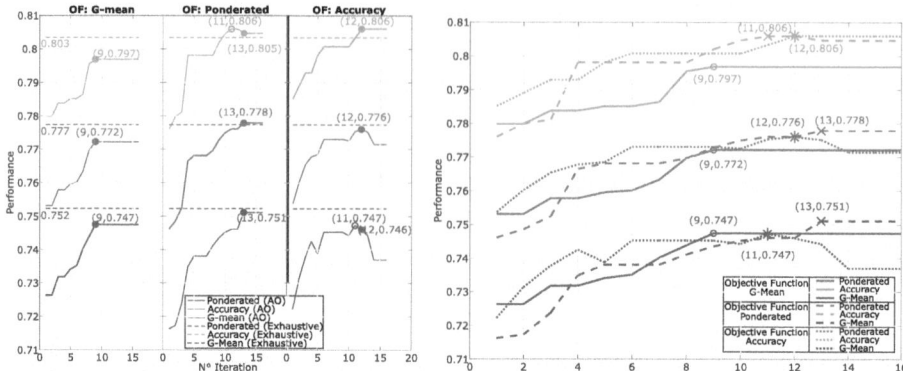

(a) Results for the AO by comparing with the exhaustive search values.

(b) Summary for the nine results among the three objective functions.

Fig. 5. Results and experiments summarization of AO for Diabetes Dataset.

most intriguing finding is that the best G-Mean value was obtained when the Ponderated metric was used as the objective function.

4.3 UCI Cardiotocography 3-Class Dataset Results

This dataset has 2126 samples, 22 attributes, and three classes. The class represents the fetal phase, where N is normal with 1655 samples, S is suspicious with 295 samples, and P is pathologic with 176, representing an unbalanced classification problem. For these experiments, $EB = 1071$, $T = 43$, and $T \times N = 430$, implying that AO performs 25.28% of the ELM runs in the exhaustive search.

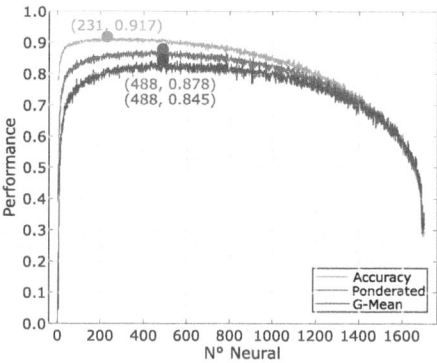

Fig. 6. Performance in terms of the ELM hidden neurons for the UCI Cardiotocography 3-Class Dataset and the exhaustive search approach.

Figure 6 shows a steady increase in their performance until they reach their optimal value and then begin to decay. It should be emphasized that the metrics have similar behavior, but with different numbers of neurons at the optimal value for each metric. Figure 7(a) shows that in this unbalanced multiclass problem, it is essential to consider G-Mean in the objective function, either when it is used or part of another, as in the Ponderated metric.

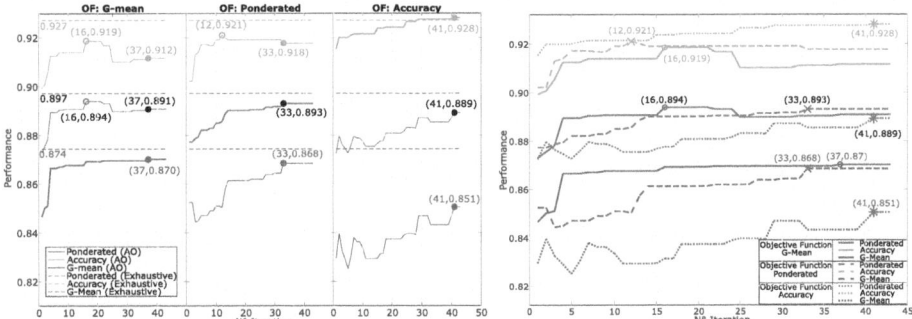

(a) Results for the AO by comparing with the exhaustive search values.

(b) Summary for the nine results among the three objective functions.

Fig. 7. Results and experiments summarization of AO for UCI Cardiotocography 3-Class Dataset.

Figure 7(b) shows that the results for the same metric in the different objective functions tend to cluster together, just as in the previous dataset. In particular, the effect of using Accuracy as the objective function is seen, being itself at the top of the graph, while G-Mean is at the bottom.

The results show the need to not neglect G-Mean in unbalanced multiclass problems. When Accuracy is used as the objective function, the network tends to be optimized based on the majority class. Finally, using Ponderated as the objective function generated a good balance in the values of G-Mean and Accuracy, maintaining the performance of G-Mean just 0.002 lower than when G-Mean is used as the objective function, while the value of Accuracy is still good.

4.4 Performance of AO Respect to the Exhaustive Search

Table 1 shows a summary of the results obtained categorized by dataset. The Dataset column shows the name of the dataset with the type of experiment and the used metric, the Time (seconds) columns is the mean of time and the respective standard deviation (STD). Finally, the Performance column show the mean of the results of each experiment and they STD.

Based on Table 2, it can be concluded that the proposed method allows efficient and effective optimization of the hyperparameters of an ELM. The results obtained always remained above 99% of the average maximum value found by

Table 1. Experiment summary table.

Dataset	Time (seconds)	Performance
UCI Heart Disease Exh(G-Mean)	61.081 ± 4.560	1.000 ± 0.000
UCI Heart Disease AO(G-Mean)	19.627 ± 3.344	1.000 ± 0.000
UCI Heart Disease Exh(Ponderated)	61.081 ± 4.560	1.000 ± 0.000
UCI Heart Disease AO(Ponderated)	21.985 ± 1.620	1.000 ± 0.000
UCI Heart Disease Exh(Accuracy)	61.081 ± 4.560	1.000 ± 0.000
UCI Heart Disease AO(Accuracy)	20.934 ± 4.869	1.000 ± 0.000
Diabetes Exh(G-Mean)	28.407 ± 0.565	0.752 ± 0.019
Diabetes AO(G-Mean)	3.128 ± 1.787	0.747 ± 0.021
Diabetes Exh(Ponderated)	28.407 ± 0.565	0.777 ± 0.014
Diabetes AO(Ponderated)	3.429 ± 1.751	0.778 ± 0.006
Diabetes Exh(Accuracy)	28.407 ± 0.565	0.803 ± 0.013
Diabetes AO(Accuracy)	4.113 ± 0.864	0.806 ± 0.021
UCI Cardiotocography 3-Class Exh(G-Mean)	987.997 ± 9.863	0.874 ± 0.012
UCI Cardiotocography 3-Class AO(G-Mean)	88.915 ± 38.111	0.870 ± 0.019
UCI Cardiotocography 3-Class Exh(Ponderated)	987.997 ± 9.863	0.897 ± 0.009
UCI Cardiotocography 3-Class AO(Ponderated)	95.299 ± 47.952	0.893 ± 0.010
UCI Cardiotocography 3-Class Exh(Accuracy)	987.997 ± 9.863	0.927 ± 0.009
UCI Cardiotocography 3-Class AO(Accuracy)	114.438 ± 34.707	0.928 ± 0.011

the exhaustive search, with average time percentages of 10.08% in the UCI Cardiotocography 3-Class Dataset, 12.52% in the Diabetes Dataset, and 34.13% in the UCI Heart Disease Dataset, enabling the application of the method in situations where time is a scarce resource to the detriment of a percentage of performance.

Table 2. Performance in percentage of the AO with respect to exhaustive search.

Dataset	Objective function	Time % (seconds)	Performance %
UCI Heart Disease	G-Mean	32.13%	100%
	Ponderated	35.99%	100%
	Accuracy	34.27%	100%
Diabetes	G-Mean	11.01%	99.34%
	Ponderated	12.07%	100.13%
	Accuracy	14.48%	100.37%
UCI Cardiotocography 3-Class	G-Mean	9.00%	99.54%
	Ponderated	9.65%	99.55%
	Accuracy	11.58%	100.11%

5 Conclusions

In this paper, an adaptation of the AO algorithm applied to the metaheuristic optimization of the hyperparameter $EB_{N\underline{o}Neural}$ of the ELM, was introduced. The results show that the algorithm, inspired by the behavior of aquilas while they hunt their prey to survive, allows the optimization of objective functions in a reduced time, while its value is close to the optimal global when compared to an Exhaustive Search approach.

Although the ELM used in this research can be considered a low-complexity model compared to other neural networks with more hyperparameters, its true potential lies in its training speed and high generalization ability despite associated randomness. For more complex ELM models, the hyperparameter search space can be so extensive that traditional optimization methods may not be feasible. The AO algorithm shows excellent potential in optimizing processing times, future research should consider this potential.

Compared to more complex ELM variations, such as Regularized ELM or Multi-layer ELM, which have a higher number of hyperparameters, using the AO algorithm may significantly reduce the number of trains without significantly affecting performance. This means that there will be greater availability of hardware resources and energy savings, which should not be underestimated.

It's clear from the graphs that maximizing the Accuracy metric in unbalanced multiclass problems can negatively impact the G-Mean. This underscores the need for new objective functions to optimize, such as the average of both. By doing so, we can avoid the situation where improving one metric comes at the cost of reducing other important metrics, potentially leading to overfitting and other model problems.

Applying soft computing algorithms, particularly the AO metaheuristic, to this problem is highly advantageous, especially when time is critical. The AO metaheuristic, with its proven robustness, low computational cost, and linear algorithmic complexity ($O(n)$), instills confidence in its ability to promptly deliver close to the best global optimal solutions.

Acknowledgements. The author Amelia E. Pizarro thanks the funding from ANID-Subdirección de Capital Humano/Doctorado Nacional/2024-21242342.

References

1. Abualigah, L., Yousri, D., Abd Elaziz, M., Ewees, A.A., Al-Qaness, M.A., Gandomi, A.H.: Aquila optimizer: a novel meta-heuristic optimization algorithm. Comput. Ind. Eng. **157**, 107250 (2021)
2. Curry, H.B.: The method of steepest descent for non-linear minimization problems. Q. Appl. Math. **2**(3), 258–261 (1944)
3. Dorigo, M., Birattari, M., Stutzle, T.: Ant colony optimization. IEEE Comput. Intell. Mag. **1**(4), 28–39 (2006). https://doi.org/10.1109/MCI.2006.329691

4. Gonzalez, P., Iglesias, P., Silva, E.: Restricted particle swarm optimization metaheuristic method. In: 2023 42nd IEEE International Conference of the Chilean Computer Science Society (SCCC), pp. 1–5 (2023). https://doi.org/10.1109/SCCC59417.2023.10315753
5. Gonzalez, P., et al.: Heuristic parametrization of anisotropic diffusion filtering. In: 2018 37th International Conference of the Chilean Computer Science Society (SCCC), pp. 1–5 (2018). https://doi.org/10.1109/SCCC.2018.8705235
6. Holland, J.H.: Genetic algorithms. Sci. Am. **267**(1), 66–73 (1992)
7. Huang, G.B., Zhu, Q.Y., Siew, C.K.: Extreme learning machine: theory and applications. Neurocomputing **70**(1), 489–501 (2006). https://doi.org/10.1016/j.neucom.2005.12.126. https://www.sciencedirect.com/science/article/pii/S0925231206000385. Neural Networks
8. Karaboga, D.: An idea based on honey bee swarm for numerical optimization. Technical report, Technical report-tr06, Erciyes University, Engineering Faculty, Computer Engineering Department (2005)
9. Kennedy, J., Eberhart, R.: Particle swarm optimization. In: Proceedings of ICNN 1995-International Conference on Neural Networks, vol. 4, pp. 1942–1948. IEEE (1995)
10. Pao, Y.H., Park, G.H., Sobajic, D.J.: Learning and generalization characteristics of the random vector functional-link net. Neurocomputing **6**(2), 163–180 (1994). https://doi.org/10.1016/0925-2312(94)90053-1. https://www.sciencedirect.com/science/article/pii/0925231294900531. Backpropagation, Part IV
11. Paz, K.: Media aritmética simple. Facultad de Ingeniería **7**, 1–13 (2007)
12. Perona, P., Malik, J.: Scale-space and edge detection using anisotropic diffusion. IEEE Trans. Pattern Anal. Mach. Intell. **12**(7), 629–639 (1990)
13. Schmidt, W., Kraaijveld, M., Duin, R.: Feedforward neural networks with random weights. In: Proceedings., 11th IAPR International Conference on Pattern Recognition. Vol. II. Conference B: Pattern Recognition Methodology and Systems, pp. 1–4 (1992). https://doi.org/10.1109/ICPR.1992.201708

Mixture of LSTM Experts for Sales Prediction with Diverse Features

Matías Soto, Felipe Cortés, Tímar Contreras, and Billy Peralta(✉)

Facultad de Ingeniería, Universidad Andres Bello, Antonio Varas 810, 7500735
Providencia, Región Metropolitana, Chile
{m.sotomejas,f.cortsarenas,t.contreraslvares}@uandresbello.edu,
billy.peralta@unab.cl

Abstract. Sales prediction is crucial for business intelligence, aiding in workforce management or resource allocation. Accurate sales forecasting is vital for financial planning and predicting both short-term and long-term company performance. In this work, we propose the use of adaptive ensembles of classification models to accommodate different trends within the data, unlike typically used machine learning models. Our approach is based on a Mixture of Experts (MoE) model using LSTM networks, with block cross-validation. We compare our proposal to various standard models in prediction tasks. Experiments show that our model achieves greater generalization on unseen stores compared to other models. As future work, we plan to extend this model to Transformer models.

Keywords: Forecasting · Mixture-of-Experts · LSTM

1 Introduction

Sales prediction plays a relevant role in business development, serving as a core component of business intelligence. By leveraging past and current sales data, it provides critical insights into workforce management, cash flow, and resource allocation. Accurate sales forecasting is key for financial planning, enabling companies to anticipate both short-term and long-term performance. This facilitates informed decision-making, enhancing supply chain management, profitability, and customer satisfaction [2]. On the other hand, sales prediction aids in understanding consumer behavior, optimizing marketing strategies, and efficiently managing resources. As a benchmark, it supports strategic planning and demand-supply alignment [16].

In the current era, data analytics considering both statistical and machine learning methods play an essential role in business management. These technologies enable organizations to get the most out of their data, making informed decisions and optimizing processes. In a nutshell, data analytics enables companies to understand their efficiency and performance, and ultimately helps the company make more informed decisions [23].

Particularly in this paper, we address the prediction of weekly sales in multiple stores of a national chain. The data comes from the publicly available Rossman database. This problem is challenging because of their changing behavior over time and because sales behavior is unique to each store (Fig. 1). Therefore, prediction is addressed using different computational techniques, especially considering deep learning including the proposed MoE-based models.

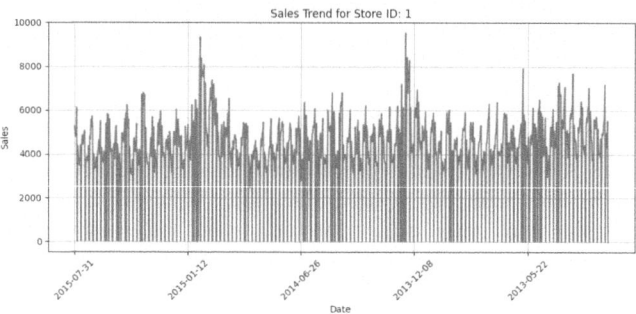

Fig. 1. Time series in one store

In general, among the computational models most commonly applied in sales prediction is the Autoregressive Integrated Moving Average (ARIMA) model, which is a classical autoregressive statistical model that requires a large number of observations in order to fit the model to a series of data [7]. In machine learning, the linear regression model is usually applied to understand the linear relationships between variables and sales [10]. In order to capture nonlinear patterns, neural networks such as Multi-Layer Perceptron (MLP) or Recurrent Neural Networks (RNN), which on the other hand are known for their black box [3] nature, are usually applied. Recently, the Prophet [27] model has been applied, which is oriented to be robust in the presence of missing data, outliers and to capture trend changes given by nonlinear growth curves.

In general, the methods used are based on models that do not explicitly consider the presence of multiple latent patterns within the sales data. We believe that these patterns are feasible, for example there may be rainy and sunny days where the patterns are highly different from each other, which makes the task difficult for a single model. In this work we propose the use of the Mixture-of-Experts (MoE) model where each expert is given by a LSTM neural network, which specializes in temporal forecasting which is suitable for the sales forecasting task where there are non-linear patterns in sales. On the other hand, the gating of the MoE model allows to identify the multiple latent patterns within its framework achieving a model that is able to adapt to multiple patterns unlike typical models.

2 Antecedents

2.1 Related Work

There are multiple works that have applied machine learning models to retail sales prediction. Below we list some relevant works in both sales prediction and business event prediction with similar patterns.

"At a general level, according to Kumar et al. [14], neural networks (NN) are among the most prominent models for various forecasting tasks. However, for time series forecasting, models like recurrent neural networks (RNN), particularly Long Short-Term Memory (LSTM) networks, are often preferred due to their ability to capture temporal dependencies. While convolutional neural networks (CNN) have proven effective in capturing spatial patterns, their application in time series forecasting is more limited and typically involves hybrid models."

Guo et al. [9] present the application of a Support Vector Machine Regression (SVR) for sales datasets with high data frequency and significant noise. This approach demonstrates the versatility of SVR to mitigate noise in datasets, thus contributing to improved prediction accuracy.

Shahi et al. [24] introduce LSTM (Long Short-Term Memory) models for forecasting in the stock market. This study highlights how recurrent neural networks, such as LSTM models, have elevated the performance in temporal predictions. The incorporation of attention mechanisms in these models has proven to be especially effective on temporally ordered data sets, improving the ability to capture complex relationships and nonlinear patterns.

Pavlyshenko [19] designs an MLP layer-based neural network considering a trend correction block to improve the accuracy of nonstationary sales predictions. This trend correction block helps to optimize predictions by combining predicted values and trend terms in the loss function of the model.

Dai and Huang [6] propose the implementation of LSTM models on the Rossmann database considering the use of hyperparameters. This study only takes a part of the dataset, studying weekly sales of stores during 2013 and validating with 2014 sales up to June 30.

A machine learning model widely used in practice is XGBoost. In the context of sales prediction, Jayakumaran et al. [13] looks at the application of boosting methods, comparing the performance between an ANN model and an XGBoost model. In this study, the dataset is used filtering on Sundays.

On the other hand, Malik et al. [17] perform a 60-day projection using the XGBoost models, as well as ARIMA and Prophet using the tabular data from the stores. In this study, XGBoost excels with better R^2, RMSE and MAE metrics than the other two models.

2.2 Mixture-of-Experts (MoE)

The Mixture-of-Experts model divides an artificial intelligence model into specialized sub-networks, or "experts", where each of these will focus on a subset of the input data, to jointly perform a task [11,20]. This is a supervised learning architecture that provides us with an efficient representation of complex nonlinear relationships in observed pairs on heterogeneous data (X, Y) [4].

By considering this explicit separation of the data, the representation of segmented or non-stationary data is facilitated by detecting non-linear patterns. On the other hand, this model can integrate the selection of variables [21] which can improve the interpretability of these models.

Within the temporal prediction task, the MoE model was applied in a spatio-temporal analysis for time series [15] in order to predict future values of multiple interrelated variables over time in a traffic context within large cities.

3 Proposed Methodology

We propose a variant KDD methodology, where we consider exploratory data analysis, ML model building and model evaluation. Moreover this methodology is applicable to other dataset.

3.1 Exploratory Data Analysis (EDA)

Exploratory data analysis is a fundamental step in our approach. For this purpose, we performed the following steps:

- Data Cleaning: We will identify and address null values, outliers or inconsistencies in sales and store data sets.
- Data Exploration: We will use various statistics such as graphs to create visualizations that allow us to understand key variables as well as their relationships.

3.2 ML Models

This work is based on the implementation of several ML models, including Multilayer Perceptron, Gated Recurrent Network and LSTM. In addition we consider the creation of a MoE model with LSTM experts to capture the multiple latent patterns within the data. These models are detailed in Sect. 4. The list of models tested in this work is:

- Multiple linear regression (MLR) [18]
- Multilayer Perceptron (MLP) [25]
- Long-Short Term Memory Network (LSTM) [26]
- Gated Recurrent Unit Network (GRU) [5]
- LSTM with Mixture of Experts (LSTM-MoE
- Mixture of LSTM Experts (MoE of LSTM)

It is worth mentioning that the neural models were trained considering the Huber Loss function known for its resistance to outliers [12]. This function was chosen because of its better results in the models considering a validation set.

3.3 Validation of ML Models

In this work we considered the block cross-validation approach to measure the performance of our models. This validation method provides an accurate and efficient evaluation by dividing the data into sequential blocks. By doing so, it avoids repetition of observations in the training and validation sets, avoids information leakage, preserves temporality and makes testing more realistic [1].

Metrics. The metrics used are: the Mean Squared Error (MSE $= \frac{1}{n}\sum_{i=1}^{n}(y_i - \hat{y}_i)^2$) calculates the average of the squared errors between the predicted values ($\hat{y}_i$) and the actual values (y_i). The MSE formula is calculated as the mean of the squared differences between the actual values and the predicted values (($y_i - \hat{y}_i)^2$), where n is the total number of observations available. On the other hand, the Mean Absolute Error (MAE $= \frac{1}{n}\sum_{i=1}^{n}|y_i - \hat{y}_i|$) represents the average magnitude of the errors between the predicted values ($\hat{y}_i$) and the actual values (y_i). The MAE formula is calculated as the mean of the absolute values of the differences between the actual values and the predicted values ($|y_i - \hat{y}_i|$). The Root Mean Squared Error (RMSE $= \sqrt{\frac{1}{n}\sum_{i=1}^{n}(y_i - \hat{y}_i)^2}$) is the square root of the resulting MSE equation and provides a measure of the dispersion of the residuals. Finally we have the Coefficient of Determination ($R^2 = 1 - \frac{\sum_{i=1}^{n}(y_i - \hat{y}_i)^2}{\sum_{i=1}^{n}(y_i - \bar{y})^2}$)) measures the proportion of the total variation in the dependent values that is explained by the model. The R^2 formula includes the variables y_i (real values), $\hat{y}_i$ (predicted values) and $\bar{y}$ (mean of the actual values), a higher R^2 indicates a better fit of the model to the data, the busy loss function is based on Huber Loss, which provides a better fit to the outliers [12].

4 Proposed LSTM-Based MoE

We propose two new models based on MoE and LSTM network. First, we propose a model where an LSTM network feeds a classical linear expert mixture model. As a second model, we propose a model where a MoE weights the output of different LSTM-based experts. We detail each one in the following.

4.1 Model LSTM + MOE

This model receives an input to the MOE under the parameters of each LSTM, where, depending is this last output triggers the weighting of the expert. The latter is composed each of a dense layer of 7 units and its respective dropout (see Fig. 2).

For this model, an LSTM layer with 50 units and 5 experts was used, giving a total of 15636 parameters.

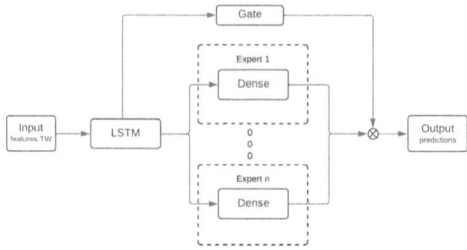

Fig. 2. LSTM structure with MOE

4.2 Model MOE of LSTM

The LSTM-based Mixture of Experts (MoE) model works with a number of experts, where each expert contains an LSTM layer (with a certain number of units) concatenated with a dense layer with outputs for forecasting, as can be seen in Fig. 3. The number of experts and the assigned units were adjusted by different combinations, which are presented in the next section.

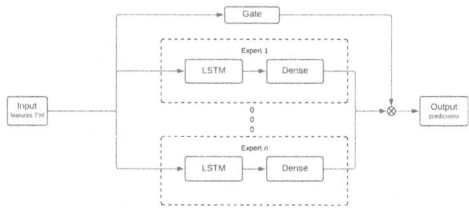

Fig. 3. Proposed structure in the LSTM-based MoE model

Each expert i receives an input $\mathbf{X} \in \mathbb{R}^{N \times T \times F}$, where N is the number of samples, T is the length of the time window, and F is the number of features in each time step. The expert's LSTM layer i processes this input and produces an output $\mathbf{h}^{(i)} \in \mathbb{R}^{N \times U}$, where U is the number of LSTM units in the expert i. This output is concatenated with a dense layer with y outputs, $\mathbf{z}^{(i)} \in \mathbb{R}^{N \times y}$, to generate the predictions:

$$\mathbf{h}^{(i)} = \text{LSTM}_i(\mathbf{X}) \qquad \mathbf{z}^{(i)} = \text{Dense}_i(\mathbf{h}^{(i)}) \qquad (1)$$

The gate input is composed in two ways for the experiments. The first one is composed of the flattened input $\mathbf{X}$, while the second proposed variant uses only the final features of the time window, denoted as $\mathbf{x}_T^{(i)} \in \mathbb{R}^F$, where $\mathbf{x}_T^{(i)}$ represents the last row of the input $\mathbf{x}^{(i)} \in \mathbb{R}^{T \times F}$. These gate configurations are represented as:

$$\mathbf{G}_1 = \text{Gate}_1(\text{Flatten}(\mathbf{X})) \qquad \mathbf{G}_2 = \text{Gate}_2(\mathbf{x}_T^{(i)}) \qquad (2)$$

Finally, the expert outputs are combined using the gate weights to generate the final prediction:

$$\mathbf{y}_{\text{pred}} = \sum_{i=1}^{N} \mathbf{G}_i \odot \mathbf{z}^{(i)} \qquad (3)$$

where $\odot$ denotes the Hadamard product.

5 Data Preprocessing and Analysis

5.1 Database

The data used in this project comes from the Rossman Store Sales [8] dataset, hosted by Kaggle. This data contains detailed information about stores and their daily sales. It is divided into Stores, which contains external information that affects stores, and Sales, which contains external information that affects stores.

Next, we will see the table with the breakdown of the dataset variable (Table 1).

Table 1. Description of variables

Variable name	Description	Data type
Store	The store where the data were recorded	Char
DayOfWeek	Day of the week	Char
Date	The date of registration	Date
Sales	Sales on the day	Numeric
Customers	The number of customers in the day	Numeric
Open	Indicates whether the store was open (1) or closed (0)	Numeric
Promo	Indicates if a store is running a promotion that day	Numeric
StateHoliday	Indicates a state holiday. Options include a (public holiday), b (Easter holiday), c (Christmas) and 0 (None)	Char
SchoolHoliday	Indicates if (Store, Date) was affected by the closing of public schools	Numeric

5.2 Data Analysis

In this section, we perform a detailed data analysis to better understand the information and distribution in our data sets. In the following, we show the most relevant observations found.

266   M. Soto et al.

- Figure 4a: In the following image, there is a graph showing the average sales per day of the week. There is a slight drop in sales from day 0 to day 6, and no sales are recorded on day 7. This information is essential to understand sales trends on different days.
- Figure 4b: The image shows a correlation matrix highlighting the relationship between key variables. The positive correlation between the number of customers and sales suggests that customer traffic is directly related to sales. In addition, the presence of promotions is also positively correlated with sales, supporting the importance of promotions in driving sales.

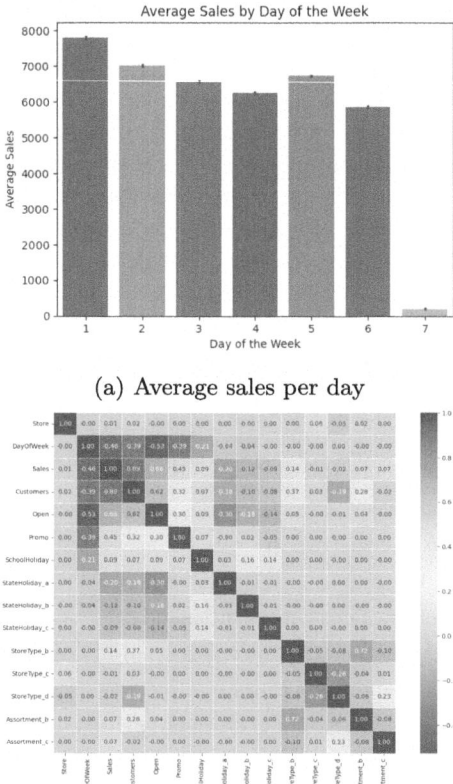

(a) Average sales per day

(b) Correlation matrix daily sales parameters with store information

Fig. 4. Data exploration and analysis

5.3 Preprocessing

Data processing was performed as follows:

1. The values with the strongest correlation were chosen ("DayOfWeek", "State-Holiday", "Promo", "Open", "SchoolHoliday", "Customers").
2. The value to predict is "Sales".
3. One hot encoding in variables of "StateHoliday", "DayOfWeek".
4. Added the variable "PreviusDaySales", to have more incoming information from the stores.
5. The data was normalized between 0 and 1, by means of a scaling with the respective minimum and maximum values.

This gives us an input of 16 characteristics for each day, in order to predict Sales.

6 Experiments

6.1 Division of the Dataset

The project does not use a traditional "holdout" method, which typically refers to dividing the data set into training and test sets and retaining a fixed percentage of the data for final evaluation (See Fig. 5). Instead, a temporal cross-validation technique is employed. The study by [22] delves into cross-validation techniques for data with temporal structures, where we can find the "cross block validation".

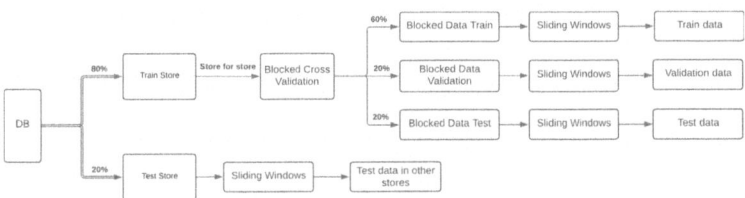

Fig. 5. Dataset splitting

Temporal cross-validation, often used in time series, splits the data into training and test sets sequentially, ensuring that the test data are always in the future with respect to the training data.

Since our dataset is based on different stores, it was decided to make subsamples with data from each of the stores. The block-cross-validation method was performed on each store sample. Our approach divides into 3 k-folds, which are distributed in a ratio of 60% for training, 20% for validation and 20% for testing (see Fig. 1), which are divided temporally by a window of 35 data (days), in order to predict 7 (days).

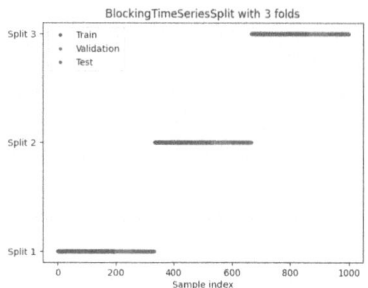
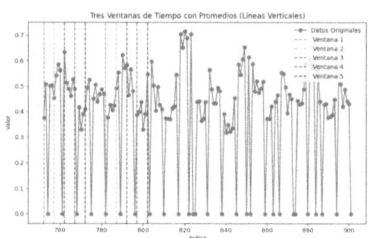

(a) Blocked cross validation implemented

(b) Sliding window for the creation of our input and output

Fig. 6. Comparison of techniques

The process of extracting the training data is The process of extracting the training data is carried out using a sliding window technique and temporal cross-validation. The function **"sliding window"** implements the sliding window in the Fig. 6b, dividing the time series into segments to train and evaluate the model. This is performed on the data obtained in the data separated by "blocked cross validation".

– Input (samples, 35, 16)
– Output (samples, 7)

7 Experiments

In this section we will perform a quantitative comparison of different ML models for the sales prediction task. Then we will perform a plotting within the test set to qualitatively analyze the predictions.

The ML models tested are linear regression, MLP, LSTM, GRU, LSTM+MoE and MoE of LSTM. In the case of MLP, a hidden layer with 100 neurons was considered. In the case of GRU, a hidden layer with 50 GRU units, including dropout, was considered. In the case of LSTM, a hidden layer with 50 LSTM units was also considered. In the case of MoE+LSTM, an LSTM layer of 50 units and 5 experts was considered. Finally, in the case of MoE of LSTM, 2 to 8 experts and 128 to 256 LSTM units per expert were considered. All parameter values were chosen based on validation data.

7.1 Quantitative Results

In Table 2 presents the different configurations of the MoE of LSTM model, we can see represented the performance during the validation, considering the number of experts and LSTM units. The results show that a model with 2 experts and 128 LSTM units has a Loss of 0.000806 and a MAE of 0.027514. When increasing the LSTM units to 256, the loss is 0.000810 and the MAE 0.027729. With 3 experts and 256 LSTM units, the loss is 0.000815 and the MAE 0.027567. A model with 6 experts and 256 LSTM units improves the MAE to 0.026834, although the loss is 0.000811. Finally, the model with 8 experts and 256 LSTM units has the best loss (0.000796) and the best MAE (0.026286), indicating that this configuration optimizes model performance. Based on the observations indicate that increasing the number of experts and the appropriate use of LSTM units allows the model to better capture the complexity and diversity of the sequential data, reducing model bias and error, which is reflected in better performance metrics.

Table 2. Table comparison LSTM-MoE with Gate 1 in Validation

N° Experts	Units LSTM	Loss	MAE
2	128	0.000806	0.027514
2	256	0.000810	0.027729
3	256	0.000815	0.027567
6	256	0.000811	0.026834
8	256	0.000796[a]	0.026286[b]

General note: This table summarizes the comparison of MoEs in validation between different numbers of experts and LSTM units. The outputs are normalized between 0 and 1.
[a] The best model in terms of Loss.
[b] The best model in terms of MAE.

It is possible to appreciate in the Table 3 that as the number of experts increases the loss values look relatively similar, but it is in the MAE metric where the clear improvement of the different variations of experts is seen, being the configuration with 9 experts and 64 LSTM units stands out as the most promising, with relatively low values of loss and MAE.

In relation to the Table 4, this focuses on observing the results of the metrics obtained in each case of the different models evaluated. The first thing to highlight is the good performance present in the multiple linear regression model demonstrating in terms of Mean Absolute Error (MAE) and Mean Squared Error (MSE), it still achieves a reasonable ability to explain the variability in the data with a coefficient of determination (R^2) of 0.8734. On the other hand, the Short-Term Memory Recurrent Neural Networks (LSTM) and the Gated Recurrent Units (GRU) stand out, similarly, the values obtained by changing the structure of gate 2 present in the MoE of the LSTM are shown to be competitive, the latter being the one with the best performance.

The analysis of the results presented in the comparison table (Table 5) for unseen store sales prediction models indicates that the MOE of LSTM G2 model is the most effective, demonstrating superior performance with the lowest values of MAE (840.7876), MSE (2.0878), RMSE (1444.9380), and the highest R^2

Table 3. Table comparison LSTM-MoE with Gate 2 in Validation

N.° Experts	Units LSTM	Loss	MAE
9	64	0.000702[a]	0.023837[b]
4	128	0.000724	0.024210
6	128	0.000748	0.024367
4	64	0.000755	0.026771
6	64	0.000757	0.025146

General Note: This table summarizes the comparison of MoEs in validation between different numbers of experts and LSTM units. The outputs are normalized between 0 and 1.
[a] The best model in terms of Loss.
[b] The best model in terms of MAE.

Table 4. Model comparison table on test data

Modelos	MAE	MSE	RMSE	R^2
GRU	847.38	1637790	1279.76	0.8680
MLP	890.46	1652763	1285.59	0.8673
LSTM	849.66	1670914	1292.63	0.8656
LSTM MOE	986.70	1993877	1412.04	0.8395
MOE of LSTM G1	881.03	1719629	1311.34	0.8615
MOE of LSTM G2[a]	798.08	1619457	1272.57	0.8697
MLR	872.52	1575676	1255.25	0.8734

General note: The test data are composed of 55254 samples, each with 7 days
[a] It presents the best values compared to the other models.

(0.8759). These metrics reflect its superior ability to capture nonlinear relationships and manage variability within the data. In contrast, models like GRU and LSTM MOE underperform, with higher errors and lower R^2 values, making them less suitable for this prediction task. The results underscore the efficiency of advanced deep learning architectures like MOE of LSTM, especially when variations in network structure, such as those seen between the G1 and G2 models, can lead to significantly more efficient outcomes with fewer parameters. For example, this structural adjustment led to a reduction from 2,255,296 parameters in model G1 to just 190,872 in model G2, highlighting how targeted changes in network design can achieve more efficient and accurate predictions (Fig. 7).

Table 5. Model comparison table on unseen store data

Modelos	MAE	MSE	RMSE	R^2	MAPE	RMSPE
GRU	1129.2427	3.3032	1817.4740	0.8037	15.4837	22.9600
MLP	1116.5714	2.9812	1726.6252	0.8228	15.1483	22.5534
LSTM	1036.5689	2.7454	1656.9296	0.8369	14.4831	21.3011
LSTM MOE	1209.0065	3.2692	1808.1065	0.8057	15.5142	22.3284
MOE of LSTM G1	980.0356	2.4736	1572.7974	0.8530	14.5304	22.2312
MOE of LSTM G2[a]	840.7876	2.0878	1444.9380	0.8759	12.8293	20.6505
RLM	1120.1481	3.0685	1751.7209	0.8177	15.9657	23.6257
XGB	983.5273	2.7862	1669.2108	0.8344	13.9995	20.9817

General note: The unseen stores data is composed of 223 unseen stores in training, which are generated by the "sliding Windows", 200700 samples, each with 7 days.

[a] It presents the best values compared to the other models.

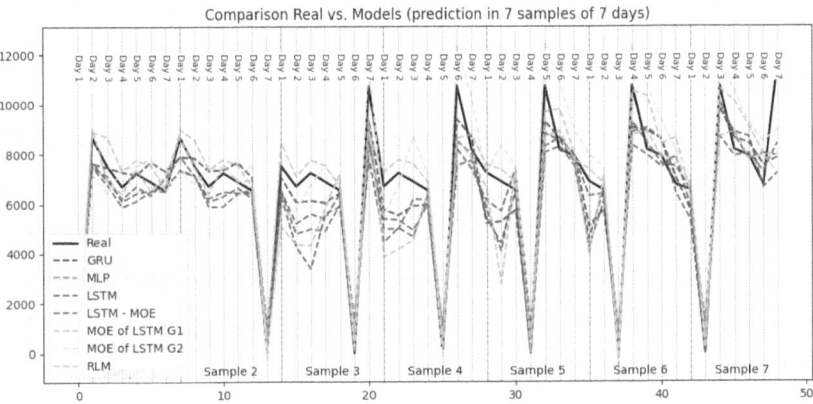

Fig. 7. Prediction in unseen stores

8 Conclusions

We have observed outstanding results with the use of the MoE assembly technique, using an expert structure, in the context of data analysis. MoE technique, using an expert framework, in the context of sequential data analysis for the sequential data analysis for the prediction of unseen store sales. Specifically, specifically, the MoE of LSTM model was used in this research, which allowed us to achieve greater generalization in the predictions. This allowed us to achieve greater generalization in the predictions of unseen store data. ted. In addition, a series of experiments were conducted during the research, which demonstrated that the MoE of LSTM model not only offers the best performance in terms of accuracy and metrics, but also in terms of the most accurate and reliable data. performance in terms of accuracy and metrics, but also maintains its generalization capabilities. generalization capabilities. These results highlight the effectiveness of the MoE approach with variation of its gate and how these can improve our obtained metrics in comparison to other metrics compared to other more traditional models or even more basic versions of mixture of expert, leaving of expert, showing its potential for applications in complex and varied data environments. complex and varied data environments. Finally, it is important to mention Finally, it is important to mention that future approaches could consider the incorporation of more complex data inputs, broader temporal more complex data inputs, more extensive temporal outputs, or variations in the combi- MoE could be considered for future approaches. This will allow further exploration of the model's capabilities and its adaptability to different contexts and types of data.

Acknowledgments. M. Soto, F Cortés and T. Contreras thank our study house, Universidad Andres Bello, and especially the faculty of the master's program in computer science for their support and guidance. B. Peralta appreciate the support of the National Center for Artificial Intelligence CENIA FB210017, Basal ANID.

References

1. Arlot, S., Celisse, A.: A survey of cross-validation procedures for model selection. Stat. Surv. **4**(none) (2010). https://doi.org/10.1214/09-ss054
2. Armstrong, J.S.: Principles of Forecasting: A Handbook for Researchers and Practitioners, vol. 30. Springer, New York (2001). https://doi.org/10.1007/978-0-306-47630-3
3. Bhadouria, S., Jayant, A.: Development of ANN models for demand forecasting. Am. J. Eng. Res **6**(12), 142–147 (2017)
4. Chamroukhi, F., Pham, N.T., Hoang, V.H., McLachlan, G.J.: Functional mixtures-of-experts (2023)
5. Cho, K., et al.: Learning phrase representations using RNN encoder-decoder for statistical machine translation. arXiv preprint arXiv:1406.1078 (2014)
6. Dai, Y., Huang, J.: A sales prediction method based on LSTM with hyperparameter search. J. Phys. Conf. Ser. **1756**, 012015 (2021)
7. Fattah, J., Ezzine, L., Aman, Z., El Moussami, H., Lachhab, A.: Forecasting of demand using ARIMA model. Int. J. Eng. Bus. Manag. **10**, 1847979018808673 (2018)
8. FlorianKnauer, W.C.: Rossmann store sales (2015). https://kaggle.com/competitions/rossmann-store-sales
9. Guo, Y., Han, S., Shen, C., Li, Y., Yin, X., Bai, Y.: An adaptive SVR for high-frequency stock price forecasting. IEEE Access **6** (2018). https://doi.org/10.1109/ACCESS.2018.2806180
10. Hoffmann, J.P.: Linear Regression Models: Applications in R. Chapman and Hall/CRC (2021)
11. Jacobs, R.A., Jordan, M.I., Nowlan, S.J., Hinton, G.E.: Adaptive mixtures of local experts. Neural Comput. **3**(1), 79–87 (1991). https://doi.org/10.1162/neco.1991.3.1.79

12. Jadon, A., Patil, A., Jadon, S.: A comprehensive survey of regression based loss functions for time series forecasting. arXiv preprint arXiv:2211.02989 (2022)
13. Jayakumaran, C., Merlin, S., Kulkarni, V.R., Stephan, T.: Entity embedding in artificial neural networks: a novel approach to sales data analysis and forecasting (2024)
14. Kumar, D., Sarangi, P.K., Verma, R.: A systematic review of stock market prediction using machine learning and statistical techniques, vol. 49 (2020). https://doi.org/10.1016/j.matpr.2020.11.399
15. Liu, H., Zhang, Y., Wang, X., Wang, B., Yu, Y.: ST-MoE: spatio-temporal mixture of experts for multivariate time series forecasting, pp. 562–567 (2023). https://doi.org/10.1109/ISKE60036.2023.10480934
16. Ma, S., Fildes, R.: Retail sales forecasting with meta-learning. Eur. J. Oper. Res. **288**(1), 111–128 (2021)
17. Malik, S., Khan, M., Abid, M.K., Aslam, N.: Sales forecasting using machine learning algorithm in the retail sector. J. Comput. Biomed. Inf. **6**(02), 282–294 (2024)
18. Montgomery, D.C., Peck, E.A., Vining, G.G.: Introduction to Linear Regression Analysis. Wiley (2021)
19. Pavlyshenko, B.M.: Forecasting of non-stationary sales time series using deep learning. arXiv preprint arXiv:2205.11636 (2022)
20. Peralta, B., Saavedra, A., Caro, L., Soto, A.: Mixture of experts with entropic regularization for data classification. Entropy **21**(2), 190 (2019)
21. Peralta, B., Soto, A.: Embedded local feature selection within mixture of experts. Inf. Sci. **269**, 176–187 (2014). https://doi.org/10.1016/j.ins.2014.01.008. https://www.sciencedirect.com/science/article/pii/S0020025514000140
22. Roberts, D.R., et al.: Cross-validation strategies for data with temporal, spatial, hierarchical, or phylogenetic structure. Ecography **40**(8), 913–929 (2017)
23. Sahoo, K., Samal, A.K., Pramanik, J., Pani, S.K.: Exploratory data analysis using Python. Int. J. Innov. Technol. Exploring Eng. **8**(12), 4727–4735 (2019)
24. Shahi, T.B., Shrestha, A., Neupane, A., Guo, W.: Stock price forecasting with deep learning: a comparative study. Mathematics **8** (2020). https://doi.org/10.3390/math8091441
25. Udaiyakumar, S., Chinnadurrai, C., Anandhakumar, C., Ravindran, S.: Electricity price forecasting using multilayer perceptron optimized by particle swarm optimization. In: 2022 Smart Technologies, Communication and Robotics (STCR), pp. 1–6 (2022). https://doi.org/10.1109/STCR55312.2022.10009414
26. Van Houdt, G., Mosquera, C., Nápoles, G.: A review on the long short-term memory model. Artif. Intell. Rev. **53**(8), 5929–5955 (2020). https://doi.org/10.1007/s10462-020-09838-1
27. Yenidoğan, I., Çayir, A., Kozan, O., Dağ, T., Arslan, Ç.: Bitcoin forecasting using ARIMA and prophet. In: 2018 3rd International Conference on Computer Science and Engineering (UBMK), pp. 621–624. IEEE (2018)

Correction to: Impact of Agricultural Production on Climate Change in South America: Comparative Analysis Between 1990 and 2020

Carlos Miguel Aizaga and Rafael Melgarejo-Heredia

Correction to:
Chapter 8 in: R. J. Barrientos and S. A. Velastin (Eds.):
Progress in Pattern Recognition, Image Analysis, Computer Vision, and Applications, LNCS 15369,
https://doi.org/10.1007/978-3-031-76604-6_8

In the originally published version of chapter 8, one of the co-author's name was incorrect.

In addition, author was unable to link his article to his account. These errors are corrected.

The updated version of this chapter can be found at
https://doi.org/10.1007/978-3-031-76604-6_8

Author Index

A
Aguilar, Eduardo I-1, II-74
Aizaga, Carlos Miguel II-104
Allende, Héctor I-198
Almeida, Jurandy II-205
Ángel González-Ordiano, Jorge II-1
Antunes, Francisco II-45
Araya, Mauricio I-92
Astudillo, César II-233
Astudillo, César A. I-178, II-151

B
Barcelos, Isabela Borlido I-162
Belém, Felipe C. I-148
Bellon de Carvalho, Gabriel II-205
Beltran, Tommy D. I-46
Beurton-Aimar, Marie II-120
Brady, Beth II-30
Bravo-Diaz, Alejandra I-63
Bustio-Martínez, Lázaro II-1

C
Calle, Roger II-74
Canales, Claudio II-219
Caro, Luis I-118
Castro, Carlos I-16
Castro, Juan Sebastian II-175
Castro-Azofeifa, César II-30
Chabert, Steren II-175
Chuquimarca, Luis E. I-46
Clarke, Colton II-60
Contreras, Tímar II-259
Cortés, Felipe II-259
Corvalán, Diego I-63
Csaholczi, Szabolcs II-191
Cubero-Pardo, Priscilla II-30

D
Delcourt, Cecile II-120
Dulau, Idris II-120

E
Encina-Chacana, Felipe II-161
Espinoza, Paulo A. II-16

F
Falcão, Alexandre X. I-148
Falcão, Alexandre I-162
Faúndez-Lizama, Thalía I-178, II-151
Frediani, João Otávio Rodrigues Ferreira I-244
Fuentes-Concha, Juan II-244

G
Gajardo-Sepúlveda, Nicolás I-178, II-151
Garcia, Eduardo I-228
Garcia, Gabriel Lino I-213, I-228, I-244
García, José I-16
Gonzalez, Paulo II-244
Gonzalez, Sergio II-135
Goyo, Manuel Alejandro I-133
Guimarães, Silvio J. F. I-148
Guimarães, Silvio Jamil F. I-162
Gutiérrez-Bahamondes, Jimmy H. I-178, II-151
Györfi, Ágnes II-191

H
Helmer, Catherine II-120
Hernández-García, Ruber II-60
Herrera-Semenets, Vitali II-1
Hidalgo, Mauricio I-133

J
Jacobs, Hanno I-31
Jodas, Danilo Samuel I-213
Júnior, Zenilton Kleber Gonçalves do Patrocínio I-148

K
Kovács, Levente II-191

L
Lacerda, Lucca S. P.   I-148
Lazcano, Vanel   I-187
Lira, Andrea   I-78
Lopatin, Javier   I-63

M
Maldonado, Camilo   I-198
Maldonado-Quispe, Percy   I-104, I-256
Mallea, Mario   I-92
Marana, Aparecido Nilceu   I-244
Martínez, Ignacio   II-233
Mauro, Jorge   I-78
Melgarejo-Heredia, Rafael   II-104
Mora, Marco   I-118
Mora-Melia, Daniel   I-178, II-151
Mora-Ramírez, Sebastián   II-30
Moreno, Sebastián   I-63
Muñoz, Bastián   I-1

N
Ñanculef, Ricardo   I-92
Negri, Pablo   II-135
Nicolis, Orietta   I-118
Núñez, Daniel   II-233
Nuñez, Felipe   I-118
Nyathi, Thambo   I-31

P
Paiola, Pedro Henrique   I-213, I-228, I-244
Papa, João Paulo   I-213, I-228, I-244
Passos, Leandro Aparecido   I-244
Patrocínio Jr, Zenilton K. G.   I-162
Pedrini, Helio   I-104, I-256
Peralta, Billy   I-118, II-259
Perdigão, Dylan   II-45
Pérez-Guadarramas, Yamel   II-1
Pizarro, Amelia E.   II-244
Prieto, Claudia   II-175

Q
Querales, Marvin   II-175
Quirós-Corella, Fabricio   II-30

R
Recur, Benoit   II-120
Remeseiro, Beatriz   I-1
Ribeiro Manesco, João Renato   I-213, I-228
Ribeiro, Bernardete   II-45
Roberts, Ian   I-92
Robles, Diego   I-78
Roudergue, César   II-219
Ruiz, Juan   II-135
Ruz, Gonzalo A.   II-16, II-161
Rycyk, Athena   II-30

S
Saavedra, Carolina   II-175
Salas, Rodrigo   II-175, II-219
Salazar-Jurado, Edwin H.   II-60
Serratosa, Francesc   II-90
Severín, Antonia   II-219
Silva, Catarina   II-45
Silvarrey, Alejo   II-135
Soto, Matías   II-259
Szilágyi, László   II-191

T
Taramasco, Carla   I-78
Torres, Romina   II-219

U
Ureña-Madrigal, Juan Pablo   II-30

V
Valle, Carlos   I-16, I-198
van den Berg, Jan   II-1
Vasquez-Iglesias, Philip   II-244
Velastin, Sergio A.   I-46
Vidal, Luciano   II-135
Vieira, Danielle   I-162
Villao, Raul J.   I-46
Vintimilla, Boris X.   I-46

Z
Zabala-Blanco, David   II-244

The manufacturer's authorised representative in the EU is Springer Nature Customer Service Centre GmbH, Europaplatz 3, 69115 Heidelberg, Germany. If you have any concerns regarding our products, please contact ProductSafety@springernature.com

Printed and bound by CPI Group (UK) Ltd, Croydon, CR0 4YY

26/03/2026

02078963-0004